Environmental Contamination Remediation and Management

There are many global environmental issues that are directly related to varying levels of contamination from both inorganic and organic contaminants. These affect the quality of drinking water, food, soil, aquatic ecosystems, urban systems, agricultural systems and natural habitats. This has led to the development of assessment methods and remediation strategies to identify, reduce, remove or contain contaminant loadings from these systems using various natural or engineered technologies. In most cases, these strategies utilize interdisciplinary approaches that rely on chemistry, ecology, toxicology, hydrology, modeling and engineering.

This book series provides an outlet to summarize environmental contamination related topics that provide a path forward in understanding the current state and mitigation, both regionally and globally.

Topic areas may include, but are not limited to, Environmental Fate and Effects, Environmental Effects Monitoring, Water Re-use, Waste Management, Food Safety, Ecological Restoration, Remediation of Contaminated Sites, Analytical Methodology, and Climate Change.

Nitish Kumar · Amrit Kumar Jha
Editors

Lead Toxicity Mitigation: Sustainable Nexus Approaches

Springer

Editors
Nitish Kumar
Department of Biotechnology
Central University of South Bihar
Gaya, Bihar, India

Amrit Kumar Jha
Krishi Vigyan Kendra Sahibganj
Birsa Agricultural University
Ranchi, Jharkhand, India

ISSN 2522-5847 ISSN 2522-5855 (electronic)
Environmental Contamination Remediation and Management
ISBN 978-3-031-46145-3 ISBN 978-3-031-46146-0 (eBook)
https://doi.org/10.1007/978-3-031-46146-0

This Springer imprint is published by the registered company Springer Nature Switzerland AG
The registered company address is: Gewerbestrasse 11, 6330 Cham, Switzerland

Preface

Lead is highly persistent in the environment, and because of its continuous use, its levels rise in almost every country, posing serious threats. Lead toxicity is an important environmental issue. One of the most deadly heavy metals, lead, plays no part in biological systems. All living creatures may be adversely affected by its traces in the air, soil, water, and biological systems, and its bioaccumulation in the food chain is especially dangerous for the health of people and animals. In trace amounts, lead is a bluish-gray metal that occurs naturally in the earth's crust. The available research shows that lead accumulates in the environment due to its non-biodegradable nature and ongoing use, which has a number of negative impacts including neurotoxicity and altered psychological and behavioral development in many organisms. Lead's speciation in soil has a significant impact on its bioavailability and, consequently, its toxicity to plants and microorganisms. In order to counteract the poisonous effects of lead, numerous plants and microorganisms have evolved detoxifying processes.

This global environmental problem is well discussed in the book, which also suggests interdisciplinary methodology for the mitigation of contamination. There are three sections in this book. The first section discusses the various sources and locations of lead in soil and plant ecosystems. The second section describes the health dangers of lead toxicity. The third section discusses methods for reducing lead toxicity and possible uses of current biological technology to address problems.

We give a general review of lead-polluted areas' potential for bioremediation using fungi, bacteria, or plants. These restoration techniques benefit from being economical and environmentally beneficial because they use plants to absorb and immobilize pollutants from soil and water, and fungus and bacteria to break them down. Phytoremediation is a well-established, well-researched practice with multiple in-field applications that make use of a wide variety of plant species. Students, educators, researchers, and environmental specialists working on lead contamination around the world will find this book to be a useful resource.

Gaya, Bihar, India — Nitish Kumar
Sahibganj, Jharkhand, India — Amrit Kumar Jha

Acknowledgments

Thanks to all the authors of the various chapters for their contributions. It had been a bit of a long process from the initial outlines to developing the full chapters and then revising them in light of the reviewer's comments. We sincerely acknowledge the author's willingness to go through this process. We also acknowledge the work and knowledge of the members of our review panels, many of which had to be done at short notice. Thanks to all the people at Springer Nature especially Nel van der Werf, and Mr. Karthikeyan Krishnan with whom we corresponded for their advice and facilitation in the production of this book.

Gaya, Bihar, India — Nitish Kumar
Sahibganj, Jharkhand, India — Amrit Kumar Jha

Acknowledgements

Contents

Contributors

Richa Aeri Survival for Women and Children (SWACH NGO), Panchkula, Haryana, India

Jawaid Ahsan Department of Biotechnology, Central University of South Bihar (CUSB), Panchanpur, Gaya, Bihar, India

Jasbir Arora Department of Anthropology, Panjab University, Chandigarh, India

Camelia Bețianu Department of Environmental Engineering and Management, "Cristofor Simionescu" Faculty of Chemical Engineering and Environmental Protection, Gheorghe Asachi Technical University of Iasi, Iasi, Romania

Petronela Cozma Department of Environmental Engineering and Management, "Cristofor Simionescu" Faculty of Chemical Engineering and Environmental Protection, Gheorghe Asachi Technical University of Iasi, Iasi, Romania

Swarup Debroy Department of Veterinary Anatomy, ICAR-IVRI, Izatnagar, Bareilly, Uttar Pradesh, India

Nitish Dhingra Electron Microscopy and Nanoscience Laboratory, Department of Soil Science, Punjab Agricultural University, Ludhiana, Punjab, India

Shallu Garg Bangalore Institute of Dental Sciences, Bangalore, India

Maria Gavrilescu Department of Environmental Engineering and Management, "Cristofor Simionescu" Faculty of Chemical Engineering and Environmental Protection, Gheorghe Asachi Technical University of Iasi, Iasi, Romania;
Academy of Romanian Scientists, Bucharest, Romania

Simmi Goel Department of Biotechnology, Mata Gujri College, Punjab, India

Shubham Gudadhe Department of Biotechnology, Central University of South Bihar (CUSB), Panchanpur, Gaya, Bihar, India

Saurabh Gupta Department of Microbiology, Mata Gujri College, Fatehgarh Sahib, Punjab, India

Alejandro Hernández-Morales Facultad de Estudios Profesionales Zona Huasteca, Universidad Autónoma de San Luis Potosí, Ciudad Valles, San Luis Potosí, Mexico;
Facultad de Ciencias Químicas, Centro de Investigación y Estudios de Posgrados, Universidad Autónoma de San Luis Potosí, San Luis Potosí, Mexico

Justin Jacob International Centre for Genetic Engineering and Biotechnology, Trieste, Italy

Neetu Jagota Department of Botany and Environment Science, DAV University, Jalandhar, Punjab, India

Ankush Kamboj Department of Soil Science, CCS Haryana Agricultural University, Hisar, India

Amanpreet Kaur Department of Microbiology, Mata Gujri College, Fatehgarh Sahib, Punjab, India

Amrit Kaur Department of Microbiology, Mata Gujri College, Punjab, India

Harleen Kaur Department of Botany and Environment Science, DAV University, Jalandhar, Punjab, India

Manjot Kaur Department of Microbiology, Mata Gujri College, Punjab, India

Ravneet Kaur Department of Botany and Environment Science, DAV University, Jalandhar, Punjab, India

Sumanveer Kaur Department of Microbiology, Mata Gujri College, Fatehgarh Sahib, Punjab, India

Ravindra Kumar Department of Agriculture, Mata Gujri College, Fatehgarh Sahib, Punjab, India

Rohtas Kumar Department of Soil Science, CCS Haryana Agricultural University, Hisar, India

Mamta Lathwal Department of Botany, Panjab University, Chandigarh, India

Amitava Paul Department of Veterinary Pathology, IIVER, Bahu-Akbarpur, Rohtak, Haryana, India

Bhairav Prasad Department of Biotechnology, Chandigarh Group of Colleges, Landran, Mohali, Punjab, India

Mamta Rani Department of Botany, Panjab University, Chandigarh, India

Gisela Adelina Rolón-Cárdenas Facultad de Estudios Profesionales Zona Huasteca, Universidad Autónoma de San Luis Potosí, Ciudad Valles, San Luis Potosí, Mexico

Moon Roy Department of Pediatric and Preventive Dentistry, TIDSHRC, Bilaspur, Chhattisgarh, India

Ashish Sharma Department of Botany and Environment Science, DAV University, Jalandhar, Punjab, India

Apoorva Shekhar Department of Veterinary Physiology and Biochemistry, College of Veterinary Science, Guru Angad Dev Veterinary and Animal Sciences University, Ludhiana, India

Deep Shikha Department of Veterinary Microbiology, IIVER, Bahu-Akbarpur, Rohtak, Haryana, India

Anjali Singal All India Institute of Medical Sciences, Bathinda, India

Chanchal Singh Department of Veterinary Physiology and Biochemistry, College of Veterinary Science, Guru Angad Dev Veterinary and Animal Sciences University, Ludhiana, India

Raghubir Singh Department of Veterinary Public Health, College of Veterinary Science and Animal Husbandry, Jalukie, Nagaland, India;
Central Agricultural University, Imphal, Manipur, India

Sushma Kumari Singh Department of Biotechnology, Central University of South Bihar (CUSB), Panchanpur, Gaya, Bihar, India

Swapnil Singh Department of Botany and Environment Science, DAV University, Jalandhar, Punjab, India

Vijay Singh Department of Botany, Mata Gujri College, Fatehgarh Sahib, Punjab, India

Sneha Department of Microbiology, Mata Gujri College, Fatehgarh Sahib, Punjab, India

Manish Kumar Verma Department of Veterinary Pharmacology and Toxicology, International Institute of Veterinary Education and Research, Bahu-Akbarpur, Rohtak, Haryana, India

Vikas Department of Soil Science, CCS Haryana Agricultural University, Hisar, India

Part I
Source and Distribution of Lead in Soil and Plant Ecosystem

Chapter 1
Environmental Lead Exposure—A Continuing Challenge

Swarup Debroy, Amitava Paul, and Deep Shikha

Abstract Lead is one the most abundantly present heavy metal on the earth crest, use of which in many can be traced back to 7000–6500 B.C. A low concentration of lead can be seen in an optimum range but when the concentration reaches up to 150–300 ppm in the environment, it can pose a serious threat to individual health. In the environment, lead can be found in both organic and inorganic forms with inorganic lead being the most predominant. The majority of lead pollution is caused by human activity to harvest and exploit the metal. In the early twentieth century, industrial workers who were working in painting, smelting, printing, plumbing, and other industries were heavily exposed to lead. Due to the use of lead in petrol after the invention of motor vehicles at the beginning of the twentieth century, there was a significant rise in ambient lead contamination. Brain and spinal cord is the most prominent organ among those harmed by lead. Young individual's intellectual development suffers long-lasting negative impacts from low-level chronic Pb exposure. Apart from that exposure to lead can have detrimental effects on different organ systems of the body. Chronic low-level lead exposure can have a long-lasting effect on the well-being of a new generation. The main focus of this chapter is to study the distribution, toxicology, and remediation of lead toxicity throughout the decades.

Keywords Lead · Environment · Metal · Exposure · Contamination

---

S. Debroy
Department of Veterinary Anatomy, ICAR-IVRI, Izatnagar, Bareilly, Uttar Pradesh 243122, India

A. Paul (✉)
Department of Veterinary Pathology, IIVER, Bahu-Akbarpur, Rohtak, Haryana 124001, India
e-mail: amit01paul@gmail.com

D. Shikha
Department of Veterinary Microbiology, IIVER, Bahu-Akbarpur, Rohtak, Haryana 124001, India

N. Kumar and A. K. Jha (eds.), *Lead Toxicity Mitigation: Sustainable Nexus Approaches*, Environmental Contamination Remediation and Management,
https://doi.org/10.1007/978-3-031-46146-0_1

1.1 Introduction

Ever growing development of industries comes with huge drawbacks, including water and air pollution and contamination of soil with toxic heavy metals emitted from those industries. This significantly deteriorates the quality of life for animals as well as for humans. All farm animals get affected by this contamination of the environment with heavy metals, because of their natural habit of pasturing in fields in the neighbourhood of the emission sources. These elements can be present on the body hair and on the skin of an animal and act as exogenous reserves, which can go inside of an animal and can get absorbed in the alimentary canal while licking their hair and also while grazing on the field. These endogenous reserves of ingested heavy metals get distributed to different organs and tissue by blood and get taken up by the hair during its growth phase, which subsequently increases the number of trace elements in an animal's body (Kabata-Pendias and Mukherjee 2007). Due to their toxicity, endurance, and capability to get absorbed into the body tissues via different modes of exposure, heavy metals are among the most harmful pollutants in the natural environment. When these toxins accumulate in body tissue more quickly than the body can eliminate them, a progressive build-up of the toxins takes place (Khudzariet al. 2013). Chronic low-level exposure to heavy metals also can have serious health effects just as much as excessive exposure to them. Even though some of these exposures and their negative effects are frequently subtle, especially on an individual level, the damage can be significant on a population level (Reis et al. 2007).

Lead is a silver-grey heavy metal with a melting temperature of 327.5 °C and a molecular mass of 207.19. Although this soft metal has good corrosion resistance, it is soluble in hot sulfuric and nitric acids. For inorganic lead compounds, the typical valence state is +2. Lead sulphide and lead oxides are not very soluble in water, but nitrate, chlorate, and chloride salts are very soluble in cold water. In addition, stable organic molecules like tetraethyllead and tetramethyllead, as well as organic acids like lactic and acetic acids, form salts with lead (WHO 1995). Despite having four electrons in its valence shell, lead only readily ionizes two of the electrons. Therefore, instead of +4, the typical oxidation state of lead in inorganic compounds is +2. The chloride, nitrate, and, to a much lesser extent chlorate, are water-soluble. Although the acetate is rather soluble, some of the salts produced with organic acids, such as lead oxalate, are similarly insoluble. Physio-chemical properties of lead salts.

Lead comes under non-essential heavy metals for living organisms, which is hazardous to biota even at a minute concentration. Traces of this metal can generally be found in the soil with an optimum concentration of 15–40 ppm. But with the increase in the concentration of lead up to 150–300 ppm in the environment, it can pose a serious threat to individual health (Dikmen et al. 2023). Because of the excessive emission from human activities over millennia, it is now impossible to accurately estimate the natural concentration of lead in the environment. According to multiple researches, natural air-lead concentrations were four to five folds lower than today's atmospheric concentration. Because of it, lead is now considered one of the most life-threatening heavy metal presents in the environment by many countries (Kazantzis et al. 1989). In the environment, lead can be found in both organic and inorganic forms with inorganic lead being the most predominant. The physical and chemical form of lead, as well as the size of lead particles, has a direct influence on its distribution, absorption into a living organism, sedimentation, and toxicity.

The first case of occupational lead poisoning was documented around 370 B.C. In the early twentieth century, industrial workers who were working in painting, smelting, printing, plumbing, and other industries were heavily exposed to lead. In 1767, several patients were hospitalized at La Charite Hospital in Paris with symptoms, which were not recognized then are now showing similarities with those of lead poisoning. Evidence showed that all those workers to some extent were exposed to lead from their occupational environment (Tong et al. 2000). Lead has been classified as a possible human and animal carcinogen and has well-documented effects on every organ system, including the immunological, reproductive, cardiovascular, and renal as well as on teeth and bones. However, the neurological system is particularly susceptible to lead's effects (White et al. 2007). Lead exposure is thought to be harmful and is linked to cognitive impairment, neuromuscular weakness, behavioural abnormalities, and hearing deficiencies in both people and animals. No "safe" level of lead exposure has been found, nor is there any level of lead that appears to be required or advantageous for the body (Flora et al. 2012). In many nations, exposure has declined as a result of the elimination of lead from gasoline. Lead has been extensively employed in industries nevertheless and levels are still high in many places because of its malleability, resilience to corrosion, and low melting point (Wang et al. 2012). Deposition of lead in the body tissue differs with the type of tissue and with the age group. 80–90% of lead deposition can be seen in the bone of adults, whereas about 70% of total lead deposition can be seen in the bone of children. A maximum amount of exposed organic lead get absorbed in the body and then may be present in different body fluid whereas inorganic lead remain unchanged and excreted through urine.

Children have a particularly high risk of lead poisoning since they absorb 4–5 times more ingested lead than an adult from a similar source. In addition, due to their natural curiosity and age-appropriate hand-to-mouth activity, children often swallow lead-containing or lead-coated particles, such as dust, flakes, and contaminated soil. Individuals who have pica, a psychological condition that causes excessive and persistent cravings for non-food objects, are more likely to pick at and consume lead paint from furniture, doorframes, and walls, which increases the risk of exposure. Children in Nigeria, Senegal, and other nations have experienced widespread lead poisoning and several deaths as a result of exposure to lead-contaminated soil and dust brought on by battery recycling and mining (WHO 2022).

Recently, the focus on lead poisoning has shifted from adults, highly exposed to this heavy metal from industrial effluent to asymptomatic children with minimal chronic lead exposure. Since chronic low-level lead exposure can have a long-lasting effect on the well-being of a new generation. In this chapter, we are going to discuss the development of knowledge on the distribution, toxicology, and remediation of lead toxicity throughout the decades of lead study.

1.2 History of Lead Uses

Being the first metal to get melted and discovered by humans, a trace of lead was found in different ancient ornaments dating back from 7000 to 6500 B.C. (Kazantzis et al. 1989). Lead was extensively used during Roman Empire, in their construction, cooking utensils, and other day-to-day objects. Lead was utilized by the ancient Romans to make water pipes and line baths. Due to its sweet flavour, lead was useful in winemaking to balance out the astringent taste of grape tannic acid. Roman upper classes consumed a lot of lead-sweetened wine, which can have up to 20 mg of lead per litre (Needleman 2004). The use of lead was started in mediaeval times for statues, ornaments, cisterns, tanks, gutters, roofs, and coffins. Lead was also used in the past to make the strips that connected the pieces of coloured glass in church windows. A little statue from Turkey that dates back to 6500 B.C. is the oldest known object manufactured by humans that contains lead. Between 3000 and 4000 B.C. Egyptian pharaohs utilized lead to glaze ceramics. Chinese, ancient Greek, and Roman coinage were made of lead 4000 years ago (Smith 1984). Lead poisoning has been linked to theories linking the fall of Rome to the Roman aristocracy's concurrent decline in fecundity and rise in psychosis (Gilfillan 1965). Smith (1882) reported cases of lead poisoning in the eighteenth century among weavers working with lead dichromate-containing dye in a cotton mill with symptoms of Jaundice and a blue line on the gums. A stricter regulation of the dyeing of the yarns and the use of personal protective equipment by mill workers resulted from an investigation into the poisonings, which ultimately eliminated all occurrences of lead poisoning. Due to their flexibility and capacity to be moulded into different diameters, in the nineteenth century lead has previously been utilized to relieve blockages in the lacrimal and nasal ducts. Burridge cites multiple cases of using the consumption of lead acetate in the treatment of dysentery and other diarrheal disease (Burridge 1851). Due to the use of lead in petrol after the invention of motor vehicles at the beginning of the twentieth century, there was a significant rise in ambient lead contamination. Throughout much of the century, this led to an increase in the community's exposure to environmental lead (Tong et al. 2000).

1.3 Lead in the Environment

The removal of gases and particles from the atmosphere is accomplished by atmospheric deposition. However, it is also a serious environmental issue in numerous regions of the world due to worries about natural ecosystem acidification and eutrophication, bioaccumulation of hazardous compounds and metals, effects on biodiversity, animal health, and global climate change. Increased pollutant concentrations in the atmosphere caused by human activities result in increased pollutant deposition, which has a negative impact on human health, crop yields, and terrestrial and marine ecosystems (Pan and Wang 2015). Lead exists in the earth's crust and is naturally found in the environment via a variety of mechanisms such as volcanic emissions and geochemical weathering. However, the majority of lead pollution is caused by human activity to harvest and exploit the metal (Fewtrell et al. 2003). Lead emissions from human activity into the environment can occur directly in the air, water, and soil. There is a constant flow of lead between these compartments even though emissions into these media may be easily monitored. Particle size has a significant impact on where atmospheric lead is found geographically concerning the source of emission. A surface, such as plants, soil, bodies of water, man-made surfaces, or the respiratory tracts of animals, eventually receives the majority of airborne lead through dry or wet deposition processes. Dry deposition occurs either through the impaction of all sizes of particles, mainly smaller particles, or the gravitational settling of bigger particles (> 10 m). Wet deposition is the outcome of either the build-up of particles by falling precipitation or the integration of particles into water droplets within clouds. Most of the lead in water is caused by industrial discharges, highway runoff, and sewage effluent, with some wet atmospheric lead deposition and direct dry deposition, which is more relevant for big bodies of water. The chemical nature of the lead affects how it disperses in water. The main causes of lead deposition in soils are the wet and dry deposition of atmospheric lead, especially close to the sources of emissions, and the discharge of sewage sludge, frequently onto agricultural land (Pattee and Pain 2003).

Soils are not excellent natural historical archives of contamination because metals are dispersed between anthropogenic and geogenic sources, and younger anthropogenic depositions cannot be separated exactly from older depositions. As a result, tree rings, peat deposits, and lake/marine sediments, in particular, are better recorders of pollution history, frequently dating back thousands of years (Savard et al. 2006). Although the historical development of Pb isotopic composition in sediments and tree rings is usually comparable, the process of metal acquisition differs. Nonetheless, soil humic layers, together with lake and bay sediments and trees, acted as effective receptor media for detecting cumulative metal pollution (particularly when Pb isotope studies were used), even at sites located a significant distance (N100 km) from the contamination source (Komárek et al. 2008). The lead concentration of various meals varies greatly, with plant-based foods being the primary source. Total diet studies in industrialized countries show a lead intake of 200–300 g per day, while values ranging from less than 100 to more than 400 g/day have been recorded. Lead solder in cans, dust ingestion by young children, and lead plumbing in places with soft-water sources all contribute significantly to daily lead intake.

Lead as well as compounds can enter the environment at any time throughout the mining, smelting, processing, usage, recycling, or disposal processes. Extensive uses of lead can be seen in batteries, gasoline additives, cables, solder, pigments, and steel products are among the many applications. In countries where leaded gasoline is still used, mobile and stationary sources of gasoline combustion account for the majority of air pollution. Air pollution is particularly severe in areas near lead mines and smelters (WHO 1995). Before the industrial revolution, environmental lead exposure to human and animal populations was comparatively minimal, but industrialization and large-scale mining have increased this heavy metal exposure. Compared to other non-essential elements, lead contamination in the environment significantly affects an organism's livelihood (Tong et al. 2000). As of 2022, Australia had the greatest lead deposits in the world, totalling 37 million metric tonnes. Despite having the second-largest lead reserves in the world, China was the world's top lead producer in 2022. They generated over two million metric tonnes of lead in that one year. Approximately, 12.3 million metric tonnes of refined lead were consumed globally in 2021 (Statista 2023a, b).

In the majority of developed countries, deliberate efforts have resulted in a decrease in the ambient lead concentration in recent years, reflecting a decline in lead's commercial use, particularly in petrol. Due to the phase-out of lead in petrol and the decrease in ambient exposure to the metal over the past 20 years, blood lead levels in the general population in these countries have decreased significantly (Tong et al. 2000). In developing nations where there are wide variations in exposure sources and pathways, lead continues to be a serious public health issue.

1.4 Toxicology and Effects of Lead (Pb) Exposure

A divalent cation, lead, has a considerable affinity for the sulfhydryl groups on proteins. Brain and spinal cord are the most prominent organ among those harmed by lead. Lead is a diverse toxin that has a variety of targets, but the deformation of enzymes and structural proteins is thought to be a major contributor to its toxicity. The endogenous opiate system's development is hampered by lead. There is no sign of a threshold as it catalytically and effectively cleaves the ribophosphate backbone of tRNA at particular places. Because of its capacity to imitate or compete with calcium, lead exhibits several hazardous qualities (Needleman 2004). As per Bailey and Kitchen (1985) lead competes with calcium for binding sites on cerebellar phosphokinase C at picomolar doses, which alters neural signalling. Because of the high lead sensitivity of astrocytes and olegodendrocytes, lead has a significant effect on blood–brain barrier and myelin sheath formation. Lead interferes with vascular permeability by interfering with collagen formation.

According to WHO's 2021 report on the impact of chemicals on Public health, lead exposure cost around a million of lives from all the over the world. According to estimates, chronic lead exposure is estimated to cause 30% of the total intellectual disability, 4.6% of the total cardiovascular diseases and 3% of kidney diseases worldwide due to its chronic effect on the health. Children's health may suffer severely from lead exposure. Lead damages the brain and spinal cord at high exposure levels, resulting in unconsciousness, convulsions, and even death. Children who recover from severe lead exposure may nevertheless have behavioural and intellectual problems. Lead is now understood to induce a spectrum of harm across numerous physiological systems at lower exposure levels that don't immediately manifest any symptoms. Lead, in particular, can have an impact on how children's brains develop, which can lower IQ, change behaviour in the form of increased antisocial behaviour and decreased attention span, as well as lower scholastic achievement. Anaemia, renal impairment, hypertension, toxicity to the reproductive organs, and immunotoxicity are further effects of lead exposure. Lead is thought to have permanent impacts on the brain and behaviour (WHO 2023). Numerous instances of anaemia have been linked to lead poisoning because lead inhibits the enzymes ferrochelatase and porphobilinogen synthase, inhibiting the production of porphobilinogen and the integration of iron into protoporphyrin IX, which blocks the synthesis of heme in blood or causes defective heme synthesis, leads to microcytic anaemia (Ara and Usmani 2015). Lead act as a calcium analogue, which interacts with ion channels, which is one of the processes by which it impairs cognition. Lead can disrupt the ultrastructure of mitochondrion and cell membrane permeability; replace essential elements like Zn, Ca and Fe, and increase the synthesis of reactive oxygen species (ROS), in addition to activating some enzyme and non-enzymatic antioxidants. At very low dose also lead can have a detrimental effect on living cell. In addition to oxidizing intracellular proteins, lipids, and nucleic acids, ROS (H_2O_2, hydroxyl radical, superoxide anion) also cause membrane damage, enzyme deactivation, and lipid peroxidation (Zhang et al. 2023). Acute and chronic exposure to lead can have a huge impact on the reproductive organs of an individual. In a study comparing infertile and fertile males, lead levels in the blood of infertile men were found to be higher (12.5 μ/dl and 6 μg/dl, respectively) (Pant et al. 2003). Epidemiological studies also demonstrate elevated blood lead levels in male employees, ranging from 10 to 40 μg/dl, as well as an increased risk of infertility as a result of lead exposure. Another study of 4000 male workers with elevated blood levels of lead more than 25 g/dl revealed that these individuals had fewer children than the control group (Ganesh 2023). According to Oehninger (2000), infertility cases involving men account for about half of all cases; environmental exposure, particularly occupational exposure in developing countries, as well as a lack of awareness of safety precautions while working in hazardous environments are the main causes of male-related infertility in men.

Young individual's intellectual development suffers long-lasting negative impacts from low-level chronic Pb exposure (Bellinger and Bellinger, 2006). In their study, Bailey and Kitchen (1985) found out that monkeys fed with lead acetate @ new born to 200 days of age, their blood lead levels ranged from 3 to 25 μg/dl. They underwent a delayed alternation test at the age of 7–8 years, in which the crucial positive stimulus was switched. The ability to learn was compromised in treated monkeys, especially at longer periods of delay. Epidemiological data show that lead exposure during early childhood results in a noticeable loss in cognitive development during the subsequent childhood years. Children are more likely to experience negative effects from lead exposure than adults because: they consume more lead per unit of body weight; they frequently put things in their mouths when they are young, possibly increasing their intake of lead; they consume more lead per unit of body weight than adults; children have higher physiological uptake rates of lead than do adults; young children are developing quickly and have underdeveloped systems, making them more susceptible to the effects of lead than do adults (Tong et al. 2000).

Lead exposure at high levels may result in renal impairment. The same issue could arise even from very little lead exposure. Acute and chronic nephropathies are the two different forms of impaired renal function. Nuclear enclosing bodies, which contain lead protein complexes, and degenerative alterations in the tubular epithelium can be used to classify acute nephropathy both visually and functionally, as can a mechanism of decreased tubular transport. It may enhance an abnormal secretion of amino acids, phosphates, and glucose, a combination known as Fanconi's syndrome, although it is not the cause of protein appearing in the urine. Chronic nephropathy, on the other hand, is easier to treat and can result in permanent morphological and functional abnormalities characterized by, hypertension, hyperuricemia and renal breakdown caused by tubulointerstitial and glomerular abnormalities (Baranowska et al. 2012). According to Carmignani et al. (2000)'s review, lead exposure has a detrimental effect on both human and animal kidneys, leading to the development of renal toxicity due to the stress on the body's oxidative system that it creates. The earlier study, however, revealed that such an impact primarily affects the kidney in chronic exposure that becomes clinically significant and that kidney injury does not typically occur in asymptomatic/acute situations. Rarely do we find information in the literature about how acute Pb exposure causes oxidative stress in an animal's kidney. When compared to other groups, the injection of lead acetate resulted in a substantial rise in urea and creatinine levels. According to a recent study by Sharma and Singh (2014), exposure to Pb acetate at doses of 10 and 150 mg/kg BW for 24 h increased the amount of thiobarbituric acid reactive substances (TBARS) in the kidneys, which is a sign of lipid peroxidation. In the bones, lead is meant to be stored in two compartments. The exchangeable pool situated at the bone surface and the non-exchangeable pool found deep within the bone cortex. Lead could move to the surface after leaving the non-exchangeable pool because it can easily reach plasma from the exchangeable pool and is actively being reabsorbed. Adults' bones contribute between 40 and 70% of the released lead in the blood, according to stable lead isotope analysis. Adults keep roughly 85–95% of their lead in their bones, but children's soft tissues contain about 70% of their high quantity of lead (Patrick 2006). Age, pregnancy dosage and rate, lead exposure, race, and gestation are only a few of the variables that affect how much lead is mobilized and stored in bones. According to Al Naimi et al. (2011), administering lead acetate at a dose of 75 mg/kg BW at 20 and 40 days results in a mild hyperplasia of haemopoietin tissue with megakaryocyte proliferation and the appearance of thin trabeculae of calcified cartilage coated by a thin coating of bone. In comparison to normal, healthy bones, the mineralized cartilage bars that developed as a result of defective osteoclast resorption are wider and extend further into the metaphyseal marrow cavity.

If the damage is too severe, especially to the nervous system's cells or tissue, treatment could not be effective. Following lead exposure, cattle is given calcium disodium edentate (Ca-ethylenediaminetetraacetic acid [EDTA]) subcutaneously or intravenously for three days. 5% dextrose and a similar amount compartmentalized to 4 treatments per day were administered subcutaneously to dogs for 2–5 days. After a one-week break from the end of the therapy, a second 5-day treatment may be required if the clinical indicators don't go away. There isn't a suitable veterinary product with Ca-EDTA on the market at the moment. Thiamine reduces the amount of lead that accumulates in tissues, which can help lessen the clinical symptoms. Thiamine and Ca-EDTA therapy appeared to have the most beneficial response (Payne and Livesey 2010). It has been demonstrated that the chelating agent succorer (meso 2, 3-dimercaptosuccinic acid [DMSA]) is effective in both dogs and birds. Compared to Ca-EDTA, DMSA has a lot fewer adverse effects. Lead removal from the GI tract may benefit from cathartics such a rumenotomy magnesium sulphate. For cases exhibiting convulsion episodes, tranquillizers or barbiturates may be administered as supportive therapy. The oxidative damage caused by severe lead poisoning may be reduced by antioxidant therapy paired with a chelating agent. However, DMSA has been utilized in conjunction with antioxidants such N-acetylcysteine. Using endoscopically guided forceps, it is feasible to extract the swallowed lead pieces from the stomach in chelonian. This should be followed by two weeks of Ca-EDTA therapy (Kaneko et al. 2008).

1.5 Steps to Prevent Lead Exposure

WHO has listed lead as one of the most 10 hazardous elements presents in the environment. Through its website, WHO has made a variety of lead information accessible, including resources for advocacy, technical advice, and information for policy makers. In order to provide policymakers, public health authorities, and health professionals with evidence-based guidance on the steps they can take to protect the health of young and adult individuals from lead exposure, WHO has developed guidelines on clinical management of lead exposure and is currently preparing guidelines on prevention of lead exposure. The Centres for Disease Control and Prevention (CDC) has a long-standing obligation to safeguard children against lead poisoning, with the elimination of lead exposure among young children as its main objective. The CDC has assisted regional health agencies in creating lead poisoning prevention initiatives since the early 1970s. Public health organizations have long depended on blood lead screening tests to detect exposed individuals because lead exposure does not manifest evident symptoms until after serious harm has already been done. An integrated programme to identify and limit sources of exposure and offer case management for kids with elevated blood lead levels must include blood lead screening for primary or secondary prevention (Ettinger et al. 2019). The Global Alliance to Eliminate Lead Paint was established by WHO and the United Nations Environment Programme due to the ongoing exposure risk posed by leaded paint in many nations.

Humanity has long been aware of lead poisoning, which first came to light in the eighteenth century during the industrial revolutions. Lead has no known biological role in the body, so when it gets inside of you, you risk major health problems that could have a deadly outcome.

References

Al-Naimi RA, Abdul-Hadi D, Zahroon OS, Al-Taae EH (2011) Toxicopathological study of lead acetate poisoning in growing rats and the protective effect of cysteine or calcium. Al-Anbar J Vet Sci 4:26–39

Ara A, Usmani JA (2015) Lead toxicity: a review. Interdisc Toxicol 8(2):55–64

Bailey C, Kitchen I (1985) Ontogenesis of proenkephalin products in rat striatum and the inhibitory effects of low-level lead exposure. Dev Brain Res 22(1):75–79

Baranowska-Bosiacka I, Gutowska I, Rybicka M, Nowacki P, Chlubek D (2012) Neurotoxicity of lead. Hypothetical molecular mechanisms of synaptic function disorders. Neurologia i neurochirurgia polska 46(6):569–578

Bellinger DC, Bellinger AM (2006) Childhood lead poisoning: the tortuous path from science to policy. J Clin Invest 116:853–857

Burridge R (1851) On the acetate of lead. Prov Med Surg J 15:347

Carmignani M, Volpe AR, Boscolo P, Qiao N, Di Gioacchino M, Grilli A, Felaco M (2000) Catcholamine and nitric oxide systems as targets of chronic lead exposure in inducing selective functional impairment. Life Sci 68(4):401–415

Dikmen F, seymenoğlu VC, ergönül mB The effect of lead (Pb) on the growth rates of two aquatic macrophyte species grown in different experimental media. Commun Faculty Sci Univ Ankara Ser C Biol 32(1):59–69

Ettinger AS, Leonard ML, Mason J (2019) CDC's lead poisoning prevention program: a long-standing responsibility and commitment to protect children from lead exposure. J Public Health Manage Pract: JPHMP 25 (Suppl 1 lead poisoning prevention)

Fewtrell L, Kaufman R, Prüss-Üstün A (2003) Lead: assessing the environmental burden of diseases at national and local levels

Flora G, Gupta D, Tiwari A (2012) Toxicity of lead: a review with recent updates. Interdiscip Toxicol 5(2):47–58

Ganesh MK (2023) Efficacy of Withaferin a on lead acetate induced testicular toxicity in Wistar rats-a biochemical analysis. Eur Chem Bull 12(5):2370–2395

Gilfillan SC (1965) Lead poisoning and the fall of Rome. J Occup Med 7:53–60

Kabata-Pendias A, Mukherjee AB (2007) Trace elements from soil to human. Springer Science & Business Media, Berlin, Heidelberg, pp 257–271, 283–293, 294–308, 381–391

Kaneko JJ, Harvey JW, Bruss ML (eds) (2008) Clinical biochemistry of domestic animals. Academic press

Kazantzis G () Lead: ancient metal—modern menace? In: Smith MA, Grant LD, Sors AI (eds) Lead exposure and child development: an international assessment. Lancaster, England, MTP Press, pp 119–128

Khudzari JM, Wagiran H, Hossain I, Ibrahim N (2013) Screening heavy metals levels in hair of sanitation workers by X-ray fluorescence analysis. J Environ Radioact 115:1–5

Komárek M, Ettler V, Chrastný V, Mihaljevič M (2008) Lead isotopes in environmental sciences: a review. Environ Int 34(4):562–577

Needleman H (2004) Lead poisoning. Annu Rev Med 55:209–222

Oehninger S (2000) Clinical and laboratory management of male infertility: an opinion on its current status. J Androl 21(6):814–821

Pan YP, Wang YS (2015) Atmospheric wet and dry deposition of trace elements at 10 sites in Northern China. Atmos Chem Phys 15(2):951–972

Pant N, Upadhyay G, Pandey S, Mathur N, Saxena DK, Srivastava SP (2003) Lead and cadmium concentration in the seminal plasma of men in the general population: correlation with sperm quality. Reprod Toxicol 17(4):447–450

Patrick L (2006) Lead toxicity part II: the role of free radical damage and the use of antioxidants in the pathology and treatment of lead toxicity. Altern Med Rev 11(2)

Pattee OH, Pain DJ (2003) Lead in the environment. Handbook Ecotoxicol 2:373–399

Payne J, Livesey C (2010) Lead poisoning in cattle and sheep. Pract 32(2):64–69

Reis MF, Sampaio C, Brantes A, Aniceto P, Melim M, Cardoso L, Gabriel C, Simao F, Miuel JP (2007) Human exposure to heavy metals in the vicinity of Portuguese solid waste incinerators e part 3: biomonitoring of Pb in blood of children under the age of 6 years. Int J Hyg Environ Health 210:455e459

Savard MM, Bonham-Carter GF, Banic CM (2006) A geoscientific perspective on airborne smelter emissions of metals in the environment: an overview. Geochemist Explor Environ, Anal 6(2–3):99–109

Sharma S, Singh B (2014) Effects of acute and chronic lead exposure on kidney lipid peroxidation and antioxidant enzyme activities in BALB-C mice (Mus musculus). Int J Sci Res 3(9):1564–1566

Smith RC (1882) Poisoning by chromate of lead in weavers. BMJ 1:8

Smith MA (1984) Lead in history. In: Lansdown R, Yule W (eds) The lead debate: the environmental toxicology and child health. Croom Helm, London, pp 7–24

Statista (2023) Distribution of lead mine production worldwide in 2019 https://www.statista.com/statistics/603544/distribution-of-world-mine-production-of-lead-by-select-country/

Statista (2023) Lead reserves worldwide by country in 2022 (in million metric tons). https://www.statista.com/statistics/273652/global-lead-reserves-by-selected-countries

Tong S, Schirnding YEV, Prapamontol T (2000) Environmental lead exposure: a public health problem of global dimensions. Bull World Health Organ 78(9):1068–1077

Wang J, Yang Z, Lin L, Zhao Z, Liu Z, Liu X (2012) Protective effect of naringenin against lead-induced oxidative stress in rats. Biol Trace Elem Res 146:354–359

White LD, Cory-Slechta DA, Gilbert ME, Tiffany-Castiglioni E, Zawia NH, Virgolini M, Basha MR (2007) New and evolving concepts in the neurotoxicology of lead. Toxicol Appl Pharmacol 225(1):1–27

World Health Organization (1995) Inorganic lead. World Health Organization.

World Health Organization (2022) Lead poisoning. World Health Organization. https://www.who.int/news-room/fact-sheets/detail/lead-poisoning-and-health

Zhang H, Hu L, Du X, Sun X, Wang T, Mu Z (2023) Physiological and molecular response and tolerance of Macleaya cordata to lead toxicity. BMC Genomics 24(1):1–13

Part II
Health Risks Linked to Lead Toxicity

Chapter 2
Effects of Lead: Neurological and Cellular Perspective

Chanchal Singh, Raghubir Singh, and Apoorva Shekhar

Abstract Lead exposure is a serious public health concern with significant neurological and cellular effects. This chapter examines the effects of lead on brain development, neurotransmitter function, and cellular processes from a neurological and cellular perspective. Lead exposure during critical periods of brain development can result in structural and functional changes in the brain, leading to cognitive and behavioral deficits. Alterations in neurotransmitter function, such as dopamine, serotonin, and glutamate, can contribute to the development of neurological conditions. At the cellular level, lead can interfere with mitochondrial function and oxidative stress, leading to cell death and inflammation. In addition, lead exposure can have long-term effects, contributing to the development of neurological disorders such as Parkinson's disease and Alzheimer's disease. While the exact mechanisms of lead toxicity are still being investigated, effective strategies to prevent lead exposure are critical, including reducing lead in the environment, improving screening and remediation efforts, and increasing public awareness of the risks associated with lead exposure.

Keywords Lead · Neurological effects · Cellular effects · Cognitive · Oxidative stress

C. Singh (✉) · A. Shekhar
Department of Veterinary Physiology and Biochemistry, College of Veterinary Science, Guru Angad Dev Veterinary and Animal Sciences University, Ludhiana, India
e-mail: drchanchal833@gmail.com

R. Singh
Department of Veterinary Public Health, College of Veterinary Science and Animal Husbandry, Jalukie, Nagaland, India

Central Agricultural University, Imphal, Manipur, India

N. Kumar and A. K. Jha (eds.), *Lead Toxicity Mitigation: Sustainable Nexus Approaches*, Environmental Contamination Remediation and Management,
https://doi.org/10.1007/978-3-031-46146-0_2

2.1 Introduction

Lead is a toxic heavy metal that has been used for various industrial purposes for many years. It has been used in various industrial applications, including paint, gasoline, batteries, and solder. Due to its extensive use, it has become a widespread environmental pollutant, posing a significant threat to human health and the ecosystem. It is affecting both humans and animals due to its ability to interfere with biological processes, particularly those related to the nervous system, the immune system, and the cardiovascular system.

In humans, its exposure can cause a range of health problems, depending on the level and duration of exposure. The most significant effects are seen in children, who are more vulnerable to lead's toxic effects than adults (Lidsky and Schneider 2003). Lead exposure in children can cause developmental delays, behavioral problems, decreased IQ, and an increased risk of neurological disorders such as Parkinson's disease, Alzheimer's disease, and multiple sclerosis (Bellinger 2004; Wu et al. 2020). In adults, lead exposure can cause anemia, high blood pressure, kidney damage, and reproductive problems.

In animals, its exposure has been observed in a variety of animal species, including birds, fish, and mammals. In birds, exposure is causing decreased reproductive success, impaired immune function, and neurological damage. In fish, exposure is a cause for developmental abnormalities, impaired growth, and reduced survival. In mammals, lead exposure has similar effects as those seen in humans, including neurological damage, reproductive problems, and decreased immune function.

The lead exposure to multicellular organisms causes widespread systemic changes in the body ranging from cellular dysfunctions to molecular alterations. The various biochemical processes required for metabolism of carbohydrate, lipids, and proteins are significantly affected. In the present, the chapter cellular and neurological alterations caused by the exposure of lead has been discussed.

2.2 Lead Exposure in the Environment

Lead is present in the environment in various forms, including soil, air, water, and food. Human exposure to lead occurs through multiple routes, including inhalation of contaminated air, ingestion of contaminated soil or water, and consumption of lead-contaminated food. Industrial activities such as mining, smelting, and battery manufacturing are the primary sources of lead pollution. Additionally, lead-based paints used in houses and buildings can also be a significant source of lead exposure.

2.3 Adverse Effects of Lead Exposure

Lead toxicity affects many biological systems, including the cellular and neurological systems. Lead toxicity can cause oxidative stress, inflammation, and damage to cellular components such as DNA, proteins, and lipids. The nervous system is particularly vulnerable to lead toxicity, as it can interfere with neurotransmitter signaling, neuronal development, and synaptic function (Atchison 1988) The long-term effects of lead exposure on the nervous system include decreased IQ, developmental delays, and an increased risk of neurological disorders such as Parkinson's disease and Alzheimer's disease (Liu et al. 2013; Raj et al. 2021).

2.4 Mechanisms of Cellular and Neurological Effects

The mechanism by which lead exerts its toxic effects is complex and involves multiple mechanisms, including interference with enzymatic activity, oxidative stress, inflammation, and disruption of ion channels and membrane transporters. The exact mechanisms by which lead causes its toxic effects can vary depending on the specific biological system or process that is affected. Lead also bind to and inhibit enzymes such as delta-aminolevulinic acid dehydratase (ALAD), which plays a critical role in heme synthesis. It generates reactive oxygen species (ROS), which may leads to oxidative stress and damage to cellular components. In the nervous system, lead (Pb) interfere with synaptic transmission and disrupt the balance of calcium ions, lead to excitotoxicity and neuronal death (Mason et al. 2014).

2.4.1 Lead Effect: Cellular Perspective

2.4.1.1 Cellular Effects

The cellular perspective of lead effects involves understanding how lead interferes with cellular functions and signaling pathways, leading to cellular dysfunction and damage. Lead can enter cells through various mechanisms, including ion channels, transporters, and receptors. Once inside the cell, lead can bind to and disrupt many cellular components, including enzymes, proteins, and DNA. One of the primary mechanisms by which lead exerts its toxic effects is through the generation of reactive oxygen species (ROS), which can cause oxidative stress and damage to cellular structures (Jomova and Valko 2011). Lead can also interfere with calcium signaling, which plays a crucial role in many cellular processes, including cell growth, differentiation, and apoptosis. Lead can bind to and inhibit calcium channels, leading to altered calcium homeostasis and impaired cellular signaling. This can affect many cellular processes, including the activation of signaling pathways, gene expression, and apoptosis. Another mechanism by which lead exerts its toxic effects is through the disruption of the cytoskeleton, which provides structural support and maintains cellular shape. Lead can interfere with the assembly and stability of microtubules and actin filaments, leading to altered cellular morphology, impaired cellular migration, and altered cellular function. Furthermore, lead can alter cellular signaling pathways, leading to aberrant cellular proliferation, differentiation, and apoptosis. Lead exposure has been associated with the activation of many signaling pathways, including MAPK/ERK, PI3K/Akt, and JAK/STAT pathways, which can contribute to cellular dysfunction and damage.

2.4.1.2 Cellular Proteins Affected by Lead

Proteins are large, complex molecules that perform various functions within cells, such as catalyzing chemical reactions, transporting molecules, and providing structural support. Lead exposure can affect many different cellular proteins, leading to a range of adverse health effects (Chasapis 2018; GoERING 1993). Here are some cellular proteins that can be affected by lead:

Metallothioneins

Metallothioneins are small, cysteine-rich proteins that are found in a variety of organisms, including humans. They are particularly abundant in the liver and kidneys, where they play an important role in detoxifying heavy metals. Metallothioneins bind to metals like lead and cadmium, sequestering them and preventing them from causing damage to cells and tissues (Bruno et al. 2016).

In humans, there are four different metallothionein isoforms: MT-1, MT-2, MT-3, and MT-4. MT-1 and MT-2 are the most abundant isoforms and are found in most tissues, while MT-3 is primarily expressed in the brain and MT-4 is found in stratified squamous epithelia. Lead exposure can increase the expression of metallothioneins in the body, as a protective mechanism against the toxic effects of lead. However, chronic exposure to lead can deplete the levels of metallothioneins, making the body more susceptible to lead toxicity. Metallothioneins are also used as biomarkers of heavy metal exposure and toxicity, as their levels can be measured in blood and urine samples.

Enzymes: Lead can bind to enzymes and alter their structure and function, inhibiting their ability to catalyze chemical reactions. This can disrupt various cellular processes, such as energy metabolism and DNA synthesis.

Ion channels: Lead can interact with ion channels, which are specialized membrane proteins that allow ions to flow in and out of cells. This can interfere with the normal flow of ions across the cell membrane, disrupting various cellular processes such as signaling and muscle contraction.

Transporters: Lead can interact with transporters, which are membrane proteins that move molecules in and out of cells. This can interfere with the normal transport of essential molecules such as nutrients and neurotransmitters, leading to cellular dysfunction.

Receptors: Lead can interact with receptors, which are proteins on the cell surface that bind to specific molecules and trigger signaling pathways. This can interfere with normal cell signaling and contribute to the development of various diseases.

Structural proteins: Structural proteins provide support and shape to cells, and they also play important roles in cell division, movement, and signaling. Lead can interact with structural proteins, such as microtubules and intermediate filaments, which provide support and shape to cells. This can disrupt the normal structure and function of cells, leading to cellular dysfunction. It can affect various structural proteins in cells, which can lead to cellular dysfunction and adverse health effects. Here are some examples of structural proteins that can be affected by lead:

Microtubules: Microtubules are long, hollow tubes made of protein subunits called tubulin. They play a critical role in maintaining cell shape and supporting cell division, as well as in intracellular transport and signaling. Lead exposure can disrupt microtubule structure and function, leading to impaired cell division and transport.

Intermediate filaments: Intermediate filaments are a diverse group of fibrous proteins that provide mechanical strength to cells and tissues. They are particularly important in cells that are subjected to mechanical stress, such as skin cells and muscle cells. Lead exposure can disrupt intermediate filament structure and function, leading to cellular dysfunction and tissue damage.

Extracellular matrix proteins: The extracellular matrix (ECM) is a complex network of proteins and carbohydrates that surrounds cells and provides structural support. ECM proteins, such as collagen and fibronectin, are important in cell adhesion, migration, and signaling. Lead exposure can disrupt ECM protein synthesis and organization, leading to impaired cell function and tissue integrity.

Cytoskeletal proteins: The cytoskeleton is a dynamic network of protein fibers that provides mechanical support to cells and helps maintain their shape and organization. Cytoskeletal proteins, such as actin and myosin, are particularly important in muscle cells and other cells that require movement. Lead exposure can disrupt cytoskeletal organization and function, leading to impaired cell movement and function.

2.4.1.3 Interfering with Enzymes

Lead can bind to enzymes and disrupt their normal function. This can lead to metabolic disruptions and affect cellular processes. Lead can bind to the active site of enzymes and inhibit their function, e.g., lead inhibits in vitro creatine kinase and pyruvate kinase activity in brain cortex of rats (Lepper et al. 2010). This can disrupt normal cellular processes that depend on enzymatic activity, such as metabolism and protein synthesis. It can cause changes in the structure of enzymes, which can affect their function. This can lead to the formation of misfolded or dysfunctional enzymes that can be harmful to the cell. Some enzymes require cofactors, such as metal ions or vitamins, to function properly. Lead can interfere with the binding of these cofactors to enzymes, leading to decreased enzymatic activity. It can cause irreversible damage to enzymes, leading to their inactivation. This can result in a loss of enzymatic activity that can have severe consequences for cellular processes (Nemsadze et al. 2009).

2.4.1.4 Disrupting Ion Channels

Lead can interfere with the function of ion channels, which play a key role in the regulation of cellular processes such as signaling and ion transport. It can disrupt ion channels in several ways, leading to cellular dysfunction. Ion channels are specialized membrane proteins that allow ions to move in and out of cells, playing a critical role in various cellular processes such as signaling and ion homeostasis. Ion channels can be opened or closed by different mechanisms, and lead can interfere with these mechanisms. For example, lead can disrupt the voltage-gating mechanism of ion channels, preventing them from opening or closing properly in response to changes in membrane potential. It can bind to ion channel proteins, altering their structure and function. This can cause the ion channels to become less selective or to conduct ions more slowly, leading to cellular dysfunction. The expression of ion channels could be altered either by increasing or decreasing their expression levels. This can disrupt ion homeostasis and signaling pathways, leading to cellular dysfunction.

Lead can enter cells through a variety of mechanisms, including passive diffusion and active transport. Once inside cells, lead can bind to proteins and interfere with cellular processes. For example, lead can bind to calcium-binding proteins and disrupt intracellular calcium signaling, which can affect cell growth and differentiation (Wani et al. 2015). Lead can also interfere with the transport of other essential metals, such as iron and zinc, leading to further cellular dysfunction.

2.4.1.5 Disruption of Calcium Signaling

Calcium ions play a critical role in many cellular processes, and lead can interfere with calcium signaling by disrupting the function of calcium channels. This can lead to a range of cellular dysfunctions, including impaired cell signaling and mitochondrial dysfunction (Yang et al. 2020). Calcium ions are involved in the regulation of gene expression, and disruption of calcium channels can alter the expression of genes within cells. This can lead to abnormal cellular function and contribute to the development of various diseases. They play a key role in regulating mitochondrial function, and disruption of calcium channels can lead to impaired mitochondrial function. This can lead to cellular dysfunction and contribute to the development of various diseases. Disruption of calcium channels can lead to excessive accumulation of calcium ions within cells, which can trigger cell death. This can contribute to the development of various diseases, including neurodegenerative diseases (Lee and Freeman 2014) and cardiovascular disease.

2.4.1.6 Altering Gene Expression

Gene expression refers to the process by which genes are transcribed into RNA and translated into proteins, which play a critical role in various cellular processes (Yang et al. 2018). Lead can alter gene expression within cells, which can affect the synthesis of proteins and other cellular components, leading to a range of health effects. Lead exposure can alter gene expression in various ways, leading to a range of adverse health effects. For example, perinatal exposure to lead (Pb) promotes Tau phosphorylation in the rat brain in a GSK-3β and CDK5 dependent manner (Gąssowska et al. 2016). Here are some ways in which lead can alter gene expression:

Epigenetic modifications: Lead exposure can cause epigenetic modifications, which refer to changes in gene expression that are not caused by changes in the DNA sequence itself. These modifications can include DNA methylation and histone modifications and can lead to changes in gene expression that can persist over time. For example, combined exposure of lead and cadmium leads to the aggravated neurotoxicity through regulating the expression of histone deacetylase (Zhou et al. 2020).

Alteration of transcription factors: Lead exposure can alter the activity of transcription factors, which are proteins that regulate gene expression by binding to specific DNA sequences. This can lead to changes in gene expression that can contribute to the development of various diseases.

DNA damage: Lead exposure can cause DNA damage, which can lead to changes in gene expression. This can occur through direct interaction with DNA or indirectly through the generation of reactive oxygen species (ROS), which can cause oxidative damage to DNA.

2.4.1.7 Disruption of Signaling Pathways

Cell signaling refers to the process by which cells communicate with each other to regulate various cellular processes, such as growth, differentiation, and apoptosis. Here are some ways in which lead can disrupt cell signaling: Lead exposure can disrupt signaling pathways within cells, which can lead to changes in gene expression. For example, lead can interfere with calcium signaling, which plays a critical role in regulating gene expression. Lead exposure can disrupt cell signaling in various ways, leading to a range of adverse health effects.

Disruption of receptor-ligand interactions: Lead exposure can disrupt the interaction between receptors on the cell surface and their ligands, which are molecules that bind to the receptors and trigger signaling pathways. This can interfere with normal cell signaling and contribute to the development of various diseases.

Interference with intracellular signaling pathways : Lead exposure can interfere with intracellular signaling pathways, which are triggered by receptor-ligand interactions and relay information within the cell. This can disrupt normal cellular function and lead to cellular dysfunctions.

Alteration of second messenger signaling: Second messengers are molecules that relay signals from the cell surface to the interior of the cell, where they trigger signaling pathways. Lead exposure can interfere with the production or function of second messengers, which can disrupt normal cell signaling.

Disruption of calcium signaling: Calcium ions play a critical role in many signaling pathways within cells, and lead exposure can disrupt calcium signaling. This can interfere with normal cellular function and contribute to the development of various diseases.

2.5 Lead Toxicity and Blood Cells

Lead exposure may cause vascular dysfunction in the brain (Olung et al. 2021). Lead toxicity can affect various types of blood cells, including red blood cells, white blood cells, and platelets, which can contribute to a range of adverse health effects. Lead exposure can inhibit the synthesis of heme, a component of hemoglobin, which can lead to decreased production of red blood cells and anemia. Lead can also cause the formation of abnormal hemoglobin molecules, which are less efficient at carrying oxygen than normal hemoglobin. This can further contribute to decreased oxygen delivery to the body's tissues and the development of anemia. Lead exposure can suppress the production and function of white blood cells, which play a critical role in the immune system's defense against infections and diseases. This can lead to an increased risk of infections and impaired immune function. It can reduce the number and function of platelets, which are responsible for blood clotting, which can lead to an increased risk of bleeding and bruising. In addition to these effects on blood cells, lead toxicity can also lead to other adverse health effects, including neurological and developmental effects, reproductive and fertility problems, and cardiovascular disease.

2.6 Lead Toxicity and Hemoglobin

Lead toxicity can affect hemoglobin, the protein in red blood cells that carries oxygen from the lungs to the body's tissues. Lead exposure can lead to anemia, a condition characterized by a decrease in the number of red blood cells or a decrease in the amount of hemoglobin in the blood. One way that lead exposure can affect hemoglobin is by inhibiting the activity of the enzyme delta-aminolevulinic acid dehydratase (ALAD), which is required for the synthesis of heme, a component of hemoglobin. Without heme, the production of hemoglobin is impaired, leading to decreased oxygen-carrying capacity in the blood and anemia. Lead exposure can also lead to the formation of abnormal hemoglobin molecules, which are less efficient at carrying oxygen than normal hemoglobin. This can further contribute to decreased oxygen delivery to the body's tissues and the development of anemia.

2.7 Oxidative Stress

One of the primary cellular effects of lead exposure is oxidative stress. Lead exposure can increase the production of reactive oxygen species (ROS), which can damage cellular components such as lipids, proteins, and DNA. Studies have shown that lead exposure can reduce the activity of antioxidant enzymes, such as superoxide dismutase (SOD) and catalase, which help protect cells against oxidative damage (Flora 2009). In addition, lead exposure can lead to the depletion of glutathione, an important cellular antioxidant (Patrick 2006a, b, c; Gasmi et al. 2022; Paithankar et al. 2021).

2.8 Inflammation

Lead exposure has also been shown to cause inflammation at the cellular level. Studies have shown that lead exposure can increase the production of pro-inflammatory cytokines, such as tumor necrosis factor-alpha (TNF-α) and interleukin-6 (IL-6) (3). This can lead to the activation of immune cells and the recruitment of inflammatory cells to affected tissues. Chronic inflammation can contribute to a variety of diseases, including cardiovascular disease and cancer (Navas-Acien et al. 2007; Taiwo et al. 2018).

2.9 DNA Damage

Lead exposure has also been linked to DNA damage at the cellular level. Studies have shown that lead exposure can cause single-strand breaks and oxidative damage to DNA (Taiwo et al. 2018). In addition, lead exposure can interfere with DNA repair mechanisms, leading to the accumulation of DNA damage over time. This can increase the risk of mutations and cancer.

2.10 Suppression of Immune Function

Lead exposure has been shown to have significant effects on immune cells, including both innate and adaptive immune responses. Studies have shown that lead exposure can reduce the number and function of immune cells, such as neutrophils, natural killer cells, and T cells (Dietert et al. 2004; Wang et al. 2015). Lead exposure has also been shown to decrease the production of cytokines, such as interleukin-2 (IL-2), interferon-gamma (IFN-γ), and tumor necrosis factor-alpha (TNF-α), which are important for immune function (Koller and Exon 2001; McElvania Tekippe et al. 2018).

Here are some of the specific effects:

(a) **Decreased number and function of immune cells**: Lead exposure has been shown to decrease the number and function of various immune cells, such as neutrophils, natural killer cells, and T cells (Dietert et al. 2004; Wang et al. 2015).
(b) **Suppressed cytokine production**: Lead exposure has also been shown to reduce the production of cytokines, such as interleukin-2 (IL-2), interferon-gamma (IFN-γ), and tumor necrosis factor-alpha (TNF-α), which are important for immune function (Koller and Exon 2001; McElvania Tekippe et al. 2018).
(c) **Altered immune cell signaling**: Lead exposure can interfere with the signaling pathways involved in the activation and differentiation of immune cells (Ahamed and Siddiqui 2007; Patrick 2006a, b, c). For example, lead exposure can inhibit the activity of the transcription factor nuclear factor-kappa B (NF-κB), which is involved in the regulation of cytokine production and immune cell activation (Luster et al. 1998; Almutairi et al. 2022).
(d) **Disrupted immune cell communication**: Lead exposure can affect the expression and function of cell surface receptors and signaling molecules, such as toll-like receptors (TLRs), which are involved in the recognition of pathogens and the activation of immune cells (Heo et al. 1996; Rosati et al. 2021). Lead exposure can also affect the production and secretion of signaling molecules, such as chemokines and cytokines, which can disrupt the communication between immune cells.

2.11 Lead Effect: Neurological Perspective

The nervous system is particularly vulnerable to lead toxicity, as it can cross the blood–brain barrier and accumulate in the brain, where it interferes with neurotransmitter signaling, neuronal development, and synaptic function. Neurological perspective of lead effects include its impact on cognitive and behavioral functions also.

2.11.1 Neurodevelopmental Effects in Children

Public health officials are concerned about lead exposure, particularly in young children who are more at risk due to greater hand-to-mouth activity and who only absorb about half of an oral dosage of water-soluble lead. The adverse effects of organic lead are far greater than that of inorganic lead since it is lipid soluble and impacts the cell rapidly. According to meta-analyses, children's IQ scores drop by 2–3 points for every 10 μg/dl increase in blood lead levels, and there is no threshold for lead's negative effects on IQ. A recent study that focuses on the negative impact of lead on the executive and attention domains of neurobehavioral function has found findings that are similar to these. Complex mechanisms underlie lead's ability to cause neurotoxicity (Rocha and Trujillo 2019).

2.11.2 Behavioral and Cognitive Effects in Children and Adults

Lead exposure in early life has long been associated with aggressive, disruptive, and erratic conduct that can lead to scholastic failure and expulsion from school (Byers and Lord 1943).Children who are exposed to lead may experience cognitive and behavioral problems, including hyperactivity, as well as problems with fine motor function, hand–eye coordination, and reaction speed. They may also perform less well on IQ tests.

Lead levels in children's dentin have been linked to unhelpful classroom behavior (Needleman et al. 1979). Boys aged 7–11 exhibit self-reported correlations between aggression, attentiveness, and delinquency and K-shell X-ray fluorescence (KXRF) measurements of lead in the tibia in addition to teacher and parent reports. A cross-sectional study of 15–24 year olds found that those with blood lead levels between 1.5 and 10 μg/dL were over 8 times more likely to meet the DSM-IV criteria for conduct disorder than those with levels in the lowest detectable range of less than 0.7 μg/dL. (Braun et al. 2008). Children with attention deficit hyperactivity disorder (ADHD) and those exposed to lead have behavioral similarities that are noteworthy (Nigg et al. 2008; Rice 2000). In discrimination reversal measures such the Wisconsin Card Sorting Test, spatial delayed alternation, go-no-go task, distractibility task, and serial reaction tasks, children with ADHD and those exposed to lead show severe impairments (Winneke 2011). Low scores on a variety of achievement tests, impulsivity, deficits in verbal processing, non-verbal thinking, reading, and arithmetic were found to be positively correlated with blood lead levels below 5 μg/dL found in children (Canfield et al. 2003). While comparable low amounts of other hazardous heavy metals, such as mercury and aluminum, are not associated with ADHD-like effects, these effects are visible when blood lead levels are below 10 μg/dL (Ha et al. 2009). Blood lead levels below 5 μg/dL are linked to mixed hyperactive-inattentive ADHD symptoms, according to the DSM-IV. When compared to kids whose lead levels are undetectable, children with lead levels below 5 μg/dL have a more than two times higher chance of being diagnosed with ADHD (Froehlich et al. 2009).

The cumulative nature of lead poisoning in adults showed detrimental effects later in life owing to the leaching of lead from bones overtime. Cognitive diseases such as Alzheimer's were found to be positively correlated with lead exposure early in life. Research support the idea that early lead exposure has latent cognitive effects that manifest later in life as Alzheimer's disease (Shih et al. 2007). It has been reported that older persons with blood lead levels of 3.46 μg/dL had tibia lead levels averaging 18.7 μg/g, which was a substantially higher cumulative lead level.

It is important that declines in a variety of cognitive abilities, including as language, processing speed, eye-hand coordination, executive functioning, verbal memory and learning, visual memory, and visuoconstruction, were strongly connected with tibia lead levels but not blood lead levels. Despite the fact that blood lead levels were low, steady state and peak blood lead levels were likely high in order to induce the elevated amounts of lead in bone yet were not recorded.

Adults exposed to increased environmental levels of lead were found to possess lower cognitive test performances like verbal and visual memory, visuospatial ability, attention, and executive functioning (Shih et al. 2007).

2.11.3 Neuropsychiatric Disorders Associated with Lead Exposure

Various studies have associated lead exposure with multiple psychiatric illnesses like schizophrenia, major depressive disorder, mood, anxiety, and general distress and cognitive development, which is adversely correlated with BLLs (Bouchard et al. 2009).People with blood lead levels > 2.1 g/dL had a 2.3-fold increased chance of meeting DSM-IV criteria for major depressive disorder and a 4.9-fold increased risk of panic disorder compared to people with blood lead levels < 0.7 g/dL. BLLs may be low or zero if assessed over the course of 30–40 days after exposure or absorption because of dispersion to multiple organs, even though bone lead levels may be higher.

2.11.3.1 Cognitive Effects

Lead exposure can cause cognitive deficits, particularly in children, who are more susceptible than adults. The cognitive effects of lead exposure include decreased IQ scores, impaired attention, memory, and learning abilities. Children with high lead levels may also exhibit behavior problems, including hyperactivity and aggression. These cognitive deficits are thought to result from lead-induced alterations in the developing brain, particularly in the prefrontal cortex, which is responsible for executive functions, attention, and decision-making (Lanphear et al. 2005; Finkelstein et al. 1998).

2.11.3.2 Behavioral Effects

Lead exposure has also been associated with behavioral abnormalities, including aggression, delinquency, and attention deficit hyperactivity disorder (ADHD). The behavioral effects of lead exposure are thought to result from the disruption of the dopaminergic system, which plays a critical role in reward, motivation, and mood regulation. Lead-induced alterations in the dopaminergic system may contribute to the development of behavioral problems by altering the balance between reward and punishment signals, leading to impulsive and aggressive behavior (Needleman 2004).

2.11.4 Neurological Mechanisms

Lead exerts its neurotoxic effects by interfering with multiple mechanisms in the nervous system, including neurotransmitter signaling, neuronal development, and synaptic function (Green and Planchart 2018). Lead can bind to and inhibit calcium channels, which are critical for synaptic transmission, leading to impaired neurotransmitter release and altered synaptic plasticity (Maeda et al. 2022). Lead can also disrupt the development of neurons, leading to abnormal neuronal migration and dendritic growth, which can affect the formation of synapses and neuronal circuits. Additionally, lead can generate reactive oxygen species (ROS), which can lead to oxidative stress and damage to cellular components. Oxidative stress can also contribute to the disruption of the blood–brain barrier, leading to the infiltration of immune cells and the production of pro-inflammatory cytokines, which can further exacerbate the neurotoxic effects of lead.

2.11.4.1 Neurotoxic Effects

The primary mechanisms by which lead causes neurological damage is through its ability to disrupt the function of neurotransmitters, including dopamine, serotonin, and norepinephrine (Savolainen et al. 1998). Lead can interfere with the release, uptake, and metabolism of these neurotransmitters, leading to dysregulation of the nervous system (Guilarte and Miceli 1992; Guilarte et al. 2003).

Another mechanism by which lead causes neurological damage is through its ability to disrupt the development of the nervous system. During critical periods of brain development, lead exposure can interfere with the formation and differentiation of neurons, resulting in permanent structural and functional changes in the brain (Cory-Slechta et al. 1997). Lead can also cause oxidative stress and inflammation in the brain, which can lead to damage to neurons and glial cells. Oxidative stress occurs when there is an imbalance between the production of reactive oxygen species (ROS) and the ability of the body's antioxidant defense mechanisms to neutralize them. Lead exposure has been shown to increase ROS production and decrease antioxidant capacity in the brain, leading to oxidative damage (Flora 2009). In addition to these mechanisms, lead can also disrupt the blood–brain barrier, which is a protective barrier that prevents harmful substances from entering the brain. Lead exposure can weaken the blood–brain barrier, allowing lead and other toxins to enter the brain and cause damage (Guilarte and McGlothan 1998).

2.12 Interventions to Mitigate Lead Toxicity

Several interventions can be implemented to mitigate the adverse effects of lead toxicity. Primary prevention strategies include reducing or eliminating exposure to lead by reducing the use of lead-containing products, implementing environmental regulations, and providing education and awareness programs. Secondary prevention strategies involve identifying and treating individuals with elevated blood lead levels through chelation therapy or other medical interventions. Additionally, various antioxidants and neuroprotective agents have been investigated for their potential to mitigate the effects of lead toxicity on the cellular and neurological systems.

2.13 Conclusion

Lead toxicity can cause a range of neurological disorders that can have lifelong consequences. These disorders can affect cognitive and behavioral functioning, as well as cause more serious conditions such as seizures and encephalopathy. It is crucial that steps are taken to reduce exposure to lead, particularly in young children who are most vulnerable to its effects. This can include measures such as reducing lead in water and soil, removing lead-based paint from homes, and increasing public awareness of the dangers of lead exposure. By taking these steps, we can work to prevent the neurological disorders caused by lead toxicity and promote healthier outcomes for individuals and communities.

References

Ahamed M, Siddiqui MK (2007) Environmental lead toxicity and nutritional factors. Clin Nutr 26(4):400–408

Almutairi MM, Nadeem A, Ansari MA, Bakheet SA, Attia SM, Albekairi TH, Ahmad SF (2022) Lead (Pb) exposure exacerbates behavioral and immune abnormalities by upregulating Th17 and NF-κB-related signaling in BTBR T+ Itpr3tf/J autistic mouse model. Neurotoxicology 91:340–348

Atchison WD (1988) Effects of neurotoxicants on synaptic transmission: lessons learned from electrophysiological studies. Neurotoxicol Teratol 10(5):393–416

Bellinger DC (2004) Lead. Pediatrics 113(4 Suppl):1016–1022

Bjorklund G, Stejskal V, Urbina MA, Dadar M, Chirumbolo S, Mutter J (2018) Metals and Parkinson's disease: mechanisms and biochemical processes. Curr Med Chem 25(19):2198–2214

Bouchard MF, Bellinger DC, Weuve J, Matthews-Bellinger J, Gilman SE, Wright RO, Weisskopf MG (2009) Blood lead levels and major depressive disorder, panic disorder, and generalized anxiety disorder in US young adults. Arch Gen Psych 66(12):1313–1319

Braun JM, Froehlich TE, Daniels JL, Dietrich KN, Hornung R, Auinger P, Lanphear BP (2008) Association of environmental toxicants and conduct disorder in US children: NHANES 2001–2004. Environ Health Perspect 116(7):956–962

Bruno CRN, Risso WE, dos Reis Martinez CB (2016) Lead accumulation and metallothionein content in female rats of different ages and generations after daily intake of Pb-contaminated food. Environ Toxicol Pharmacol 48:272–277

Byers RK, Lord EE (1943) Late effects of lead poisoning on mental development. Am J Dis Child 66(5):471–494

Canfield RL, Henderson CR, Cory-Slechta DA, Cox C, Jusko TA, Lanphear BP et al (2003) Intellectual impairment in children with blood lead concentrations below 10 microg per deciliter. N Engl J Med 348:1517–1526

Chasapis CT (2018) Shared gene-network signatures between the human heavy metal proteome and neurological disorders and cancer types. Metallomics 10(11):1678–1686

Cory-Slechta DA, Brockel BJ, O'Mara DJ (1997) Lead exposure and the central nervous system: neural mechanisms of lead-induced neurotoxicity. Curr Opin Neurol 10(3):193–198

Dietert RR, Lee JE, Hussain I, Piepenbrink MS (2004) Developmental immunotoxicology of lead. Toxicol Appl Pharmacol 198(2):86–94

Finkelstein Y, Markowitz ME, Rosen JF (1998) Low-level lead-induced neurotoxicity in children: an update on central nervous system effects. Brain Res Rev 27(2):168–176

Flora SJ (2009) Structural, chemical and biological aspects of antioxidants for strategies against metal and metalloid exposure. Oxid Med Cell Longev 2(4):191–206

Gasmi A, Noor S, Piscopo S, Menzel A (2022) Toxic metal–mediated neurodegradation: a focus on glutathione and GST gene variants. Arch Razi Instit 77(2):525–536

Gąssowska M, Baranowska-Bosiacka I, Moczydłowska J, Tarnowski M, Pilutin A, Gutowska I, Adamczyk A (2016) Perinatal exposure to lead (Pb) promotes Tau phosphorylation in the rat brain in a GSK-3β and CDK5 dependent manner: relevance to neurological disorders. Toxicology 347:17–28

GoERING PL (1993) Lead-protein interactions as a basis for lead toxicity. Neurotoxicology 14(2):45–60

Green AJ, Planchart A (2018) The neurological toxicity of heavy metals: a fish perspective. Compar Biochem Physiol C Toxicol Pharmacol 208:12–19

Guilarte TR, McGlothan JL (1998) Selective decrease in NR1 subunit splice variant mRNA in the hippocampus of Pb^{2+}-exposed rats: implications for synaptic targeting and cell surface expression of NMDAR complexes. Brain Res Mol Brain Res 56(1–2):151–157

Guilarte TR, Miceli RC (1992) Age-dependent effects of lead on the dopamine system and behavior in the rat. Brain Res 586(2):224–232

Guilarte TR, Toscano CD, McGlothan JL, Weaver SA (2003) Environmental enrichment reverses cognitive and molecular deficits induced by developmental lead exposure. Ann Neurol 53(1):50–56

Ha M, Kwon HJ, Lim MH, Jee YK, Hong YC, Leem JH, Jo SJ (2009) Low blood levels of lead and mercury and symptoms of attention deficit hyperactivity in children: a report of the children's health and environment research (CHEER). Neurotoxicology 30(1):31–36

Heo Y, Parsons PJ, Lawrence DA (1996) Lead differentially modifies cytokine production in vitro and in vivo. Toxicol Appl Pharmacol 137(1):73–81

Jomova K, Valko M (2011) Advances in metal-induced oxidative stress and human disease. Toxicology 283(2–3):65–87

Koller LD, Exon JH (2001) The immunotoxicity of environmental lead. Environ Health Perspect 109(Suppl 1):21–45

Lanphear BP, Hornung R, Khoury J et al (2005) Low-level environmental lead exposure and children's intellectual function: an international pooled analysis. Environ Health Perspect 113(7):894–899

Lee J, Freeman JL (2014) Zebrafish as a model for investigating developmental lead (Pb) neurotoxicity as a risk factor in adult neurodegenerative disease: a mini-review. Neurotoxicology 43:57–64

Lee J, Freeman JL (2020) Exposure to the heavy-metal lead induces DNA copy number alterations in zebrafish cells. Chem Res Toxicol 33(8):2047–2053

Lepper TW, Oliveira E, Koch GDW, Berlese DB, Feksa LR (2010) Lead inhibits in vitro creatine kinase and pyruvate kinase activity in brain cortex of rats. Toxicol Vitro 24(3):1045–1051
Lidsky TI, Schneider JS (2003) Lead neurotoxicity in children: basic mechanisms and clinical correlates. Brain 126(1):5–19
Liu J et al (2013) Impact of low blood lead concentrations on IQ and school performance in Chinese children. PLoS ONE 8(5):e65230
Luster MI, Simeonova PP, Gallucci RM et al (1998) Role of inflammation in chemical-induced hepatotoxicity. Toxicol Appl Pharmacol 153(2):136
Maeda N, Shimizu S, Takahashi Y et al (2022) Oral exposure to lead acetate for 28 days reduces the number of neural progenitor cells but increases the number and synaptic plasticity of newborn granule cells in adult hippocampal neurogenesis of young-adult rats. Neurotox Res 40:2203–2220
Mason LH, Harp JP, Han DY (2014) Pb neurotoxicity: neuropsychological effects of lead toxicity. BioMed Res Int 2014:1–8
McElvania Tekippe E, Allen IC, Hulseberg PD et al (2018) Granulocyte/macrophage colony-stimulating factor (CSF) and CSF-1 receptor pathways cooperate in the differentiation of microglia into a unique intermediate stage. J Neuroinflamm 15(1):268
Navas-Acien A, Guallar E, Silbergeld EK, Rothenberg SJ (2007) Lead exposure and cardiovascular disease: a systematic review. Environ Health Perspect 115(3):472–482
Needleman HL (2004) Lead poisoning. Annu Rev Med 55:209–222
Needleman HL, Gunnoe C, Leviton A, Reed R, Peresie H, Maher C, Barrett P (1979) Deficits in psychologic and classroom performance of children with elevated dentine lead levels. New Engl J Med 300(13):689–695
Nemsadze K, Sanikidze T, Ratiani L, Gabunia L, Sharashenidze T (2009) Mechanisms of lead-induced poisoning. Georg Med News 172–173:92–96
Nigg JT, Knottnerus GM, Martel MM, Nikolas M, Cavanagh K, Karmaus W, Rappley MD (2008) Low blood lead levels associated with clinically diagnosed attention-deficit/hyperactivity disorder and mediated by weak cognitive control. Biol Psych 63(3):325–331
Olung NF, Aluko OM, Jeje SO, Adeagbo AS, Ijomone OM (2021) Vascular dysfunction in the brain; implications for heavy metal exposures. Curr Hypertens Rev 17(1):5–13
Paithankar JG, Saini S, Dwivedi S, Sharma A, Chowdhuri DK (2021) Heavy metal associated health hazards: an interplay of oxidative stress and signal transduction. Chemosphere 262:128350
Patrick L (2006a) Lead toxicity, a review of the literature. Part 1: exposure, evaluation, and treatment. Altern Med Rev. 11(1):2–22
Patrick L (2006b) Lead toxicity, a review of the literature Part 2: the clinical implications. Altern Med Rev 11(2):114–127
Patrick L (2006c) Lead toxicity part II: the role of free radical damage and the use of antioxidants in the pathology and treatment of lead toxicity. Altern Med Rev 11(2):1287
Raj K, Kaur P, Gupta GD, Singh S (2021) Metals associated neurodegeneration in Parkinson's disease: insight to physiological, pathological mechanisms and management. Neurosci Lett 753:135873
Rice DC (2000) Parallels between attention deficit hyperactivity disorder and behavioral deficits produced by neurotoxic exposure in monkeys. Environ Health Perspect 108(suppl 3):405–408
Rocha A, Trujillo KA (2019) Neurotoxicity of low-level lead exposure: History, mechanisms of action, and behavioral effects in humans and preclinical models. Neurotoxicology 73:58–80
Rosati E, Sabatini R, Rampino T et al (2021) Involvement of TLR4 and PPAR α receptors in hippocampal inflammation and oxidative stress following exposure to a high-fat diet and air pollution. Environ Pollut 268(Pt A):115918
Sanders T, Liu Y, Buchner V, Tchounwou PB (2009) Neurotoxic effects and biomarkers of lead exposure: a review. Rev Environ Health 24(1):15–46
Savolainen KM, Loikkanen J, Eerikäinen S, Naarala J (1998) Interactions of excitatory neurotransmitters and xenobiotics in excitotoxicity and oxidative stress: glutamate and lead. Toxicol Lett 102:363–367

Shih RA, Hu H, Weisskopf MG, Schwartz BS (2007) Cumulative lead dose and cognitive function in adults: a review of studies that measured both blood lead and bone lead. Environ Health Perspect 115(3):483–492

Taiwo OA, Maroufi A, Bello OS (2018) Genotoxicity and potential carcinogenicity of lead: current perspectives and future directions. Front Public Health 6:254

Wang J, Feng W, Zhao Y et al (2015) Effects of lead exposure on T lymphocyte subsets and serum cytokines in workers. Biol Trace Elem Res 168(1):66–72

Wang SS, Xu X, Lu AX, Li WH, Liu JX, Liu C, Yan CH (2022) Amelioration of cholesterol sulfate for lead-induced CTX cell apoptosis based on BDNF signaling pathway mediated cholesterol metabolism. Ecotoxicol Environ Saf 248:114307

Wani AL, Ara A, Usmani JA (2015) Lead toxicity: a review. Interdiscip Toxicol 8(2):55–64

Winneke G (2011) Developmental aspects of environmental neurotoxicology: lessons from lead and polychlorinated biphenyls. J Neurol Sci 308(1–2):9–15

Wu YC, Lin YC, Yu HL, Chen JH, Chen TF, Sun Y (2020) Increased risk of Alzheimer's disease in individuals with exposure to lead: a systematic review and meta-analysis. Sci Total Environ 707:135–213

Yang M, Li Y, Wang Y, Cheng N, Zhang Y, Pang S, Shen Q, Zhao L, Li G, Zhu G (2018) The effects of lead exposure on the expression of HMGB1 and HO-1 in rats and PC12 cells. Toxicol Lett 288:111–118

Yang L, Li X, Jiang A, Li X, Chang W, Chen J, Ye F (2020) Metformin alleviates lead-induced mitochondrial fragmentation via AMPK/Nrf2 activation in SH-SY5Y cells. Redox Biol 36:101626

Zhou R, Zhao J, Li D, Chen Y, Xiao Y, Fan A, Wang HL (2020) Combined exposure of lead and cadmium leads to the aggravated neurotoxicity through regulating the expression of histone deacetylase 2. Chemosphere 252:126589

Chapter 3
Lead Exposure and Poisoning in Livestock and Wildlife

Deep Shikha, Amitava Paul, Swarup Debroy, and Manish Kumar Verma

Abstract Lead poisoning in both humans and animals is a serious issue on a global scale. Lead toxicity in animals may serve as a sentinel to detect environmental lead contamination and related risks to human health. Lead poisoning is more common in cattle and dogs (pets) in veterinary medicine. Significant risk factors associated with the toxicities include young animals, pica, and higher accessibility to lead. Reduced accessibility, selective eating habits, or decreased sensitivity all serve to prevent lead poisoning in other animals. It has been recognised for more than a century that ammunition, such as bullets or gunshot, can poison wildlife with lead. The most well-known exposure methods for wildlife are by ingestion of embedded ammunition fragments in their diet or direct ingestion of environmental spent lead bullets. A possible health risk to humans exists if high tissue lead concentrations in wild animals and livestock penetrate the food chain. While there has been significant improvement in lowering lead exposure in humans from a variety of sources, improvements in lowering lead exposure in wild animals and livestock from a potential source have been inconsistent and occasionally ineffectual. A global campaign to minimize the use of lead content has arisen as a result of the broad harmful effects of lead on the growth, health, reproductive efficiency, and life span of all living organisms. Use of non-lead-based paint, switching to nontoxic shot in place of lead shot, and other practises could eventually minimize the effects of lead on the environment and health of livestock, wildlife, humans in the long run.

D. Shikha
Department of Veterinary Microbiology, IIVER, Bahu-Akbarpur, Rohtak, Haryana 124001, India

A. Paul (✉)
Department of Veterinary Pathology, IIVER, Bahu-Akbarpur, Rohtak, Haryana 124001, India
e-mail: amit01paul@gmail.com

S. Debroy
Department of Veterinary Anatomy, ICAR-IVRI, Izatnagar, Bareilly, Uttar Pradesh 243122, India

M. K. Verma
Department of Veterinary Pharmacology and Toxicology, IIVER, Bahu-Akbarpur, Rohtak, Haryana 124001, India

N. Kumar and A. K. Jha (eds.), *Lead Toxicity Mitigation: Sustainable Nexus Approaches*, Environmental Contamination Remediation and Management,
https://doi.org/10.1007/978-3-031-46146-0_3

Keywords Livestock · Wildlife · Gun shot · Lead · Poisoning · Food chain · Pets

3.1 Introduction

All living being is poisonous to lead, and there is no safe level of exposure. It is a broad toxin that can impair all biological systems, including the heart, bones, intestines, kidneys, reproductive, and nervous systems. It also interacts with a wide range of physiological and natural processes. Since at least 4200 B.C., lead poisoning has been a problem for humans. It has been blamed for contributing to the demise of the Roman Empire. Lead is a prevalent toxin of humans, animals, and pets despite today's improved understanding of lead toxicity (National Wildlife Health Center 2016; Paul et al. 2021). 80% of all child fatalities from unintentional poisoning are caused by the highly dangerous chemical lead and considered as a highly noxious element to both humans and animals. It will therefore almost certainly have a negative effect wherever it occurs. Due to the highly substantial health risks associated with lead toxicities, lead residues in animals must be properly controlled (Sharpe and Livesey 2006). Cattle are more likely to get lead poisoning due to their highly unselective eating habits. They will readily lick industrial grease, chew batteries, and drink crankcase oil (Merck Veterinary Manual 2013; Paul et al. 2022a). Cattle of all ages can become poisoned with lead, although calf poisoning is more common. Young calves are particularly susceptible to lead poisoning due to their natural curiosity, active calcium absorption, and the fact that milk and milk substitutes promote lead absorption. The most vulnerable livestock are cattle, with calves being the most common victims (Waldner et al. 2002; Paul et al. 2022b). However, lead poisoning can affect any domestic animal, including dogs, horses, and even birds and/or poultry. The least susceptible are pigs. Because cats and dogs have the propensity of licking their fur, are particularly vulnerable to lead exposure through soil and dust.

Since the 1800s, the effects of lead pollution on wildlife have been studied. Historically the most typical way for wildlife to become exposed was by consumption of used old lead shot pellets in wetlands where hunters had discharged large amounts of shot. Today's wildlife biologists are collaborating with hunters and fishermen to employ non-lead alternatives to reduce these unnatural deaths, as they are aware of a number of exposure paths via which wildlife can consume lead and become sick or even die.

In wild, animals are more likely to come into contact with lead through the remains of human activities than through natural sources in the wild. This exposure may result from a variety of things, such as contaminated sediment or water, old-building paint, mining tailings, industrial or residential usage of lead, such as in wheel weights. But most wildlife biologists agree that direct intake of lead from used ammunition or abandoned fishing gear is the main way that species become contaminated with lead.

Wildlife is not thought to be exposed to lead from consuming plants or animals that have absorbed lead because lead does not biomagnify. Lead contamination, on the other hand, results from direct exposure to lead, which almost often occurs when an animal ingests it, either unintentionally when eating something else that contains lead or deliberately after mistaking it for a natural material.

3.2 Lead Contamination in Livestock

Around the world, domestic animals are frequently poisoned by lead (Pb). Old lead acid batteries are the most typical cause of lead poisoning in animals. Battery cases degrade and grow brittle over time, making them easy pickings for intrepid cattle. They are freely accessible to stock, which readily licks or consumes the lead and lead salts that they contain. The main contributors of lead poisoning in cattle are old batteries. The main source of lead toxicity is ignorance on the part of humans. Lead toxicity is a persistent issue, and considerable effort is put into finding and managing affected animals to stop their products from getting into the human food chain (Sharpe and Livesey 2006; Waldner et al. 2002).

Additional sources of lead poisoning in animals include licking and ingesting lead-based paint from old paint cans, buildings, or other painted materials, intake of ashes left over from burning old painted objects, linoleum, and consuming sump oil. Silage contaminated with lead shot, automobile grease and oil filters, caulking, putty, and even access to leadlight windows have caused fatal lead poisoning in stock. Lead pieces from collars used at pipeline junctions (such as those on major water pipe lines) might provide a poisoning risk and are probably the source of fatal lead poisoning (Radostits et al. 2007).

3.2.1 Impacts of Lead on Livestock

Due to their indiscriminate eating habits, cattle are the species most susceptible to lead poisoning. They will happily eat crankcase oil, lick grease from machinery, and gnaw on batteries, peeling lead paint, ashes, and pretty much any other possible source of lead they might encounter. The reticulum (fore-stomach) of ruminant animals frequently becomes lodged with lead and other heavy materials. This presents a source from which the bodies of cattle, sheep, and goats can continue to absorb lead. Old paint pigments include finely ground, very soluble lead. In comparison to clean metallic lead, lead that has been exposed to acidic conditions in batteries and silage is quickly absorbed from an animal's gut and provides a larger danger of poisoning (Paul and Sujatha 2022; Waldner et al. 2002).

Lead poisoning affects cattle of all ages, but calves are more frequently affected. Dairy cattle have been found to have the highest incidence. According to estimates, each year 150,000 cattle worldwide are exposed to fatal levels of lead, which causes at least 20,000 immediate deaths (Radostits et al. 2007; Waldner et al. 2002).

The risk of lead poisoning may increase in dry weather. Particularly if they are low in trace components or minerals, hungry stock may acquire a perverted appetite (pica). Hungry animals are also more likely to enter "no-go" areas, such as the area around farm sheds or the farm trash dump, where there may be some leftover feed. The likelihood that livestock will discover dangers like out-dated batteries increase as pasture cover decreases.

It has been proven that horses can become poisoned by grazing on soil that has been tainted with lead shot from traps. The rate and quantity of lead intake, together with the age, the form of the lead, the condition of the animal, may all have an impact on the uptake and effect on the animal. Ewes that were grazing close to lead mining regions experienced spontaneously aborted. In comparison to non-pregnant ewes, the lethal lead dose seems to be lower in pregnant ewes. Horses and sheep have unintentionally consumed lead through drinking tainted spring or stream water or eating tainted grass in the vicinity of a lead smelter and mining region (Radostits et al. 2007; Payne and Livesey 2010; Pareja-Carrera et al. 2014).

3.2.2 Major Sources of Lead on the Farm Premises

Cattle are frequently involved in accidents during the planting and harvesting procedures when used oil and equipment battery disposal is done improperly. Other sources of lead include flora growing close to smelters or on the sides of roadways, paint, linoleum, grease, lead weights, lead shot, and polluted vegetation. Lead toxicity is commonly encountered in urban settings, and in young infants and small animals lead toxicities has been linked to renovation of historic homes painted with lead-based paint.

Due to the pleasant taste of some lead compounds, animals may find sump oil and other sources of lead to be appealing. Boredom from confinement and ravenous appetites brought on by hunger and phosphorus deficiency may be contributing factors. However, predisposing factors are not essential.

3.3 Lead Contamination in Wildlife

Although lead exposure in wildlife has been related to lead dispersion from mining, coal combustion, battery processing, fuel, and waste incineration, the primary exposure pathway is through direct ingestion of used ammunition or lost fishing gear (National Wildlife Health Center 2016). Lead ammunition is referred to be the "largest, largely unregulated source of lead intentionally discharged into the environment" by scientists (Pacelle 2017). Lead ammunition is referred to be the "largest, largely unregulated source of lead intentionally discharged into the environment" by scientists. For various shooting circumstances, shotgun rounds come in a variety of sizes. Depending on the size, a shotgun shell may contain hundreds of pellets. Lead pellets have a concentration of lead that is > 95%. Lead is gradually delivered into the circulatory system of an animal after a single shotgun pellet penetrates its gastrointestinal (GI) system (Moazeni et al. 2014). There is enough lead in a single shotgun pellet to result in organ failure and death. (Humane Society of the United States 2018; Harrison and Lightfoot 2006; Puschner and Poppenga 2009; Pain et al. 2014). Through either direct or indirect lead exposure, wildlife is more susceptible to lead poisoning.

3.3.1 Direct Exposure

When an animal consumes discarded lead ammunition, ammo fragments, broken fishing tackle, or any other lead source present in the environment, they are directly exposed to lead (Humane Society of the United States 2018; Harrison and Lightfoot 2006; Puschner and Poppenga, 2009; Pain et al. 2014). Since more than a century ago, lead poisoning of wildlife caused by ammunition (gunshot) has been recognised. Birds frequently mistake lead items for seed or grit when they are searching for food on the ground (Harrison and Lightfoot 2006). After ingesting spent lead ammunition from the environment or ammunition bits entrenched in their food, birds suffer lead poisoning. Among wildfowl and terrestrial game birds, particularly those with a muscular gizzard that eat grit to help grind their food, consumption of spent lead gunshot is common; while among Raptors and scavenger birds, consumption of ammo-fragments embedded birds and mammals killed by people, or their discarded remains are commonly seen. Lead pellets may be available for unintentional consumption by ducks in wetlands for up to 25 years or longer, depending on the environmental circumstances (Haig et al. 2014).

3.3.2 Indirect Exposure

When an animal feeds on another that has consumed lead or been shot with lead ammunition, it exposes itself to lead indirectly (Harrison and Lightfoot 2006; Puschner and Poppenga, 2009; Pain et al. 2014). Indirect lead exposure has an impact on carnivorous animals and scavenger or predatory birds like eagles, vultures, and condors (Haig et al. 2014; Pain et al. 2014). The most common known cause of death for the critically endangered California condor (*Gymnogyps californianus*) is lead poisoning via scavenging activities. Indirect exposure can also happen when lower species like earthworms or water that has been poisoned with lead are consumed (Haig et al. 2014; Puschner and Poppenga, 2009). In the environment, lead fishing sinkers and ammunition slowly decay and leach into the land and water. The American robin (*Turdus migratorius*) and American woodcock (*Scolopax minor*) are susceptible to lead poisoning through indirect exposure when eating earthworms in areas with high environmental lead levels.

3.3.3 Impacts of Lead on Wildlife

For millennia, lead (Pb) has been utilised in fishing gear and ammunition. Even though lead occurs naturally, it serves no vital biological purpose, and large concentrations like those found in ammunition and tackle attribute a number of direct and indirect risks to wildlife. Used lead ammo and tackle when consumed by wild animals can be harmful and have long lasting effects on the environment. After being biologically incorporated into plants and invertebrates in the soil, lead is then consumed by wildlife. Ingestion of used ammunition and lost fishing gear by reptiles, birds, and mammals can have a variety of harmful impacts on individuals. These effects on the individual may result from population-level effects in some species, including ducks, eagles, condors, doves, and loons.

Lead is a potentially dangerous non-essential metal that has no counterproductively advantageous impacts on living things (ATSDR 2007; EFSA 2010). Inorganic lead, once ingested by an animal, exhibits non-specific, accumulative metabolic effects that are unrelated to the source. The avian taxon is presumably the one that is most affected by lead poisoning from ammunition consumption as it has been the focus of the most investigation. However, lead's harmful effects are well documented from numerous laboratory and field studies and are roughly similar in all vertebrates. Clinical signs of poisoning in birds are typically associated with long-term exposure at a level that does not typically cause abrupt loss of biological function or death, though death may result. Lethargy, muscle atrophy and loss of fat reserves, anaemia, green diarrhoea staining the vent, wing droop, loss of coordination and balance, and other nervous indicators such leg paralysis or convulsions are some of the symptoms (Krone 2018; De Francisco et al. 2003; Pattee and Pain 2003). Acute exposure to high levels of lead, on the other hand, causes birds to die rapidly without any visible symptoms.

Birds that have consumed ammunition or ammunition fragments may have quick elimination of lead from the gastrointestinal system with minimal lead absorption, retention until complete erosion, solubilisation, and absorption, or any combination of these outcomes. The bloodstream carries absorbed lead, which is promptly deposited into soft tissues like the liver and kidneys, as well as into bone and a bird's developing feathers. In contrast to lead in soft tissues, which has a much shorter half-life (weeks to months), lead in bone is retained for a very long time and builds up over the length of an animal's lifetime. Blood lead levels (PbB) remain elevated for several weeks or months following exposure. The physiological effects of lead in birds have been extensively studied (Pain and Green 2015). There are a variety of biological and environmental factors that can increase or decrease a bird's susceptibility to lead poisoning, and lead sensitivity can vary between different species. Lead may have indirect effects in addition to its direct effects on life and welfare. These may include an increased risk of contracting infectious diseases, parasite infestations (due to lead's immunosuppressive effects), and a higher risk of dying from a variety of other causes, such as being shot and colliding with power lines (Kelly and Kelly 2005; Ecke et al. 2017), due to lead's effects on muscular strength and coordination (Bedrosian et al. 2012; Pain and Green 2015).

According to EFSA (2010), there are currently no established "no observed adverse effect levels" (NOAEL) or "predicted no effect concentrations" (PNEC) for lead in humans, and other vertebrates are expected to experience the same. As a result, using tolerable lead exposure criteria requires accepting a certain amount of avoidable harm. Lead poisoning has also been thoroughly researched in wildfowl in addition to terrestrial birds, including game and predatory species (Beintema 2001; Martinez-Haro et al. 2011; Newth et al., 2013; AEWA 2011, Butler et al. 2005; Potts 2005).

3.4 Effects of Lead on Aquatic Animals

Numerous studies have demonstrated that marine creatures, particularly invertebrates, have the capacity to absorb metals. Perhaps more than any other group, bivalve mollusks like clams, oysters, mussels, and scallops are renowned heavy metal concentrators. The related mussel *Mytilus galloprovincialis* showed similar lead absorption efficiency of about 60% in both the laboratory and the field, whether it was in the dissolved phase or linked with food. In contrast to dissolve lead, which was more likely to initially concentrate in the gill and eventually end up in the shell, lead in food initially concentrated in the digestive gland. Scientists found that oysters absorbed 277 ppm of lead over the course of 20 weeks in lab studies. Another study in oysters shown a concentration ratio, in the animals compared to the water, of 6600. The highest lead concentrations were detected in invertebrate organisms living in sediments in the Singapore River in Southeast Asia, such as polychaetes, whereas the lowest concentrations were found in fish (Fadl et al. 2021; Ali and Ahmad 2014).

Lead's toxic effects are easily seen in extremely contaminated places, such as mine drainage sites, where there are fewer remaining animals and plants than in other sections of their environment. Mining waste with lead that was dumped in streams has severely destroyed populations of fish and aquatic invertebrates. Low levels of lead act gradually over time with subtle detrimental effects on behaviour and/or reproduction that may result in population extinction while high levels of lead instantly kill animals. A population that depends on a certain prey species for sustenance may also perish as a result of poisoning of the prey species (Al-Balawi et al. 2013; Alkahemal-Balawi et al. 2011).

Different aquatic invertebrate species—those without backbones like worms and insects—are more or less sensitive to the effects of lead. The kind of lead (i.e.

whether it is an organic compound or in ionic form), the amount of it dissolved in water, or the water hardness or other mineral composition, all affect the uptake and toxicity of lead. Lead levels between 1 and 500 ppb caused the death of aquatic snail embryos. 41% of eggs hatched at 100 ppb, although it took 37 days longer than expected in development. All of the hatched snail had died after an extra 15 days. Typically, mature animals are far more forgiving than young ones. It seems that some larger, more complex invertebrates, such adult mollusks (clams, oysters, snails, and squid), crabs, may be able to adapt to habitats with high lead concentrations (Alkahemal-Balawi et al. 2011; Authman et al. 2015).

The metallic or inorganic forms of lead do not seem to biomagnify along food chains, unlike DDT, PCBs, methyl mercury, and organic forms of lead. Inorganic lead does not seem to accumulate at increasing amounts in higher trophic levels, according to research from Oklahoma State University and other institutions. The top layer of silt in the pond contained 529 parts per million (ppm) of lead, which is roughly 40,000 times more lead than the water, which was 13 ppb. Plankton had an average lead content of 281 parts per million (ppm) (dry weight); bottom-dwelling invertebrates had a level of 37 ppm; and mosquitoes and fish had levels of 11.5 ppm. Because of this, lead levels did not increase further up the food chain, even though lead was likely eaten by plants and animals, including plankton, which are then fed by mosquito fish. Due to their quick absorption, organic lead compounds (such as tetraethyl- or tetramethyl-lead) are very hazardous to aquatic life. In comparison to inorganic forms of lead, organic forms of lead, such methyl-mercury may be more likely to biomagnify in the food web (Alkahemal-Balawi et al. 2011).

Fish have responded to lead concentrations as low as 7 ppb in a variety of sub-lethal ways. Gross pollution may result in fish fatalities, whereas sub-lethal toxicity may inflict subtle harm to fish populations over longer times and across larger areas of the ecosystem as lead concentrations spread and were diluted out. Like invertebrates, fish vary greatly in their sensitivity to lead concentrations depending on their species, size, and life stage. Lead bioavailability which is significantly influenced by water quality also plays a substantial role (Ali and Ahmad 2014).

Even at lead concentrations at or below the 50 parts per billion previously thought to be safe in drinking water for humans, chronic lead exposure to fish may cause higher mortality rates, decreased hatching success, and indications of neurotoxicity as indicated by higher incidences of black tails (darkening of the caudal area) and spinal curvatures. Fish can develop some biochemical lesions that have also been seen in humans, such as anaemia, neurotoxicity, haemolytic anaemia, and red blood cell stippling. Lower stamina and a reduced ability to swim are observable behavioural changes. This may be due to less oxygen delivery and less oxygen exchange at the gill surfaces. Some species have been shown to have decreased ion exchange capacity, which is essential for osmoregulation (the maintenance of mineral and water balance) following exposure to lead. On the other side, damage to the brain structures or pathways that control movement may lead to a reduction in swimming ability. Fish exposed to lead have experienced prolonged periods of hyperactivity; this may be due to disruption of the neurotransmitter activity in the brain that underlies circadian rhythms (Alkahemal-Balawi et al. 2011; Authman et al. 2015).

However, lead exposure is specifically harmful to fish because it can lead to overproduction of mucus, which can obstruct the gill's ability to diffuse gases and absorb oxygen from the water and expel carbon dioxide. By interacting with the glycoproteins (proteins with sugar groups attached) in mucus, lead may really alter the physical properties of this substance. The protective and hydrodynamic resistance properties of the mucus may be impacted by these changes, which may have an effect on the fish's capacity to survive (Al-Balawi et al. 2013).

Lead's effects are influenced by water quality elements, such as hardness. Lead at a concentration of 0.48 ppm inhibited many species of fish from hatching their eggs in studies in water with a hardness of 25–40 mg/l (as calcium carbonate). Among the fish analysed were bluegill, white sucker, channel catfish, rainbow and lake trout, and catfish. Lead levels as low as 0.12 ppm after hatching caused death in the larvae. After being transferred to clean water, lead leaves fish far more quickly than it does humans, taking 3–4 weeks as opposed to years. But certain symptoms of lead exposure may endure even after switching to clean water. Concerningly, tests on blood samples from sports fisherman who frequently consume more locally obtained fish from the lead contaminated lakes revealed that their lead levels were more than 40% higher than those of those who did not routinely consume locally caught fish. (Alkahemal-Balawi et al. 2011).

3.5 Lethal Levels of Lead in Animals

i. Cattle: Chronic lead poisoning can result from intakes larger than 6 mg/kg body weight, and acute lead poisoning can result from intakes > 10 mg/kg BW.
ii. Sheep: Exclusively lambs often experience this condition, and symptoms of poisoning start to show at intakes higher than 4.5 mg/kg BW.
iii. Pigs, Goats, and Rabbits: Are more resilient than sheep or cows in this regard. At intakes of 60 mg/kg BW, very modest indications of poisoning start to appear. This is the same as 130 μg per dl of blood concentration.
iv. Horses: Respiratory "roaring" occurs at intakes of 6.4 mg/kg BW. Signs of anaemia occur at intakes of 7.4 mg/kg.
v. Birds and poultry can eat up to 100 mg/kg of feed without showing any signs of discomfort. Serious poisoning was caused at levels of 500 mg/kg.
vi. Dogs and cats: At intakes of 5 mg/kg BW/day, nervous symptoms of poisoning start to manifest.

3.6 Permissible Limits Across Different Media

The permitted levels of lead in ambient air are 0.05 mg/L for drinking water, 0.75 mg/m^3 for sensitive areas (bird sanctuaries), 1.0 mg/m^3 for residential areas, and 1.5 mg/m^3 for industrial zones (CPCB 1997). Discharge of industrial effluent into inland surface waterways is permitted at 0.10 mg/L. These are generally in accordance with global standards.

3.7 Animals as Environmental Indicators

Determination of lead concentration in animals offers a way to keep an eye on probable sources and foresee environmental hazards. Since over 60% of the body load of this metal is found in the feathers, using them to measure lead exposure in birds is highly effective.

Animals that have been poisoned in the wild should serve as a reminder of the need to limit lead emissions in the environment. It is simple and affordable to monitor blood-lead concentrations in dogs and cats. Without subjecting young children to testing, the findings help forecast human risk. To assess the danger of lead exposure, this monitoring approach might be employed even before a child is brought into the house. We as a society have much too frequently used resource and conducted our industrial activities in a casual manner. Lead's harmful effects on ecosystems, animals, and plants serve as a warning sign.

3.8 Clinical Manifestation of Lead Toxicity

Finding dead livestock, frequently close to a fence or some other obstacle, is frequently the first indication of lead poisoning. Where impacted animals have been seen, they exhibit symptoms of injury to the CNS. They typically cease grazing and become very unresponsive and dull. Animals with acute lead poisoning display symptoms such as colic, a stumbling gait, rolling eyes, slobbering, muscle spasms, blindness, clumsy attempts to scale obstacles, enhanced susceptibility to environmental stimuli, head pushing, and convulsions. This may be followed by death. Muscle spasms, which can affect any region of the body but may be particularly obvious around the face, ears, and eyelids, may occasionally accompany these symptoms. In some cases, tongue paralysis, circling, and "star-gazing" have also been recorded. Animals suffering from prolonged poisoning might display signs of recumbency, anaemia, anorexia, and wasting. There could be death and paralysis. Acute attacks can occasionally happen while a animal is suffering from prolonged poisoning. Horses exhibit dyspnoea. Additionally, gastrointestinal and neurological symptoms may coexist. Constipation or diarrhoea could result from the gastrointestinal consequences. For any animal exhibiting anxious symptoms, immediate veterinarian advice should be sought. Lead poisoning's symptoms might be mistaken for those of other conditions that have neurological effects, including plant poisonings, PE (polioencephalomalacia), and metabolic conditions like hypomagnesaemia or grass tetany. To stop further losses and choose the right course of action for sick animals, a precise early diagnosis is essential.

3.9 Lead Residues in Animal and Animal Products

The slaughter of lead-exposed animals for human consumption must wait until their tissues conform to food requirements. Even after a stock has recovered from lead poisoning, the liver and kidney may still contain unacceptable lead levels for several months. First, restrictions on slaughter apply to the affected mob as a whole since some exposed animals, even those that do not show signs of poisoning, may have absorbed enough lead to leave tissue residues. These restrictions are usually confirmed in a written agreement between the stockowner and their Local Land Services District Veterinarian.

Blood tests can be used to distinguish between healthy and ill animals. After testing, the majority of the mob is normally exempt from the slaughter ban. Blood samples should be collected within 42 days of the most recent lead exposure to ensure that all contaminated supplies are identified. The lab or the veterinarian collecting the sample may keep it for later examination.

3.10 Management, Prevention of Lead Contamination and Alternatives to Leads

Prevention is the best cure. A global movement to minimise the use of lead content has been launched in response to the increased awareness of the harmful effects of lead. In the farm's garbage can or in the machinery shed, outdated batteries, lead-based paint, sump oil, and similar materials can all pose a threat to health. Stocks are adept at recognising these hazards. Batteries for electric fences and other farm machinery must also be kept away from livestock. Where possible, take any outdated batteries off your property and dispose of them by taking them to a recycling facility. Clean up any materials that have spilled or damaged battery casings. Pica or aberrant feeding behaviour in animals will be reduced by good nutrition and consistent, regular feeding. Areas used for vehicle maintenance and machinery storage should be separate from areas used by livestock. It is best to keep livestock-only areas apart from those used for vehicle repair and equipment storage. If your cattle graze land that has an easement for a water pipe, an electricity line or a gas line, confirm with the relevant authorities that there is no lead danger associated with an installation. Look out for these risks when driving stock and "scouting ahead" before moving stock onto new territory.

It is critical that the risky location is well gated and managed to prevent stock access if a possible hazard, such as a tip site, cannot be eliminated. Lastly, be sure that none of the contractors working on your property leave any old batteries in paddocks. Following harvest, cases of lead toxicity increase. Make sure harvesting contractors don't leave used batteries lying around in the pasture where livestock will eventually discover them.

In recent years, there has been a lot of work put into developing, testing, and regulating alternatives to lead-based ammunition. Manufacturers have created non-toxic rifle bullets and ammo that can be used in all modern shotgun gauges without risk. There are many non-toxic alternatives for tackle and ammunition, including bismuth, steel, tungsten, tin, and bismuth.

Replacement of lead shot with nontoxic shot, which could decrease the impacts of lead shot on the health of animals, human, and the environment, is the only long-term option for significantly lowering migratory bird losses from lead poisoning. Waterfowl, doves, and other species of migratory birds have all been shown to be negatively impacted by lead shot. Lead shot also has an effect on upland species, such as ring-necked pheasants. Prior to the federal ban on lead shot for waterfowl hunting, alternatives to lead shot were not always easy to find. However, steel shot in particular is now accessible at a price that is equivalent to lead shot ammunition. An improved lead-free alternative is:

- The most popular lead substitute in ammunition (bullets and slugs) is copper, which doesn't fragment, has superior killing power (a wider wound channel), and better ballistics.
- Steel is a further enticing substitute for ammunition (shotgun pellets), but since it is lighter than lead, heavier loads, and larger propellant charges are required.
- Tungsten, tin, and steel are other alternative materials for sinkers and jigs. Though more expensive and 70% heavier than lead, tungsten has grown in favour, especially for ice fishing. Tin is flexible, less dense than lead, and simple to cast. Steel is more durable than lead and generates noise, which attracts fish.

In order to operate older and historical shotguns safely, nontoxic shot is now also available (Cabela's 2008). Despite costing more than lead, nontoxic alternatives to lead shot only add 1–2% to the average hunter's annual expenses (Scheuhammer and Norris 1995). For instance, Schulz et al. (2006) evaluated the waterfowl crippling rates before and after nontoxic-shot regulations were implemented in the US. After a 5-year phase-in period, they found that the application of non-toxic shot limitations led to lower crippling rates for ducks and geese.

3.11 Conclusions

There is considerable evidence that lead shot has harmful effects on the environment, humans, cattle, and wildlife. Ingesting lead shot has injured or killed more than 100 different bird species, including raptors, ducks, and upland birds. Lead shot has a variety of negative impacts on wildlife, including decreased survival, poor physical condition, behavioural changes, and impaired reproduction. Even after the shot has been removed, lead can still be present in game meat and expose people to it. The impacts of lead poisoning to humans, livestock and wildlife are entirely preventable by simply switching to non-toxic ammunition, fishing tackle, proper disposal of lead equipment like old used batteries, lead-based paint crankcase oil and use of lead free paints. There are many different kinds of lead-free ammunition and tackle available, most of which are composed of solid copper and serve as secure, accessible, and competitive substitutes. When using lead fishing tackle or ammunition, it's crucial to bury any carcasses or remnants that may contain lead to prevent wildlife from scavenging them. In addition, any lead that has been released into the environment (from ammunition, fishing line or an old battery) needs to be collected and disposed of correctly. Lead poses a very substantial risk to human health, hence lead residues in livestock and wild animal products must be scrupulously avoided to safeguard human health. To address the negative effects of lead on the health of livestock, wildlife, humans, and the environment, a comprehensive global approach is urgently needed.

References

AEWA (2011) Literature review: effects of the use of lead fishing weights on water birds and wetlands. Doc St C Inf 7(6):1–20

Al-Balawi HFA, Al-Akel AS, Al-Misned F, Suliman EAM, Al-Ghanim KA, Mahboob S, Ahmad Z (2013) Effects of sub-lethal exposure of lead acetate on histopathology of gills, liver, kidney and muscle and its accumulation in these organs of *Clarias gariepinus*. Braz Arch Biol Technol 56:293–302

Ali AS, Ahmad RSA (2014) Effect of different heavy metal pollution on fish. Res J Chem Environ Sci 2(1):74–79

Alkahemal-Balawi HF, Ahmad Z, Al-Akel AS, Al-Misned F, Suliman EAM, Al-Ghanim KA (2011) Toxicity bioassay of lead acetate and effects of its sub-lethal exposure on growth, haematological parameters and reproduction in *C. gariepinus*. Afr J Biotechnol 10(53):11039

ATSDR (2007) Toxicological profile for lead. Agency for Toxic Substances and Disease Registry, Atlanta. https://www.atsdr.cdc.gov/toxprofiles/tp13.pdf

Authman MM, Zaki MS, Khallaf EA, Abbas HH (2015) Use of fish as bio-indicator of the effects of heavy metals pollution. J Aquac Res Dev 6(4):1–13

Bedrosian B, Craighead D, Crandall R (2012) Lead exposure in bald eagles from big game hunting, the continental implications and successful mitigation efforts. PLoS ONE 7(12):e51978

Beintema NH (2001) Lead poisoning in water birds: international update report 2000. Wageningen

Butler DA, Sage RB, Draycott RA, Carroll JP, Potts D (2005) Lead exposure in ring-necked pheasants on shooting estates in Great Britain. Wildl Soc Bull 33(2):583–589

Cabela's (2008) Cabela's shooting and reloading (catalogue). Cabela's, Sidney

CPCB (1997) CPCB newsletters
De Francisco N, Ruiz Troya JD, Agüera EI (2003) Lead and lead toxicity in domestic and free living birds. Avian Pathol 32(1):3–13
Ecke F, Singh NJ, Arnemo JM, Bignert A, Helander B, Berglund ÅM, Borg H, Bröjer C, Holm K, Lanzone M, Miller T (2017) Sublethal lead exposure alters movement behavior in free-ranging golden eagles. Environ Sci Technol 51(10):5729–5736. https://doi.org/10.1021/acs.est.6b06024
EFSA (2010) Scientific opinion on lead in food. EFSA panel on contaminants in the food chain (CONTAM). EFSA J 8(4):1570. https://doi.org/10.2903/j.efsa.2010.1570
Fadl SE, El-Habashi N, Gad DM, Elkassas WM, Elbialy ZI, Abdelhady DH, Hegazi SM (2021) Effect of adding *Dunaliella algae* to fish diet on lead acetate toxicity and gene expression in the liver of *Nile tilapia.* Toxin Rev 40(4):1155–1171
Haig SM, D'Elia J, Eagles-Smith C, Fair JM, Gervais J, Herring G, Rivers JW, Schulz JH (2014) The persistent problem of lead poisoning in birds from ammunition and fishing tackle. Condor Ornithol Appl 116(3):408–428
Harrison GJ, Lightfoot TL (2006) Clinical avian medicine. Spix Publishing Inc., Palm Beach
Humane Society of the United States (2018) Lead ammunition: toxic to wildlife, people and the environment. http://www.opb.org/news/article/how-to-save-a-swanfrom-lead-poisoning/. Accessed 17 May 2018
Kelly A, Kelly S (2005) Are mute swans with elevated blood lead levels more likely to collide with overhead power lines? Water Birds 28:331–334
Krone O (2018) Lead poisoning in birds of prey. In: Birds of prey: biology and conservation in the XXI century, pp 251–272
Martinez-Haro M, Taggart MA, Martín-Doimeadiós RR, Green AJ, Mateo R (2011) Identifying sources of Pb exposure in water birds and effects on porphyrin metabolism using non-invasive fecal sampling. Environ Sci Technol 45(14):6153–6159
Merck Veterinary Manual (2013) Overview of lead poisoning. http://www.merckvetmanual.com/mvm/toxicology/lead_poisoning/overview_of_lead_poisoning.html
Moazeni M, Alibeigi FM, Sayadi M, Mofrad EP, Kheiri S, Darvishi M (2014) The serum lead level in patients with retained lead pellets. Arch Trauma Res 3(2):59
National Wildlife Health Center (2016) Concerns rise over known and potential impacts of lead on wildlife. U.S. Geological Survey. https://www.nwhc.usgs.gov/disease_information/lead_poisoning/. Accessed 17 May 2018
Newth JL, Cromie RL, Brown MJ, Delahay RJ, Meharg AA, Deacon C, Norton GJ, O'brien MF, Pain DJ (2013) Poisoning from lead gunshot: still a threat to wild waterbirds in Britain. Eur J Wildl Res 59:195–204
Pacelle W (2017) U.S. Fish and Wildlife service says it's time to ban lead ammo, fishing tackle on federal lands. In: Humane society of the United States. blog.humanesociety.org/wayne/2017/01/u-s-fishwildlife-service-says-time-ban-lead ammo-fishing-tackle-federal-lands. Accessed 17 May 2018
Pain DJ, Green RE (2015) An evaluation of the risks to wildlife in the UK from lead derived from ammunition. In: A report to the UK Government, p 120. http://www.leadammunitiongroup.org.uk/wp-content/uploads/2015/06/LAG-Report-June2015-Appendices-without-Appendix-6.pdf
Pain DJ, Cromie R, Green RE (2014) Poisoning of birds and other wildlife from ammunition-derived lead in the UK. In: Oxford Lead Symposium, p 58
Pareja-Carrera J, Mateo R, Rodríguez-Estival J (2014) Lead(Pb) in sheep exposed to mining pollution: implications for animal and human health. Ecotoxicol Environ Saf 108:210–216
Pattee O, Pain D (2003) Lead in the environment. In: Hoffman DJ, Rattner BA, Burton GA, Cairns J (eds) Handbook of ecotoxicology, 2nd edn. CRC Press, Boca Raton, pp 373–408
Paul A, Sujatha K (2022) Concurrent effect of *Linum usitatissimum* and *Emblica officinalis* on Lead induced oxidative stress and histomorphological changes in uterus of female wistar rats. Exp Anim Med Res 12(2):264–272

Paul A, Sujatha K, Srilatha CH, Kumar NV (2021) Amelioration of Lead induced toxicity on rat ovary with *L. usitatissimum* (flaxseed) and *E. officinalis* (amla). Pharma Innov J 10(8):1124–1130

Paul A, Sujatha K, Srilatha CH, Kumar NV (2022a) Effect of Lead toxicity on liver and kidney in Wistar rats and its amelioration with *L. usitatissimum* (flaxseed) and *E. officinalis* (amla). Indian J Vet Pathol 46(1):49–52

Paul A, Sujatha K, Srilatha CH, Kumar NV (2022b) Haematobiochemical study on effect of *L. usitatissimum* (flaxseed) and *E. officinalis* (amla) against lead toxicity in Wistar rats. Indian J Vet Pathol 46(2):141–149. https://doi.org/10.5958/0973-970X.2022.00023.2

Payne J, Livesey C (2010) Lead poisoning in cattle and sheep. Practice 32:64–69

Potts GR (2005) Incidence of ingested lead gunshot in wild grey partridges (*Perdix perdix*) from the UK. Eur J Wildl Res 51:31–34

Puschner B, Poppenga RH (2009) Lead and zinc intoxication in companion birds. Compendium 31(1):E1-12

Radostits OM, Gay CC, Hinchcliff KW, Constable PD (2007) Veterinary medicine: a textbook of the diseases of cattle, sheep, goats, pigs and horses, 10th edn. WB Saunders, Philadelphia, pp 1799–1808

Scheuhammer AM, Norris SL (1995) A review of the environmental impacts of lead shotshell ammunition and lead fishing weights in Canada. In: Occasional paper number 88, Canadian wildlife service. National Wildlife Research Centre, Hull, p 56

Schulz JH, Padding PI, Millspaugh JJ (2006) Will mourning dove crippling rates increase with nontoxic-shot regulations? Wildl Soc Bull 34(3):861–864

Sharpe RT, Livesey CT (2006) Lead poisoning in cattle and its implications for food safety. Vet Rec 159:71–74

Waldner C, Checkley S, Blakley B, Pollock C, Mitchell B (2002) Managing lead exposure and toxicity in cow-calf herds to minimize the potential for food residues. J Vet Diagn Invest 14:481–486

Chapter 4
A Systematic Review of Lead Exposure on Mental Health

Jasbir Arora, Anjali Singal, Justin Jacob, Shallu Garg, and Richa Aeri

Abstract Lead is the most potent and persistent toxic metal found naturally in the earth's crust. Humans are exposed to lead particles through inhalation, ingestion, or skin contact, through several sources that influence food, drinking water, soil, and air. If this heavy metal enters the body, interferes with the organ systems. US Centers for Disease Control and Prevention (CDC) recently declared that no level of lead can be considered "safe". The outcomes associated with lead exposure have become more apparent and are an ever-increasing concern across the globe, as a plethora of disorders are caused by its contact. Not only adults but young children and fetuses are also vulnerable to its toxic effects leading to neurodevelopment and kidney-related disorders. This chapter provides insight into the clinical manifestation of lead exposure especially the impact on the mental health (neurodevelopment) of the fetus, children, and adults, the mechanism of lead-induced neurotoxicity, available biomarkers, challenges, mitigation, preventive measures, and therapy for lead exposure. Future recommendations to undertake necessary steps to protect the population from environmental toxicity have also conversed.

Keywords Biomarkers · Fetus · Mental health · Lead toxicity · Neurotoxicity

J. Arora (✉)
Department of Anthropology, Panjab University, Chandigarh, India
e-mail: arora.jasbir06@gmail.com

A. Singal
All India Institute of Medical Sciences, Bathinda, India

J. Jacob
International Centre for Genetic Engineering and Biotechnology, Trieste, Italy

S. Garg
Bangalore Institute of Dental Sciences, Bangalore, India

R. Aeri
Survival for Women and Children (SWACH NGO), Panchkula, Haryana, India

N. Kumar and A. K. Jha (eds.), *Lead Toxicity Mitigation: Sustainable Nexus Approaches*, Environmental Contamination Remediation and Management,
https://doi.org/10.1007/978-3-031-46146-0_4

4.1 Introduction

Lead (Pb) is the most potent and persistent toxic metal naturally occurring element in the earth's crust. Lead in the environment in present its inorganic form. A level of 0.016 mg/dl of blood lead levels has been observed in pre-industrial humans, indicating the minimum contribution of natural sources for adding lead to the environment (Ezzati et al. 2004). Most lead exposure occurs through anthropogenic activities. Known sources of lead exposure are cigarette smoke, combustion exhaust, sewage sludge, fertilizers, mining, Pb-bearing sulfide deposits, Pb additives in petrol, Pb water pipes, Pb in paints, toys, etc. (Meyer et al. 2008). These sources influence food, drinking water, soil, and air. Drinking water is the major pathway of accumulation of Pb in humans which arises mostly due to the use of lead piping, which is still being used in some places (Neeti and Prakash 2013). Other sources of the increased level of Pb in water are landfills, electroplating, Au–Ag-Pb–Zn mining, etc. (Obeng-Gyasi 2019b). Lead exposure has also been associated positively with socio-economic status and smoking (Remy et al. 2019). Humans are exposed to lead particles through inhalation, ingestion, or via skin contact. Human fetuses are exposed to lead through the placenta (Iwai-Shimada et al. 2019; Rísová 2019; Singh et al. 2020). Ingested inorganic Pb is mainly excreted through urine or feces. It has been observed that children have a higher tendency of lead exposure than adults, which can be explained on a body weight basis (Carrington et al. 2019).

In adult blood, more than 90% of lead is absorbed by the erythrocytes with a half-life ranging from 28 to 36 days. However, some lead from the serum enters the soft tissues and is accumulated in the brain, kidney, liver, muscles, bones, and teeth over some time. In addition, as the age increases, the reserved pool of lead in the bones during calcium deficiency, pregnancy, or menopause is released back into the blood. This acts as an endogenous source of lead poisoning to various organs and is a major risk (ATSDR 2010; Klotz and Göen 2017; Organization 2019).

Lead, being a heavy metal, interferes with the organ systems of the body (Mansouri et al. 2020). Nowadays, it is very uncommon to find acute cases of lead toxicity which usually occurs through occupational exposure or accidental exposure to lead-containing agents. However, chronic cases of lead poisoning have been seen, as determined by elevated levels of blood lead concentration. The six countries Algeria, Yemen, Myanmar, North Korea, Iraq, and Afghanistan pose a higher risk of exposure because these countries did not follow the international standard of low-lead levels in the gasoline till 2013 (Cassleman et al. 2020). However, the developed nations took the initiative and banned the lead-containing gasoline and paint which drastically decreased chronic lead exposure but, low-level exposure remains a matter of concern (Flora et al. 2012).

Neurotoxicants may be defined as any substance or chemical that alters the regular operation and/or compromise adaptability in the central and/or peripheral nervous system, either during development or at maturity (Cardenas-Iniguez et al. 2022). Approximately, 200 substances can fall under neurotoxicants.

In the last two decades, lead neurotoxicity has been well recognized. Previous norms related to safe levels of lead was restricted to blood lead level < 10ug/dL, however, more recently, the US Centers for Disease Control and Prevention (CDC) updated and recommended that no level of lead can be considered "safe" (Bellinger 2018; Betts 2012). In the recent estimation of the global disease burden, lead has been ranked 17th to 30th most important contributor to the disability-adjusted life years (DALY) (Lim et al. 2012). The outcomes associated with lead exposure have become more apparent and are an ever-increasing concern across the globe. Not only adults but also young children are vulnerable to its toxic effects leading to neurodevelopment and kidney-related disorders (Murata et al. 2009). A plethora of disorders are caused by lead exposure. The adverse effects include behavioral alterations in children and neuronal developmental disorders in infants, a deficit in motor coordination, and cardiovascular disorders. In addition, its chronic exposure reduces fertility and causes renal dysfunction, convulsion, or even coma (Lustberg and Silbergeld 2002; Weaver et al. 2005).

A report by WHO (2004) states that in the year 2000, ~ 120 million people had blood lead levels between 5 and 10 mg/dl and nearly the same number had values > 10 mg/dl. This report further revealed that 90% of these children belong to underdeveloped countries. 40% of these children had blood lead concentrations of > 5 mg/dl and 50% had blood lead concentrations of > 10 mg/dl (Ezzati et al. 2004). A shocking figure of 9.8 million disability-adjusted life years (DALYs) was caused by the disease burden from mild mental retardation associated with lead exposure (Ezzati et al. 2004). Furthermore, the report of the Global Burden of disease dataset 2019 confirmed that ~ 800 million children have hazardous levels of lead in their bodies. The majority of the cases belong to Southeast Asia and India is the highest contributor to it. In India, nearly 275 million children are reported with raised lead levels and out of total deaths due to lead poisoning, 26% of deaths are reported in India GBD Compare (Metrics IFH Evaluation 2017).

This chapter provides insight into clinical manifestation of lead exposure especially the impact on the mental health (neurodevelopment) of the fetus, children, and adults, the mechanism of lead-induced neurotoxicity, available biomarkers, challenges, mitigation, preventive measures, and therapy for lead exposure. There is a need to undertake necessary steps to protect the population from environmental toxicity, future recommendations for the same have been discussed in the concluding part of the chapter.

4.2 Material and Methodology

This review paper includes a systematic search in Google scholar, PubMed, and MEDLINE. The search was conducted using different keyword strings as (("Lead exposure" OR "Lead toxicity" OR "Lead effects") AND "Mental health") on Google Scholar, and (("Lead exposure" OR "Lead toxicity" OR "Lead effects") AND ("Human brain" OR "cognitive" OR "Behavioral problems")) on PubMed and

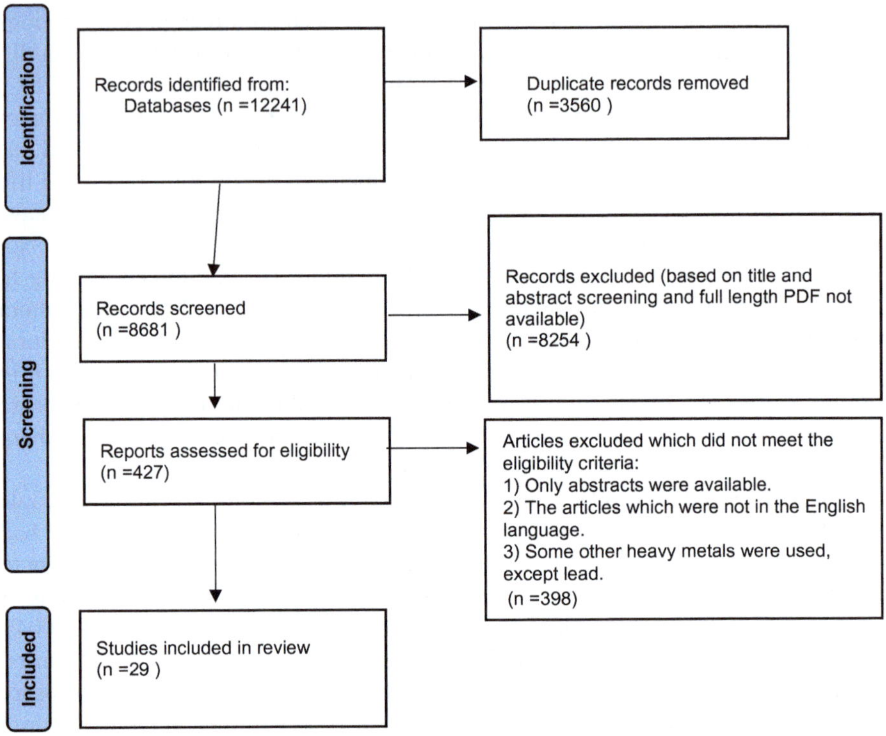

Fig. 4.1 PRISMA flow chart illustrating the selection process of the included studies. *Source* Page et al. (2021)

Medline. The keywords were searched in the title and abstracts of the research papers. No criteria was set to select population-based studies. The screening of literature was done from 2013 to 2023. The search was confined to English language and included data on humans only. Besides the database search, a manual search including the reference lists of original articles and previous reviews was also performed. The studies based on association or effects of lead exposure, toxicity on humans were included for the present paper (Fig. 4.1 PRISMA flow diagram). Out of these, only those studies which complied with the inclusion criteria were included. Blogs, private websites, and newspaper articles were excluded. A total of 29 studies were screened for the review.

4.3 Lead and Mental Health

Environmental and occupational lead exposure is a serious public health concern. The individuals at high exposure are prone to encephalopathy, anemia, and kidney damage. Even the lower doses of exposure can alter the cognitive development in children as well as in adults (Fenga et al. 2016). It is well documented that blood lead levels are inversely associated with neuropsychological development. It can also affect the mental conditions like anger, mood, general distress, schizophrenia, and violence (Cassleman et al. 2020). Because of its health risks even at low exposure, it has been suggested that there is no safe level of lead exposure.

4.3.1 Effects on the Fetus

Human embryonic and fetal growth are the foundation for healthiness and illness across the lifespan. The exposure of fetus to lead can have a deep impact on its growth, as it can cross the placental barrier and accrues in the tissues of the developing fetus (Gundacker and Hengstschläger 2012). BPb level of 10–15 μg/dL in the women of reproduction age affects the growing fetus. The exposure events of lead happening before the child birth are correlated to various long-term health and behavioral issues (Brubaker et al. 2009; Canfield et al. 2005; Cecil et al. 2011). It is assumed that intrauterine contact to lead disrupts the development of brain networks before birth. Interrupted connectivity in utero may lead to reduction in the integrity of network structure, which in turn may result in cognitive deficits linked with neurotoxic insults from lead exposure. The fetus exposed to lead deviate from characteristic patterns of neurodevelopment as per fetal resting-state functional connectivity (RSFC) MRI study. The relatively lesser age-linked rise in cross-hemispheric connections and increased connectivity of the superior frontal gyrus and the posterior cingulate cortex has been reported in prenatal lead-exposed fetuses (Thomason et al. 2019).

4.3.2 Cognitive and Behavioral Effects in Children

There are numerous studies citing the potency of lead to affect the cognition and behavior of children. Moreover, different blood lead levels show different types of cognitive and behavioral problems. Earlier it was known that blood lead levels at 10 μg/dL and above can cause cognitive and behavioral problems among children. Blood concentrations at or below 10 g/dL resulted in neurophysiological and neurobehavioral deficits. It may have an impact on academic performance (Liu et al. 2013) distractibility, memory problems, decreased verbal and quantitative scores, impaired visual-motor coordination, and longer reaction times (Bellinger 2008; Canfield et al. 2003a, b; 2004; 2005; Chiodo et al. 2004; Dietrich et al. 1993; Lanphear et al. 2005; Rocha and Trujillo 2019). However, recent studies found concrete evidence of significant damage to the brain at levels below 5 μg/dL. Studies conducted by various researchers (Canfield et al. 2003a, b; Chiodo et al. 2004; Lanphear et al. 2005) showed that children with blood lead levels of below 5 μg/dL are linked to impulsivity, deficiencies in verbal processing, non-verbal thinking, reading, and arithmetic, reaction time, attention, as well as low scores on a variety of achievement tests. Simple reaction time, teacher report cards, and neuropsychological tests have all shown that children's attention is interrupted at levels below 3 g/dL (Chiodo et al. 2004; Després et al. 2005; Min et al. 2009). Moreover, it has been proved that continuous low-level lead exposure leads to cognitive and psychological impairment instead high-level exposure results in seizures, commas, and death.

Generally, lead toxicity is asymptomatic in children. However, children below the age of 5 years may show symptoms like poor appetite, lethargy, weight loss, abdominal pain, vomiting, constipation, headache, irritability, tiredness, and nervousness (Collin et al. 2022). Anemia is one of the symptoms of lead toxicity and thus, the pale color (a result of anemia) of children may signify lead exposure. Children who have been exposed to lead may also have learning issues. Lead encephalopathy exhibits symptoms like seizures, vomiting, clumsiness, agitation, etc. The symptomatic cases of lead poisoning must be treated immediately.

Table 4.1 illustrating the few works conducted on infants and children for showing a positive association of lead exposure and mental health. It is well documented that different biomarkers including blood, urine, bone can successfully estimate the lead exposure in children. However, most of the studies used blood (venous, umbilical) as a choice of biomarker. It is clear from the table that blood lead levels even at 2.9 μg/L can affect the mental health of children. Rodríguez-Carrillo et al. (2022) did not find any association between urine lead concentration (median: 0.42 μg/L) and behavioral functions (Rodríguez-Carrillo et al. 2022).

4.3.3 Lead Exposure and Mental Health in Adults

Around the globe, usage of leaded gasoline was at its peak from 1940 to the early 1900s. Therefore, it is expected that millions of adults must have lead exposure during their childhood phase. Adult lead poisoning in adults is usually brought on by the usage of traditional medicine and occupational exposure (Kumar 2009). There is an agreement that children exposed to lead suffer from low IQ levels, inattention, hyperactivity, and tend to indulge in violence, antisocial behavior, etc., but research focused on neurobehavioral functioning in adults gives mixed opinions. However, most of the studies found a positive association between lead exposure and psychiatric problems in adults (Cassleman et al. 2020). Occupational exposure to lead can affect general intellectual performance, processing, attention, visuospatial abilities, and motor functions (Fenga et al. 2016).

Lead encephalopathy is rarely seen in adults but individuals using traditional medicines or having occupational exposure to lead may suffer from it (Kumar 2009). Generally, high level of lead exposure (occupational) in adults shows symptoms like limb and abdominal pain, headache, numbness, tingling in hands and feet, memory loss, and mood swings. Abnormal and low sperm count can be a reason for lead exposure in men. Women exposed to lead during pregnancy may experience miscarriage or deliver a premature baby. Sustained low-level exposure leads to anemia.

Table 4.1 Lead exposure and mental health in children and adolescents

References	Year	Population	Biomarker used	Levels of lead found	Whether lead exposure associated with mental health	Manifestations
Kadawathagedara et al. (2023)	2023	In cohort of ZIKV epidemic in pregnant females (*n* = 297) French Territory of America (islands of Guadeloupe, Martinique, Saint-Martin, and Saint-Barthelemy as well as the territory of French Guiana in South America)	Blood lead levels	13.5 μg/L	Yes	• Prenatal lead exposure resulted in lower ASQ scores • Some associations were male specific • Prenatal exposure not associated with the behavioral scores
Lu et al. (2023)	2023	Mother–child cohort, Shanghai, China (*n* = 275)	Umbilical cord blood	Total participants: 44.0 (24.5) μg/L (Median and IQR) Males: 44.0 (24.3) μg/L Females: 46.0 (24.0) μg/L	Yes	• Cord serum DHA positively related to fine motor scores in male children • Decreased motor function in female children
Yildiz et al. (2023)	2023	Adolescents (*n* = 228) Turkey	Blood, plasma, and urine	**Low aggression** (Blood: 5.5 μg/L Urine: 1.57 μg/L) **Normal aggression** (Blood: 6.07 μg/L Urine: 1.38 μg/L) **High aggression** (Blood: 14.8 μg/L Urine: 3.77 μg/L)	Yes	• Significant correlation between the aggression levels and lead toxicity

(continued)

Table 4.1 (continued)

References	Year	Population	Biomarker used	Levels of lead found	Whether lead exposure associated with mental health	Manifestations
Gari et al. (2022)	2022	Poland	Cord blood	Median: 9.9 μg/L Range: 2.9–70 μg/L	Yes	• Impact of prenatal Pb level on child cognitive development observed shortly after birth still persist at 7 years of age
Abd-Wahil et al. (2022)	2022	Autism spectrum disorder ASD ($n = 81$) and typical developed TD children ($n = 74$) between age 3 and 6 years from Malaysia	Urine	ASD children (0.26 μg/L) TD children (0.58 μg/L)	Unknown	• ASD children might have a decreased ability to excrete the heavy metals (including Pb) and may be considered poor detoxifiers relatively to TD children
Malavika et al. (2022)	2022	Children age 9–15 years ($n = 70$) from Rajasthan, India	Venous blood	Median blood lead level: 4.9 μg/L	Yes	• Increased tendency of social withdrawal and excessive worrying with increasing BLLs • Factor 3 and 4 relating to anxiety and depression significantly higher in high BLL group
Rodríguez-Carrillo et al. (2022)	2022	INMA-Granada Cohort Adolescent males age 15–17 years ($n = 125$), Spain	Urine	Concentrations of Pb in urine (median = 0.42 μg/L)	No	• Pb concentrations were found to be inversely related to behavioral functioning
Tlotleng et al. (2022)	2022	BT20 cohort ($n = 100$), South Africa	Bone	Males: 5–11 μg/g Females: 4–14 μg/g	Yes	• Significant association between bone lead levels and anger aggression

(continued)

Table 4.1 (continued)

References	Year	Population	Biomarker used	Levels of lead found	Whether lead exposure associated with mental health	Manifestations
Ishitsuka et al. (2020)	2020	Pregnant women, Japan (n = 17,267)	Blood	Geometric mean: 0.5 μg/L	No but pregnant females of poor economic area with high lead blood levels had symptoms of depression	• No association between blood lead exposure and psychological symptoms within the investigated blood lead levels among pregnant women in Japan
Chouhdari et al. (2020)	2020	Children younger than 13 years (n = 32)Tehra, Iran	Blood	Mean BLL 9.78 ± 3.44 μg/dl	Yes with opium poisoning	• 70% of opium-poisoned children, BLL was ≥ 5 μg/dl • Most of children poisoned with opium were boys, the BLLs were statistically higher in girls
Winter and Sampson (2017)	2017	Follow up of birth cohort of PHDCN, Chicago (n = 1255)	Blood	Average BLL when younger than 6 years: 6.14 μg/L	Yes	• 1 μg per deciliter increase in average childhood • Blood lead level increased in • Adolescent impulsivity, anxiety or depression, and body mass index
Roy et al. (2009)	2009	756 children 3–7 years of age attending pre- and elementary schools in Chennai, India	Blood	Mean blood lead level: 11.4 ± 5.3 μg/dL	Yes	• Blood lead was associated with higher anxiety, social problems, and higher scores in the ADHD index

Adults exposed to lead may show aggression (more than normal), hyperactivity, and insomnia. Delirium, coma, seizures, numbness, and motor deficits are some other symptoms.

Table 4.2 describes that individuals having blood lead levels even at a level of 0.6 μg/L can show symptoms of ill mental health in adults. Bone is also used as a biomarker for the estimation of long-term exposure of lead in adults. The adults having elevated levels of lead may experience psychiatric problems including anxiety, fatigue. They may suffer from Alzheimer's disease schizophrenia and Parkinson's disease.

4.4 Mechanism of Lead-Induced Neurotoxicity

Of all the organ affected, lead accumulation primarily affects the brain and elicits excito-toxicity of the brain, the release of various neurotransmitters, the second messenger system, as well as apoptosis. One of the major effects is its function to substitute for calcium which results in mitochondrial dysfunction and apoptosis cascade leading to neuronal death. Mitochondrial dysfunction results in increased reactive oxygen species and activation of programmed cell death (Simons 1986). Since various enzymes such as acetylcholine, dopamine, and neurotransmitters are dependent on calcium signaling, the substitution with lead alters their activity (Goldstein 1993). In addition, lead delays the differentiation of glial progenitors that can cause demyelination and hypomyelination. Indirectly, lead also dysregulates chemosynthesis with increased production of aminolevulinic acid (ALA) further inhibiting the gamma aminobutyric acid receptors (GABA). As a result of this cascade, anemia occurs along with a deficit in neurocognitive ability. These symptoms might appear immediately or may be delayed with common symptoms including reduced intellectual ability, vision loss, and behavioral issues (Hwang 2007; Lidsky and Schneider 2006).

4.4.1 Effects of Lead on Neurodevelopment

The brain is extremely sensitive to Pb during the developmental period and especially during the gestational period which is marked by neuronal proliferation, migration, and differentiation. Lead exposure adversely affects the developing brain even at minute levels (Bellinger 2008; Liu et al. 2014). During early development, hippocampal-dependent spatial learning and memory are affected as lead exposure alters the NR1 and NR2 subunits and signaling of *N*-methyl-D-aspartate (NMDA) receptors (Guilarte and McGlothan 1998). Additionally, Pb exposure alters brain-derived neurotrophic factors (BDNF) (Neal et al. 2010) as well as impairs hippocampal LTP via epigenetic modulation through DNA methyltransferases and methyl cytosine binding proteins (Schneider et al. 2013). From the animal studies on

Table 4.2 Long-term lead exposure and mental health in adults

References	Year	Population	Biomarker used	Levels of lead found	Whether lead exposure associated with mental health	Manifestations
Wen et al. (2022)	2022	Cohorts of Northern Sweden Health and Disease Study (NSHDS) included in the European FP7 EnviroGenomarkers project	Blood	–	Yes related to Alzheimer's Disease	Four miRNAs were significantly associated to lead exposure, with hsa-miR-3651, hsa-miR-150-5p and hsa-miR-664b-3p being negatively and hsa-miR-627 positively associated to lead levels
Takeuchi et al. (2021)	2021	920 healthy right-handed individuals Mena Age (20.7 years) Japan	Hair	Mean lead levels: 396.65 ppm	Yes in terms of microstructural changes	• Greater lead levels were weakly but significantly associated with increase of working memory-related activity in the areas of WM network
Ma et al. (2019)	2019	Schizophrenic patients (n = 95) and healthy subjects (n = 95), China	Blood	Schizophrenic patients = 0.626 μg/L Healthy subjects = 0.546 μg/L	Yes	• Higher concentrations of serum Pb were significantly associated with an elevated risk of schizophrenia

(continued)

Table 4.2 (continued)

References	Year	Population	Biomarker used	Levels of lead found	Whether lead exposure associated with mental health	Manifestations
Reuben et al. (2019)	2019	Children exposed to lead at the age of 11 years of age (18–38 years of age) $n = 579$	Blood	Mean (SD) blood lead level: 11.08 (4.96) μg/dL	Yes	• Each 5-μg/dL increase in childhood blood lead level was associated with a 1.34-point increase in general psychopathology
McFarlane et al. (2013)	2013	Port Pirie Cohort study, Australia ($n = 210$)	Blood	Mean BLL up to 7 years = 17.2 μg/L	Yes	• For a 10 mg/dL increase, females showed increase in anxiety problems, somatic problem and antisocial personality problems • While non-significant for males
Weisskopf et al. (2010)	2010	Mean age: 66.5 years	Bone	Tibia lead ≥ 19.1 ug/g in the fourth Quartile	Yes, increased risk of Parkinson's disease	• Increasing odds of PD with increasing tibia (cortical) bone lead, which has a half-life of decades
Coon et al. (2006)	2006	121 PD patients and 414 age-, sex-, and race-, frequency-matched controls ≥ 50 years of age	Bone	32.2%	Yes Parkinson's disease	• Pb plays a role in the etiology of PD in exposed individuals • Whole-body lifetime Pb exposure showed a statistically significant elevation of risk of PD

neonatal rats, it is clear that alterations of long-term potentiation (LTP) and paired-pulse facilitation (PPF) of hippocampal dentate gyrus occurred (Ruan et al. 1998). This hippocampal long-term potentiation is an important component of learning as well as memory. With lead exposure, the dendritic spines, the crucial structure of pyramidal neurons are severely affected. Such changes can cause a reduction or loss of learning and memory function. Any deviation in synaptic formation, maturation, or structural alterations is detrimental to the development of the brain. Post-synaptic neuroligins (NLGNs) are integral proteins that are required for synaptogenesis, dendritic spine maturation, and its stability. Upon lead exposure, NLGN proteins are disrupted and causes altered dendritic spine formation in hippocampus (Zhao et al. 2018). In addition, exposure to lead might result in abnormal alteration of the synaptic transmission that is often associated with memory impairment. This was predicted from a study, where it was observed that PB-exposed rats had decreased expression of NR2A and phosphorylated GluR1 causing synoptical morphological and functional alterations in the hippocampal CA1 pyramidal neurons that ultimately leads to behavioral changes (Wang et al. 2016). From previous studies, it has already been established that children with the blood lead concentration of 10 μg/dL are also susceptible to intellectual impairment exhibiting low intelligent quotient. Accordingly, each increase of 10 μg per deciliter in the blood results in a significant decline in the IQ by 4.6 points which is considered to be a tremendous decline (Canfield et al. 2003a, 2003b; Heidari et al. 2022; Lanphear et al. 2005). During fetal development, lead can cross the placental membrane. In an assessment of prenatal exposure to lead, it has been observed that levels of lead in the cord blood were 1.23 μg/dl indicating that during development, lead can effortlessly pass the placental barrier and accrue in the blood of the fetus (Jedrychowski et al. 2009; Vigeh et al. 2014). A study observed that levels of lead in the first-trimester maternal plasma and whole blood can be a predictor of neurodevelopmental disorders in the early stage of life (Hu et al. 2006). Interestingly, epigenetic modifications such as DNA methylation are known to play a part in hampered neurodevelopment. This was estimated by a study wherein trimester-specific maternal blood lead concentrations, DNA methylations in umbilical cord blood, and infant neurodevelopmental outcomes were measured (Rygiel et al. 2021).

Moreover, exposure to Pb makes modification in nitric oxide synthase which alters the brain vasculature, consequently affecting the serotonergic system and may intensify aggressive behavior (Martínez-Lazcano et al. 2018). So, the neurobehavioral dysfunctions and deficits in cognition can be linked to environmental Pb exposure (Santa Maria et al. 2019).

4.5 Available Biomarkers of Lead Exposure

Biological monitoring measures are important for the evaluation of the toxicological agents that might be detrimental to human health and the environment (Berlin et al. 1982). Biomarker monitors and measures the interaction with the biological system and the physical, chemical, or biological agents. Biomarkers give an index of the contamination of the biological system and identify the risk factors associated with it. A variety of biomarkers exists to determine the toxicity of an element such as biological fluids like blood, and urine, and biological tissues such as hair, bone, and nail are being tested for lead exposure. Despite this, the difficulty in assessment exists due to the complex toxico-kinetics of lead within the tissue. Blood is the most common biomarker to assess the levels of ingested/inhaled lead. In humans, more than 50% of lead is transferred to the bloodstream upon exposure (DeSilva 1981). Other biomarkers include bone, teeth, hair, and nail (Barbosa et al. 2005). Bone lead levels have been correlated with serum lead levels in adults that confers its use as a potential biomarker (Hernández-Avila et al. 1998; Hu et al. 1998). Hair is an attractive biomarker for monitoring lead exposure, as it is readily available and obtained via non-invasive procedures and is of low cost (Schuhmacher et al. 1991). Nails have been also used for the detection of chronic exposure of lead due to its several advantages. Lead in nail is considered to remain uneffected from the metabolic activities of the body and reflects the long-term exposure of lead. In this regard, toenails are considered to be superior than fingernails because toenail remains less exposed to the other environmental contaminations (Nowak and Chmielnicka 2000; Takagi et al. 1988).

4.6 Treatment to Reduce Lead Levels in Human Body

The most effective treatment available to quickly reduce the blood lead level is chelation therapy (Collin et al. 2022). In 1950, the early chelating agent, EDTA, was fetched into clinical use as an antidote for lead toxicity. In chelation therapy, which is a clinical intervention, chelating agents like calcium disodium ethylene diamine tetraacetic acid (CaNa 2 EDTA), Succimer (2,3 meso-dimercaptosuccinic acid or DMSA, oral chelating agent, for mild and asymptomatic cases), are administered, which in turn binds to Pb and removes it from the different tissues of the body (Hao et al. 2013). DMSA being an antioxidant significantly diminishes Pb-induced oxidative stress and apoptosis (Obeng-Gyasi 2019a). Though chelation therapy considerably removes the Pb ions from the body, however, because of the side effects of the chelating agents, their use is limited to severe cases of overexposure to heavy metals (Aaseth et al. 2015; Kushwaha et al. 2018). Another medical treatment for individuals with overt lead intoxication involves decontamination and supportive care (Kosnett et al. 2007).

4.7 Prevention Therapy

Lead poisoning affects many organs in adults as well as in children. The prevention of exposure to lead through the environment is the primary therapy. The most common sources of lead exposure in children and adults are air, soil, water, paint chips, candies, and toys; some of which can be easily avoided (Collin et al. 2022). Secondary prevention includes regular screening through different biomarkers. Depending upon the levels of lead found medical therapies and dietary supplements may help to lower the level of lead in the body but cannot remove it completely. The early detection of lead poisoning and monitoring of blood lead levels may avoid significant consequences. Preventing direct exposure and taking a proper diet rich in natural antioxidants, vitamins (Flavonoids), iron may prevent lead build-up in the tissues (Collin et al. 2022; Wang et al. 2021).

4.8 Suggestions and Recommendations

It has become increasingly evident from the previous data that even a low level of environment toxicity to lead can be significantly hazardous to young adolescents leading to various mental disorders and intellectual deficits. In addition, it is also responsible for various cancers, cardiovascular diseases, kidney disorders, and gastrointestinal problems.

- To prevent the lead toxicity in air, policymakers need to design/ update the laws to protect the environment effectively.
- Efforts have to be made at global levels via formation of administrative policies, promoting research in the field, and making stringent laws for monitoring pollution levels.
- Association of lead toxicity and its effects on developing brain resulting in behavior and intellectual deficits has been clearly reported. Still, the efforts to control its exposure are inadequate both at the national and international levels. To prevent various health-related issues due to potential toxins, it is important to identify these and restrict their usage before they enter the environment.
- Various programs could be created at government level to aware public about the health hazards to these toxins via distributing pamphlets, pasting posters in public places, and by advertisements on television and social media.
- Also, it is important to educate people on various sources of exposure and routes of inhalation.
- Few suggestions and recommendations for preventing environmental toxicity include

  – Industries need to certify that that the raw material used and the final product as well as factory emissions are not environment hazardous.

– It is important to assess the effect of potential toxins on the developing brain by various relevant bodies including pollution control boards and agencies
– The prevalence of mental disorders, intelligence deficits, and learning disabilities should be recorded in a national database which needs to regularly updated.

- Longitudinal studies need to be designed to study the effect of different toxins on brain at multiple exposure timings as majority of intellect issues are a cumulative effect of chemical and social exposure from fetal life to childhood. Therefore, neuroimaging at starting from the fetal period to adulthood may help to monitor the different stages of brain development and maturation, under various environmental conditions.
- It is important to assess neuropsychological and behavioral aspects of children at early ages through standardized instruments.

4.9 Conclusions

Previous studies and data's suggest that lead neurotoxicity may be a contributing factor for adverse mental health outcomes, even at levels generally considered to pose low or no risk. Thus, no blood concentration of lead is safe; its neurological and behavioral effects are thought to be irreversible. WHO has also identified lead as one of 10 chemicals of major public health concern needing immediate action by Member States to protect the health of workers, children and women. Efforts has to be made at the national and international levels to control the lead exposure. Adverse behavioral outcomes observed in children with similarly low blood lead levels emphasizes the need for considering ways to further reduce environmental lead exposures. Therefore, it is important to state the awareness and advocacy surrounding the issue of such neurotoxicant exposures in order to influence policy makers that can enact legislation to largely mitigate and eliminate these toxic environmental exposures.

References

Aaseth J, Skaug MA, Cao Y, Andersen O (2015) Chelation in metal intoxication—principles and paradigms. J Trace Elements Med Biol 31:260–266

Abd-Wahil MS, Jaafar MH, Md-Isa Z (2022) Assessment of urinary lead (Pb) and essential trace elements in autism spectrum disorder: a case-control study among preschool children in Malaysia. Biol Trace Element Res 200(1):97–121. https://doi.org/10.1007/s12011-021-02654-w

ATSDR (2010) Case studies in environmental medicine (CSEM) lead toxicity

Barbosa F, Tanus-Santos JE, Gerlach RF, Parsons PJ (2005) A critical review of biomarkers used for monitoring human exposure to lead: advantages, limitations, and future needs. Environ Health Perspect 113(12):1669–1674

Bellinger DC (2008) Very low lead exposures and children's neurodevelopment. Curr Opin Pediat 20(2):172–177

Bellinger D (2018) Tetraethyl lead, paints, pipes, and other lead exposure routes: the impact on human health

Berlin A, Yodaiken R, Logan D (1982) International seminar on the assessment of toxic agents at the workplace roles of ambient and biological monitoring, Luxembourg, 8–12 December, 1980. Int Arch Occup Environ Health 50(2):1258

Betts KS (2012) CDC updates guidelines for children's lead exposure. Environ Health Perspect 120(7):a268. https://doi.org/10.1289/ehp.120-a268

Brubaker CJ, Schmithorst VJ, Haynes EN, Dietrich KN, Egelhoff JC, Lindquist DM et al (2009) Altered myelination and axonal integrity in adults with childhood lead exposure: a diffusion tensor imaging study. Neurotoxicology 30(6):867–875

Canfield RL, Henderson CR Jr, Cory-Slechta DA, Cox C, Jusko TA, Lanphear BP (2003a) Intellectual impairment in children with blood lead concentrations below 10 μg per deciliter. N Engl J Med 348(16):1517–1526

Canfield RL, Kreher DA, Cornwell C, Henderson CR (2003b) Low-level lead exposure, executive functioning, and learning in early childhood. Child Neuropsychol 9(1):35–53

Canfield RL, Gendle MH, Cory-Slechta DA (2004) Impaired neuropsychological functioning in lead-exposed children. Develop Neuropsychol 26(1):513–540

Canfield R, Jusko T, Kordas K (2005) Environmental lead exposure and children's cognitive function. Rivista Italiana Di Pediatria Ital J Pediat 31(6):293

Cardenas-Iniguez C, Burnor E, Herting MM (2022) Neurotoxicants, the developing brain, and mental health. Biol Psych Global Open Sci 2:223–232

Carrington C, Devleesschauwer B, Gibb HJ, Bolger PM (2019) Global burden of intellectual disability resulting from dietary exposure to lead, 2015. Environ Res 172:420–429

Cassleman KL, Dorrance KA, Todd AC (2020) Neuropsychiatric implications of chronic lead exposure. Milit Med 185(5–6):e914–e918

Cecil KM, Dietrich KN, Altaye M, Egelhoff JC, Lindquist DM, Brubaker CJ, Lanphear BP (2011) Proton magnetic resonance spectroscopy in adults with childhood lead exposure. Environ Health Perspect 119(3):403–408

Chiodo LM, Jacobson SW, Jacobson JL (2004) Neurodevelopmental effects of postnatal lead exposure at very low levels. Neurotoxicol Teratol 26(3):359–371

Chouhdari A, Farnaghi F, Hassanian-Moghaddam H, Zamani N, Sabeti S, Shahrabi-Farahani H (2020) Blood lead levels in opium-poisoned children: one cross-sectional study in Iran. Addict Health 12(3):159–166

Collin MS, Kumar Venkataraman S, Vijayakumar N, Kanimozhi V, Arbaaz SM, Stacey RS et al (2022) Bioaccumulation of lead (Pb) and its effects on human: a review. J Hazard Mater Adv 7:100094

Coon S, Stark A, Peterson E, Gloi A, Kortsha G, Pounds J et al (2006) Whole-body lifetime occupational lead exposure and risk of Parkinson's disease. Environ Health Perspect 114(12):1872–1876

DeSilva P (1981) Determination of lead in plasma and studies on its relationship to lead in erythrocytes. Occup Environ Med 38(3):209–217

Després C, Beuter A, Richer F, Poitras K, Veilleux A, Ayotte P et al (2005) Neuromotor functions in Inuit preschool children exposed to Pb, PCBs, and Hg. Neurotoxicol Teratol 27(2):245–257

Dietrich KN, Berger OG, Succop PA, Hammond PB, Bornschein RL (1993) The developmental consequences of low to moderate prenatal and postnatal lead exposure: intellectual attainment in the Cincinnati Lead Study Cohort following school entry. Neurotoxicol Teratol 15(1):37–44

Ezzati M, Lopez AD, Rodgers AA, Murray CJ (2004) Comparative quantification of health risks: global and regional burden of disease attributable to selected major risk factors. World Health Organization, New York

Fenga C, Gangemi S, Alibrandi A, Costa C, Micali E (2016) Relationship between lead exposure and mild cognitive impairment. J Prevent Med Hyg 57(4):E205

Flora G, Gupta D, Tiwari A (2012) Toxicity of lead: a review with recent updates. Interdiscip Toxicol 5(2):47–58

Garí M, Grzesiak M, Krekora M, Kaczmarek P, Jankowska A, Król A et al (2022) Prenatal exposure to neurotoxic metals and micronutrients and neurodevelopmental outcomes in early school age children from Poland. Environ Res 204:112049. https://doi.org/10.1016/j.envres.2021.112049

Goldstein G (1993) Evidence that lead acts as a calcium substitute in second messenger metabolism. Neurotoxicology 14(2–3):97–101

Guilarte TR, McGlothan JL (1998) Hippocampal NMDA receptor mRNA undergoes subunit specific changes during developmental lead exposure. Brain Res 790(1–2):98–107

Gundacker C, Hengstschläger M (2012) The role of the placenta in fetal exposure to heavy metals. Wien Med Wochenschr 162(9–10):201–206

Hao P, Han SH, Liu HY, Chandni V, Cai XQ, Zhang YH (2013) Relationship of inflammation and endothelial dysfunction with risks to cardiovascular disease among people in Inner Mongolia of China. Biomed Environ Sci 26(10):792–800

Heidari S, Mostafaei S, Razazian N, Rajati M, Saeedi A, Rajati F (2022) The effect of lead exposure on IQ test scores in children under 12 years: a systematic review and meta-analysis of case-control studies. Syst Rev 11(1):106. https://doi.org/10.1186/s13643-022-01963-y

Hernández-Avila M, Smith D, Meneses F, Sanin LH, Hu H (1998) The influence of bone and blood lead on plasma lead levels in environmentally exposed adults. Environ Health Perspect 106(8):473–477

Hu H, Rabinowitz M, Smith D (1998) Bone lead as a biological marker in epidemiologic studies of chronic toxicity: conceptual paradigms. Environ Health Perspect 106(1):1–8

Hu H, Téllez-Rojo MM, Bellinger D, Smith D, Ettinger AS, Lamadrid-Figueroa H et al (2006) Fetal lead exposure at each stage of pregnancy as a predictor of infant mental development. Environ Health Perspect 114(11):1730–1735. https://doi.org/10.1289/ehp.9067

Hwang L (2007) Environmental stressors and violence: lead and polychlorinated biphenyls. Rev Environ Health 22(4):313–328

Ishitsuka K, Yamamoto-Hanada K, Yang L, Mezawa H, Konishi M, Saito-Abe M et al (2020) Association between blood lead exposure and mental health in pregnant women: results from the Japan environment and children's study. Neurotoxicology 79:191–199. https://doi.org/10.1016/j.neuro.2020.06.003

Iwai-Shimada M, Kameo S, Nakai K, Yaginuma-Sakurai K, Tatsuta N, Kurokawa N et al (2019) Exposure profile of mercury, lead, cadmium, arsenic, antimony, copper, selenium and zinc in maternal blood, cord blood and placenta: the Tohoku Study of Child Development in Japan. Environ Health Prevent Med 24:1–11

Jedrychowski W, Perera FP, Jankowski J, Mrozek-Budzyn D, Mroz E, Flak E et al (2009) Very low prenatal exposure to lead and mental development of children in infancy and early childhood. Neuroepidemiology 32(4):270–278

Kadawathagedara M, Muckle G, Quénel P, Michineau L, Le Bot B, Hoen B et al (2023) Infant neurodevelopment and behavior in Guadeloupe after lead exposure and Zika maternal infection during pregnancy. Neurotoxicology 94:135–146. https://doi.org/10.1016/j.neuro.2022.11.007

Klotz K, Göen T (2017) Human biomonitoring of lead exposure. Met Ions Life Sci 17:99–121

Kosnett MJ, Wedeen RP, Rothenberg SJ, Hipkins KL, Materna BL, Schwartz BS et al (2007) Recommendations for medical management of adult lead exposure. Environ Health Perspect 115(3):463–471

Kumar A (2009) Lead loadings in household dust in Delhi, India. Indoor Air 19(5):414–420

Kushwaha A, Hans N, Kumar S, Rani R (2018) A critical review on speciation, mobilization and toxicity of lead in soil-microbe-plant system and bioremediation strategies. Ecotoxicol Environ Saf 147:1035–1045

Lanphear BP, Hornung R, Khoury J, Yolton K, Baghurst P, Bellinger DC et al (2005) Low-level environmental lead exposure and children's intellectual function: an international pooled analysis. Environ Health Perspect 113(7):894–899

Lidsky T, Schneider J (2006) Adverse effects of childhood lead poisoning: the clinical neuropsychological perspective. Environ Res 100(2):284–293

Lim SS, Vos T, Flaxman AD, Danaei G, Shibuya K, Adair-Rohani H et al (2012) A comparative risk assessment of burden of disease and injury attributable to 67 risk factors and risk factor clusters in 21 regions, 1990–2010: a systematic analysis for the Global Burden of Disease Study 2010. Lancet 380(9859):2224–2260

Liu J, Li L, Wang Y, Yan C, Liu X (2013) Impact of low blood lead concentrations on IQ and school performance in Chinese children. PLoS ONE 8(5):e65230

Liu JA, Chen Y, Gao D, Jing J, Hu Q (2014) Prenatal and postnatal lead exposure and cognitive development of infants followed over the first three years of life: a prospective birth study in the Pearl River Delta region, China. Neurotoxicology 44:326–334

Lu A-X, Wang S-S, Xu X, Wu M-Q, Liu J-X, Xu M et al (2023) Sex-specific associations between cord blood lead and neurodevelopment in early life: the mother-child cohort (Shanghai, China). Ecotoxicol Environ Saf 249:114337. https://doi.org/10.1016/j.ecoenv.2022.114337

Lustberg M, Silbergeld E (2002) Blood lead levels and mortality. Arch Intern Med 162(21):2443–2449

Ma J, Yan L, Guo T, Yang S, Guo C, Liu Y et al (2019) Association of typical toxic heavy metals with Schizophrenia. Int J Environ Res Public Health 16(21):4200

Malavika L, Goyal T, Mitra P, Saikiran G, Sharma S, Sharma P (2022) Risk factors for lead toxicity and its effect on neurobehavior in Indian children. Indian J Clin Biochem 37(3):294–302. https://doi.org/10.1007/s12291-021-00995-w

Mansouri B, Błaszczyk M, Binkowski LJ, Sayadi MH, Azadi NA, Amirabadizadeh AR, Mehrpour O (2020) Urinary metal levels with relation to age, occupation, and smoking habits of male inhabitants of eastern Iran. Biol Trace Element Res 195(1):63–70. https://doi.org/10.1007/s12011-019-01848-7

Martínez-Lazcano JC, López-Quiroz A, Alcantar-Almaraz R, Montes S, Sánchez-Mendoza A, Alcaraz-Zubeldia M et al (2018) A hypothesis of the interaction of the nitrergic and serotonergic systems in aggressive behavior induced by exposure to lead. Front Behav Neurosci 12:202

McFarlane AC, Searle AK, Van Hooff M, Baghurst PA, Sawyer MG, Galletly C et al (2013) Prospective associations between childhood low-level lead exposure and adult mental health problems: the Port Pirie cohort study. Neurotoxicology 39:11–17

Metrics IFH Evaluation (2017) GBD compare data visualization. In: IHME. University of Washington, Seattle, WA

Meyer PA, Brown MJ, Falk H (2008) Global approach to reducing lead exposure and poisoning. Mutat Res Rev Mutat Res 659(1–2):166–175

Min MO, Singer LT, Kirchner HL, Minnes S, Short E, Hussain Z, Nelson S (2009) Cognitive development and low-level lead exposure in poly-drug exposed children. Neurotoxicol Teratol 31(4):225–231

Murata K, Iwata T, Dakeishi M, Karita K (2009) Lead toxicity: does the critical level of lead resulting in adverse effects differ between adults and children? J Occup Health 51(1):1–12

Neal AP, Stansfield KH, Worley PF, Thompson RE, Guilarte TR (2010) Lead exposure during synaptogenesis alters vesicular proteins and impairs vesicular release: potential role of NMDA receptor–dependent BDNF signaling. Toxicol Sci 116(1):249–263

Neeti K, Prakash T (2013) Effects of heavy metal poisoning during pregnancy. Int Res J Environ Sci 2(1):88–92

Nowak B, Chmielnicka J (2000) Relationship of lead and cadmium to essential elements in hair, teeth, and nails of environmentally exposed people. Ecotoxicol Environ Saf 46(3):265–274

Obeng-Gyasi E (2019a) Lead exposure and cardiovascular disease among young and middle-aged adults. Med Sci 7(11):103

Obeng-Gyasi E (2019b) Sources of lead exposure in various countries. Rev Environ Health 34(1):25–34

Organization WH (2019) Preventing disease through healthy environments: exposure to lead—a major public health concern (9240037632)

Page MJ, McKenzie JE, Bossuyt PM, Boutron I, Hoffmann TC, Mulrow CD et al (2021) The PRISMA 2020 statement: an updated guideline for reporting systematic reviews. BMJ 372:n71. https://doi.org/10.1136/bmj.n71

Remy S, Hambach R, Van Sprundel M, Teughels C, Nawrot TS, Buekers J et al (2019) Intelligence gain and social cost savings attributable to environmental lead exposure reduction strategies since the year 2000 in Flanders, Belgium. Environ Health 18:1–9

Reuben A, Schaefer JD, Moffitt TE, Broadbent J, Harrington H, Houts RM et al (2019) Association of childhood lead exposure with adult personality traits and lifelong mental health. JAMA Psych 76(4):418–425. https://doi.org/10.1001/jamapsychiatry.2018.4192

Rísová V (2019) The pathway of lead through the mother's body to the child. Interdiscip Toxicol 12(1):1–6

Rocha A, Trujillo KA (2019) Neurotoxicity of low-level lead exposure: history, mechanisms of action, and behavioral effects in humans and preclinical models. Neurotoxicology 73:58–80

Rodríguez-Carrillo A, Mustieles V, D'Cruz SC, Legoff L, Gil F, Olmedo P et al (2022) Exploring the relationship between metal exposure, BDNF, and behavior in adolescent males. Int J Hyg Environ Health 239:113877. https://doi.org/10.1016/j.ijheh.2021.113877

Roy A, Bellinger D, Hu H, Schwartz J, Ettinger AS, Wright RO et al (2009) Lead exposure and behavior among young children in Chennai, India. Environ Health Perspect 117(10):1607–1611. https://doi.org/10.1289/ehp.0900625

Ruan D-Y, Chen J-T, Zhao C, Xu Y-Z, Wang M, Zhao W-F (1998) Impairment of long-term potentiation and paired-pulse facilitation in rat hippocampal dentate gyrus following developmental lead exposure in vivo. Brain Res 806(2):196–201

Rygiel CA, Dolinoy DC, Bakulski KM, Aung MT, Perng W, Jones TR et al (2021) DNA methylation at birth potentially mediates the association between prenatal lead (Pb) exposure and infant neurodevelopmental outcomes. Environ Epigenet 7(1):5. https://doi.org/10.1093/eep/dvab005

Santa Maria MP, Hill BD, Kline J (2019) Lead (Pb) neurotoxicology and cognition. Appl Neuropsychol Child 8(3):272–293

Schneider J, Kidd S, Anderson D (2013) Influence of developmental lead exposure on expression of DNA methyltransferases and methyl cytosine-binding proteins in hippocampus. Toxicol Lett 217(1):75–81

Schuhmacher M, Domingo J, Llobet J, Corbella J (1991) Lead in children's hair, as related to exposure in Tarragona Province, Spain. Sci Total Environ 104(3):167–173

Simons T (1986) Cellular interactions between lead and calcium. Br Med Bull 42(4):431–434

Singh L, Anand M, Singh S, Taneja A (2020) Environmental toxic metals in placenta and their effects on preterm delivery-current opinion. Drug Chem Toxicol 43(5):531–538

Takagi Y, Matsuda S, Imai S, Ohmori Y, Masuda T, Vinson J et al (1988) Survey of trace elements in human nails: an international comparison. Bull Environ Contam Toxicol 41(5):1258

Takeuchi H, Taki Y, Nouchi R, Yokoyama R, Kotozaki Y, Nakagawa S et al (2021) Lead exposure is associated with functional and microstructural changes in the healthy human brain. Commun Biol 4(1):912. https://doi.org/10.1038/s42003-021-02435-0

Thomason ME, Hect JL, Rauh VA, Trentacosta C, Wheelock MD, Eggebrecht AT et al (2019) Prenatal lead exposure impacts cross-hemispheric and long-range connectivity in the human fetal brain. Neuroimage 191:186–192

Tlotleng N, Naicker N, Mathee A, Todd AC, Nkomo P, Norris SA (2022) Association between bone lead concentration and aggression in youth from a sub-cohort of the Birth to Twenty Cohort. Int J Environ Res Public Health 19(4):2200

Vigeh M, Yokoyama K, Matsukawa T, Shinohara A, Ohtani K (2014) Low level prenatal blood lead adversely affects early childhood mental development. J Child Neurol 29(10):1305–1311

Wang T, Guan R-L, Liu M-C, Shen X-F, Chen JY, Zhao M-G, Luo W-J (2016) Lead exposure impairs hippocampus related learning and memory by altering synaptic plasticity and morphology during juvenile period. Mol Neurobiol 53:3740–3752

Wang Z, Bao J, Wang T, Moryani HT, Kang W, Zheng J et al (2021) Hazardous heavy metals accumulation and health risk assessment of different vegetable species in contaminated soils from a typical mining city, central China. Int J Environ Res Public Health 18(5):2617

Weaver VM, Jaar BG, Schwartz BS, Todd AC, Ahn K-D, Lee S-S et al (2005) Associations among lead dose biomarkers, uric acid, and renal function in Korean lead workers. Environ Health Perspect 113(1):36–42

Weisskopf MG, Weuve J, Nie H, Saint-Hilaire M-H, Sudarsky L, Simon DK et al (2010) Association of cumulative lead exposure with Parkinson's disease. Environ Health Perspect 118(11):1609–1613

Wen Q, Verheijen M, Wittens MMJ, Czuryło J, Engelborghs S, Hauser D et al (2022) Lead-exposure associated miRNAs in humans and Alzheimer's disease as potential biomarkers of the disease and disease processes. Sci Rep 12(1):15966. https://doi.org/10.1038/s41598-022-20305-5

Winter AS, Sampson RJ (2017) From lead exposure in early childhood to adolescent health: a chicago birth cohort. Am J Public Health 107(9):1496–1501. https://doi.org/10.2105/AJPH.2017.303903

Yıldız S, Pirinççioğlu AG, Arıca E (2023) Evaluation of heavy metal (lead, mercury, cadmium, and manganese) levels in blood, plasma, and urine of adolescents with aggressive behavior. Cureus 15(1):157

Zhao Z-H, Zheng G, Wang T, Du K-J, Han X, Luo W-J et al (2018) Low-level gestational lead exposure alters dendritic spine plasticity in the hippocampus and reduces learning and memory in rats. Sci Rep 8(1):3533. https://doi.org/10.1038/s41598-018-21521-8

[illegible]

Chapter 5
Human Health Hazards and Risks Generated by the Bioaccumulation of Lead from the Environment in the Food Chain

Camelia Bețianu, Petronela Cozma, and Maria Gavrilescu

Abstract Taking into account the significant health concerns generated by heavy metals, the assessment of lead exposure from foodstuff consumption is a representative topic in the evaluation of human health risk. Human exposure to lead is associated to numerous harmful effects, in particular, in children and pregnant women. The transfer of lead from soil and water to plants and through food sources to humans is the main route of exposure. Bioaccumulation is a dynamic process that plays a key role in the uptake of lead from terrestrial and aquatic ecosystems and transfer to higher levels of the food chain by biomagnification. Bioaccumulation mechanism is governed by the bioavailability, uptake/absorption, bioconcentration, and biomagnification. The present study synthetically analyzes the mechanisms and processes involved in the accumulation of lead from the abiotic environment and the trophic transfer by food webs in order to identify the potential risks generated for humans. Also, the study comprehensively surveyed the literature and discusses the main methodologies applied for assessing the human health risks generated by lead-contaminated food consumption.

Keywords Bioaccumulation · Food chain · Lead dietary exposure · Human health risk (HHR)

C. Bețianu (✉) · P. Cozma · M. Gavrilescu (✉)
Department of Environmental Engineering and Management, "Cristofor Simionescu" Faculty of Chemical Engineering and Environmental Protection, Gheorghe Asachi Technical University of Iasi, Iasi, Romania
e-mail: betianuc@tuiasi.ro

M. Gavrilescu
e-mail: mgav@tuiasi.ro

M. Gavrilescu
Academy of Romanian Scientists, Bucharest, Romania

N. Kumar and A. K. Jha (eds.), *Lead Toxicity Mitigation: Sustainable Nexus Approaches*, Environmental Contamination Remediation and Management,
https://doi.org/10.1007/978-3-031-46146-0_5

5.1 Introduction

In recent years, contamination of soil, air, and water with potentially toxic elements (PTEs) especially heavy metals (HMs), is a global environmental concern with important implications on human health, mainly due to accumulation in the food chain (Tchounwou et al. 2012; Ali et al. 2019; Briffa et al. 2020; Labudda et al. 2022). The increased interest of scientist on this topic is a consequence of the fact that this group of pollutants presents toxic effects, even at low concentration, non-biodegradability, persistence in the environment, bioaccumulation in living organisms (Bharagava and Mishra 2018; Ali et al. 2019; Balali-Mood et al. 2021; Uddin et al. 2021; Mitra al. 2022); also, HMs are considered the most widespread pollutants in the environment (Wani et al. 2015; Balali-Mood et al. 2021; Swaringen et al. 2022).

Lead (Pb) has toxic properties at high concentrations, and the ubiquitous use of this element has determined contamination of the environment at global scale, generating considerable effects on human health (Nag and Cummins 2022). Human exposure to lead has been connected to different sources of industrial, agricultural, and domestic applications (Wani et al. 2015; Briffa et al. 2020; Collin et al. 2022). Lead is released into environment by both anthropogenic and natural sources (Wani et al. 2015; Collin et al. 2022; Nag et al. 2022). The main anthropogenic sources of lead are mining processes, metal processing, transport, coal burning, oil burning, painting processes, finishing operations, nuclear power plants and power lines, glass manufacturing, textile industry, industrial effluents, fertilizers, pesticides, leather tanning, car batteries, sewage sludge, lead water pipes, wood preservation, additives in pigments and gasoline, and paper processing (Tchounwou et al. 2012; Collin et al. 2022; Kumar et al. 2022a, b; Nag and Cummins 2022). According to Aslam et al. (2021), lead contamination is expected to increase over the next decade as a consequence of widespread use in the automotive industry and electric bicycles. However, some authors claim that the releasing into environment of this metal has decreased in recent years as a consequence of the reduction of lead in fuels, paints, and sanitary pipes (Nag et al. 2022).

Numerous scientific researches have focused on assessing the effects of lead on various species of microorganisms (Borgulat et al. 2021; Tang et al. 2022), plants (Sharma and Dubey 2005; Pourrut et al. 2011; Shahid et al. 2011; Aslam et al. 2021; Kumar et al. 2022a, b), animals (Zhang et al. 2014; Lee et al. 2019; Latif et al. 2022), and humans (Tchounwou et al. 2012; Wani et al. 2015; Nag and Cummins 2022).

Lead exposure is a concern for all age groups, but children's exposure is a major public health problem, mainly due to the devastating consequences of this metal has on growth and development (Wani et al. 2015). According to data reported by UNICEF around 1 in 3 children, meaning approximately 800 million, present a blood lead levels at or above 5 $\mu g/dL$ (Ress and Fuller 2020; Wilson et al. 2022).

Humans exposure to lead is associated to a large number of harmful effects and diseases, effects on nervous system as neurodevelopmental problems at children, brain injury, neurotoxicity, and cell mortality, kidney failure, cardiovascular diseases, oxidative stress, liver effects, skeleton effects, reproductive malfunctions, immune system damage, cancer, including mortality (Wani et al. 2015; Tchounwou et al. 2012; Manea et al. 2020; Briffa et al. 2020; Collin et al. 2022; Mitra et al. 2022; Swaringen et al. 2022). Thus, some authors concluded that exposure to lead generated lethal effects on around 540 000 persons at worldwide scale in 2016 (Nag and Cummins 2022).

Human exposure to lead can occur through the following pathways by inhalation, by ingestion of contaminated food and water and by dermal absorption (Wani et al. 2015; Nag et al. 2022; Collin et al. 2022). One of the main routes of exposure remains the ingestion of lead-contaminated food, which generates adverse health effects on children and adults.

In accordance with the previously presented arguments, the study performed a literature synthesis by analyzing the bioaccumulation process of lead and transfer along the length of the food chain, analysis carried out from the perspective of identifying hazards and risks pose on human health by ingestion of contaminated food products.

The dynamic processes such as the transfer, transport, and accumulation of potentially toxic elements, such as Pb, in different abiotic environments, are directly linked to potential hazards and risks pose to human health and ecological receptors. Understanding the bioaccumulation mechanisms has important implications in identifying the possibilities for human health protection, as well as other organisms in aquatic and terrestrial ecosystems exposed to HMs.

5.2 Bioaccumulation of Lead

Bioaccumulation is an active biological process, characterized by the increasing in the concentration of a toxic compound by uptake from the abiotic environment in the tissues of living organisms and then accumulates in intracellular space, where it is sequestered by proteins and peptide ligands (Proc et al. 2021; Collin et al. 2022).

Bioaccumulation is actually the result of several processes such as uptake/absorption and loss, the uptake occurs through breathing but also through the food route, while loss occurs through excretion, metabolism, passive diffusion, transfer, and growth (Borgå 2013). The accumulation process occurs when the rate of uptake of toxic compound is higher than the rate of loss by metabolism or elimination (Chojnacka and Mikulewicz 2014; Chormare and Kumar 2022), which usually follows first-order kinetics (Savoca and Pace 2021).

The rate of bioaccumulation is mainly influenced by type of exposed organism as well as the physical, chemical, and biological properties of the contaminant, environmental conditions, and the food chain (ATSDR 2007). Bioconcentration of lead in biota is directly influenced by factors such as age, size, trophic level, species, and dietary behavior. Bioaccumulation mechanism is governed by the bioavailability, uptake/absorption, bioconcentration, and biomagnification (Collin et al. 2022) (Fig. 5.1).

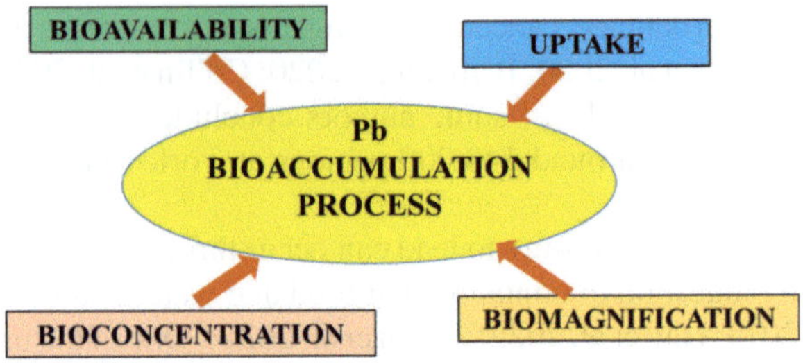

Fig. 5.1 Bioaccumulation mechanisms of lead in biota

Bioaccumulation depends on the binding stability of a chemical substance in the cellular compartments of the living organism and the half-life time. From this point of view, three categories of bioaccumulative substances are known, substances with a fast metabolic half-life, represented by chemical compounds that are degraded in a few weeks/days/hours/minutes, slow metabolic half-life, the compound is degraded in a few months and unassimilated from the gut (Chojnacka and Mikulewicz 2022). Understanding the process of bioaccumulation of metals and metalloids allows the assessment of hazards and risk related to their presence in the environment and in the food chain (Chojnacka and Mikulewicz 2014).

The hazards generated to the presence of lead in soil are represented by the transfer of the metal in the soil–plant system and bioaccumulation in organs and tissues, and the transfer in the food chains. Plants can absorb free lead ions through capillary mechanisms or through cellular respiration (Collin et al. 2022). According to the ability of superior plants to adapt to presence in their growing environment of the toxic metals, plants are categorized in three main categories—indicators, excluders, and accumulators (Uddin et al. 2021). *Metal indicator plants* are represented by sensitive species to HMs presence, which can be used as possible bioindicators of metals pollution in soil or water (Mganga et al. 2011). The concentration of lead in the above-ground tissues of these plants reveals the level of soil contamination. While *excluder plants* include species that tolerate the presence of heavy metals in their environment up to a threshold concentration, for these plants, the accumulation coefficient is always lower than unit (Labudda et al. 2022). These plants present the ability to develop defense mechanisms by blocking the accumulation and transport of metal ions through the roots or through efflux pumps. *Hyperaccumulators* are plants with the ability to accumulate high concentrations of different PTEs in their tissues without showing significant toxicity effects (Anguilano et al. 2022). Around 720 plant species are considered to be hyperaccumulators, they are able to accumulate high amounts of metal ions (Labudda et al. 2022) in the aerial parts without showing toxic effects (Suman et al. 2018). About 25% of the hyperaccumulators identified belong to the family *Brassicaceae*; and a significant number are part of the families *Steraceae, Euphorbiaceae, Rubiaceae, Fabaceae, Scrophulariacea, Myrtaceae, Proteaceae, Caryophylaceae* (Suman et al. 2018), many of these families include edible species such as vegetables, legumes.

The criteria for classification hyperaccumulator species have not been empirically defined, however, some studies (Rahman Farooqi et al. 2022) reveal that a plant can be considered hyperaccumulator for lead if shows an accumulation capacity higher than 1000 mg/kg dry weight (Suman et al. 2018). These plants show specific physiological mechanisms regarding the uptake, root-shoot translocation process, detoxifying mechanisms, sequestration, and bioconcentration of metals in different organs (Collin et al. 2022).

5.2.1 Bioaccumulation of Lead in Crops

Lead pollution in agroecosystems represents a global environmental issue (Naikoo et al. 2019), due to its high toxicity, it is classified as one of the most toxic heavy metals in the environment (Abedi et al. 2022; Zulfiqar et al. 2019; Kumar et al. 2020; Souahi et al. 2021; Collin et al. 2022). Pb and its compounds persist in soil for long periods because form stable complexes, where present for around 150–1500 years (Kanwal et al. 2020), resulting in the bioaccumulation in agricultural plants through various pathways—water, air, and soil (Kumar et al. 2020), followed by subsequent trophic transfer to other trophic levels (Kumar et al. 2022a, b).

Considering the global interest posed by metals exposure, the evaluation of Pb intake from common agricultural products, such as grains, pulses, vegetables, and fruits, which are the base of the food pyramid, is representative in order to analyze the hazards and risks to humans exposed through food consumption. Several studies had been carried out in respect with transport, translocation, the accumulation, and toxic effects of lead in various crops such as rice (*Oryza sativa*) (Ashraf et al. 2015; Al-Saleh and Abduljabbar 2017; Thakur et al. 2017), maize (*Zea mays*) (Gu et al. 2019; Metanat et al. 2019; Abedi et al. 2022; Hernández-Pitalúa et al. 2022), wheat (*Triticum aestivum*) (Wu et al. 2020; Wang et al. 2011; Souahi et al. 2021), *Avena sativa* (Souahi et al. 2021), barley (*Ordeum vulgare*) (Souahi et al. 2021), rye (*Secale cereale*), millet (*Panicum miliaceum*), potato (*Solanum tuberosum*) (Musilova et al. 2015; Khan et al. 2017; Raletsena et al. 2023), tomato (*Solanum lycopersicum*) (Baldi et al. 2021), carrot (*Daucus carota*) (Ngole 2011), broad bean, (*Vicia fava*) (Shahid et al. 2011; Saadaoui et al. 2022), common bean (*Phaseolus vulgaris*) (Baldi et al. 2021), onion (*Allium cepa*) (Jiang et al. 2014), spinach (*Spinacia oleracea*) (Ali et al. 2015; Hussain et al. 2021), which are the most representative dietary sources of lead accumulation in human body.

Rice is the most consumed cereal at global scale (Al-Saleh and Abduljabbar 2017), thus generating a major interest in this agricultural product regarding the analysis of lead accumulation from contaminated soil. In studies carried out in China, near a mining site (Huang et al. 2022), the lead concentration determined in rice was 0.11 $\pm$ 0.14 mg/kg, and 17% of the samples exceed the maximum concentration limit (MCL)—0.2 mg/kg. Norton and co-workers (2014) performed a meta-analysis determining lead concentration in rice seeds from 13 countries, investigating an impressive number of 1578 of rice seed samples. The results showed that all analyzed samples contained Pb (mean 0.04–1.85 mg/kg), high levels were determined for samples collected from China, followed by Nepal, India, and Sri Lanka, the other samples registered a mean value of lead in rice below 0.15 mg/kg. The highest concentrations of lead were recorded for samples collected from a site mine impacted, with 27.49 mg Pb/kg. However, excluding samples from areas known to be contaminated, only 0.6% of samples exceeded the standard for lead level in rice (0.2 mg/kg). Moreover, has been shown significant differences in lead accumulation between different varieties of rice, thus, the results demonstrated average concentrations of lead in grains ranged between 3.5 and 5.1 mg/kg (Liu et al. 2013).

Other studies reported high Pb accumulation in wheat grains, the registered values ranged from 0.015 to 22.6 mg/kg (Wu et al. 2020). Guo et al. (2018) analyzed 16 varieties of wheat and the results showed values of Pb in roots of 69.65 mg/kg, in stems of 4.93 mg/kg, in leaves of 1.90 mg/kg, respectively 0.24 mg/kg in grains, exceeding the MCL. Studies performed in Bangladesh showed that content of Pb in rice, wheat, and maize exceeded the maximum permissible limits in cereals, the highest values were obtained in wheat 4.04 mg/kg, while, rice and maize contained 2.22 and 1.43 mg/kg (Kumar et al. 2022a, b). Maximum values are linked with high concentrations of lead in groundwater (0.06–0.16 mg/L), used for agricultural crops irrigation in the investigated area. Sharma et al. (2018) conducted studies in India, area Punjab, and showed extremely high concentration of lead in wheat, rice, and maize, respectively 16.98, 17.13, and 18.28 mg/kg, the results demonstrating a high bioaccumulation potential of lead in crops.

El-Hassanin et al. (2020) found that Pb accumulation in maize grains was ranged from 0.01 to 0.55 mg/kg, samples cultivated in low-quality water-irrigated sites exceeded the maximum permissible limits; while, another work determined a level of 0.49 mg/kg of lead accumulated in maize grains (Lu et al. 2015). In experimental studies carried out in field, the concentration of Pb found in soil, roots, stems, leaves, and grains ranged in the following order: 172–551.34, 14.02–26.00, 8.07–14.09, 4.713–7.100, 0.3133–1.0533 mg/kg, demonstrating the bioconcentration and transfer of the toxic compound from soil and accumulation into the organs of maize plants (Chiwetalu et al. 2022).

However, Aslam et al. (2021) estimated that maize plants translocate from roots to leaves around 2–5% of the uptake Pb; also the relationship between the amount of lead in soil and its effect on maize production has been examined by several papers, the exposure at 30 ppm metal concentration recorded a decrease of 76–85% of the production (Ghani et al. 2010).

Studies have shown that the degree of accumulation and translocation in wheat is dependent on concentration and bioavailability of metal in soil (Sharma et al. 2018). Nevertheless, the mechanism of speciation, translocation, and accumulation of lead in grains is not completely understood (Wu et al. 2020).

A number of studies focused to assess lead contamination in the edible parts of plant leaves in different geographical areas, found that average concentration of lead varies on the vegetal species, soil pH, bioavailability, soil properties, metal concentration, mobile forms, chemical form, the irrigation water quality, etc. (Musilova et al. 2015; Najmi et al. 2023). It is well known that vegetables are essential components of the human diet, and cultivated on agricultural land contaminated with Pb, accumulate high metal by roots or may absorb these metal contaminants at foliar level (Ahmed et al. 2022). However, the higher metal accumulation has been observed in leafy vegetables, followed by tubers and fruits (Liu et al. 2012; Sulaiman et al. 2020). Pb concentrations in *Spinacia oleraceae* (spinach) leaves were higher than in *Daucus carota* (carrot) tubers, despite the fact that the carrot has a direct root contact with the contaminated soil, demonstrating a high translocation rate of lead to the aerial part of plant, that may indicate a selective uptake mechanism in the case of spinach (Ngole 2011). For example, study conducted by Khan et al. 2017 had investigated metal and metalloids concentrations in water, soil, and *Solanum tuberosum* (potato) at three different sites, and the result reveals a concentration of 1.504 mg/kg of Pb in tubers. Furthermore, other research (Musilova et al. 2015) describes a significant positive correlation between the content of Pb mobile forms in the soil and the rate of accumulation in potatoes, and may pose a hazard for human consumption through the high content of Pb ions in potatoes (0.244–0.855 mg/kg), exceeding the MCL (0.1 mg/kg).

In leafy vegetables, *Brassica chinensis*, *Amaranthus gangeticus,* and *Brassica rapa,* the lead concentration was ranged between 0.03 and 0.05 mg/kg (Sulaiman et al. 2020). The average contents of lead in vegetables such as lettuce, radish, mint, parsley, and jarjir (Arugula) have indicated values ranged of 0.858–1.175 mg/kg for Pb (Najmi et al. 2023).

According to Nag and Cummins, (2022) metals bioaccumulate in plants in the following order of $BAF_{root} > BAF_{stem} > BAF_{leaf} > BAF_{fruit} > BAF_{seed}$. This order of accumulation in plant organs was confirmed for lead. Several studies (Sharma and Dubey 2005; Guo et al. 2018; Usman et al. 2020; Baldi et al. 2021; Collin et al. 2022) demonstrated that roots are more able to accumulate Pb ions, their subsequent translocation to stem and leaf or other parts is highly restricted (Vasile et al. 2021). It has been reported that Pb remains predominantly in the roots level (Sharma and Dubey 2005; Wu et al. 2020), from the total amount uptake, only 5% of the ions are transported to the aerial organs of the plant, the 95% remains accumulated at the roots level (Collin et al. 2022).

Most studies show a low rate of translocation of lead between the roots and the aerial part, but taking into account the toxic properties of lead, its presence in edible parts of crops is an important factor that may induce important risks to human health through dietary exposure.

5.2.2 The Mechanism of Lead Accumulation

The mechanism of Pb accumulation in plants is based on the soil–plant relationship. Through the root system, plants take up water and nutrients from the soil and along with them absorb the free Pb (II) ions which are distributed in the top soil (Collin et al. 2022). Two main mechanisms for plant metals uptake are described here: one is passive uptake, determined by the membrane concentration gradient, and the other one is inducible uptake, determined by the specific substrate (Aslam et al. 2021; Schützendübel and Polle 2002). The transport of the absorbed ions is carried out through the xylem vessels or in the accompanying phloem cells that support the transport from the root system to the aerial parts of the plant, the stem, the leaves. Pb transport can be synthesized in this way: soil solution - epidermal cells of the root system—xylem vessels—leaves. Pb binds primarily to the root cell wall via esterified pectins. Polysaccharides, present in the walls of root cells, have functional groups such as –COOH, –OH, and –SH, involved in the binding of heavy metals to the roots; these play an important role in avoiding and tolerating metal stress (Rai et al. 2019). Under chemical stress conditions, polysaccharides from agricultural crops generate changes in the integrity of the cell membrane structure and cell organelles such as in chloroplasts and mitochondria, inactivation of enzymes by replacing integral components or binding to the sulfhydryl or carboxyl group, and changes in nucleic acid conformation (Rai et al. 2019). Transporter proteins play an important role in facilitating Pb homeostasis (Aslam et al. 2021).

Since lead is a non-essential element for plant development, there are no specific channels for the uptake and transport of these ions (Collin et al. 2022). The translocation of heavy metals occurs along the up-flowing in xylem vessels that transport the dissolved nutrients and discharge into the endoderm (Sharma and Dubey 2005).

Another pathway of lead accumulation is represented by the foliar absorption of lead ions from the contaminated air by transfer through cuticles and stomata, then through the endoderm, the ions bind to the cell wall and the plasma membrane (Shahid et al. 2017).

5.3 Trophic Transfer of Lead in the Food Chain

Trophic transfer of heavy metals in food chains is a topic discussed in relation with terms as bioconcentration, metal enrichment, bioaccumulation, biomagnification, biodilution, and trophodynamics (Ali and Khan 2019).

Trophic transfer, named also *biotransference* (Cardwell et al. 2013; Ali and Khan 2019), refers to transfer of a toxic compound through the food webs, from a low trophic level to the higher level. It is considered that trophic transfer is specific to the chemical pollutant-biological species system, meaning that trophic transfer and biomagnification can be specific to each trophic chain, and it has been demonstrated that there are differences between laboratory and field studies (Cardwell et al. 2013).

The accumulation process of heavy metals from the abiotic media—soil, sediments, and water in the body of a living organisms—depends on some factors related to the properties of toxic metal (speciation, pH, mobility, bioavailability) as well as the complex physiological mechanisms developed by the organisms for the metabolism, homeostasis, and detoxification of the contaminant (Ali et al. 2019; Ahmed et al. 2019).

The *trophodynamics* of lead in trophic chain (*biomagnification* or *biodilution*) has a major importance in assessing the risk pose by metals on biota and humans (Gao et al. 2021). *Biodilution* occurs in long trophic webs, while in short trophic chains, the phenomenon of *bioamplification* of lead occurs, which depends on biological species, community structure, environmental conditions, physiological parameters, and route of exposure (Griboff et al. 2018; Soliman et al. 2022).

Biodilution is defined as a decrease of the toxic pollutant concentration with increasing trophic position in the food chain (Ali and Khan 2019; Griboff et al. 2018). It has been established that lead trophodynamics varies in different food webs (Gao et al. 2021), in some food chain, lead concentration may biomagnify in biota (Rubio-Franchini and Rico-Martínez 2011; Tingson et al. 2019; Wei et al. 2016), while in other food chain tends to biodilute significantly (Zhang et al. 2017; Chen and Folt 2000; Gao et al. 2021; Hu et al. 2021; Gu et al. 2022). In general, terrestrial species show a strong pattern of biomagnification, while aquatic species does not follow a particular behavior pattern (Ahmed et al. 2019).

Plants, the foundation on food chain, represent the primary producers and a pathway of exposure to Pb of herbivorous animals (primary consumers), which accumulate Pb and are consumed by organisms located at higher levels (secondary or tertiary level). Primary producers represent the critical link in the trophic transfer of lead, because they ensure a permanent flow of metal ions between abiotic (soil, water) and biotic elements (Fig. 5.2). Thus, the species located on the top of the trophic web, including humans, tends to accumulate important levels of toxic substances through biomagnification. Life span is an important parameter involved in biomagnification, since organisms at higher trophic levels have a longer life span, which generates a prolonged period of exposure to toxic compounds, increasing the effects.

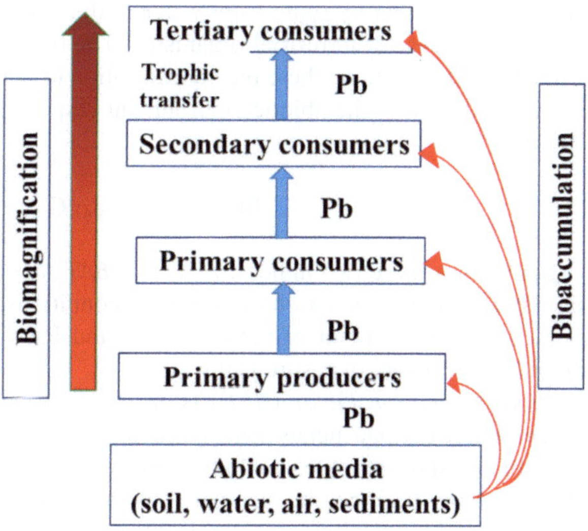

Fig. 5.2 The trophic transfer of lead in the food chain.

5.3.1 Tools for Bioaccumulative Potential Assessment

Usually, some indices such as *bioaccumulation factor* (BAF), *bioconcentration factor* (BAC), *transfer factor* (TF), *metal transfer factor* (MTF), *biota sediment accumulation factor* (BSAF), *biomagnification factor (BMF),* and *trophic magnification factor* (TMF) are applied for assessment of the accumulation, concentration, and transfer of lead into biota (van den Brink et al. 2015; Sun et al. 2020; Borgå et al. 2011; Savoca and Pace 2021). Most of these indices show the degree of accumulation of lead in living organisms in comparison to the degree of its accumulation in abiotic environment.

Transfer factor (*TF*), representing the soil-to-plant metal transfer factor, can be expressed as the ratio between the concentration of metal in plant and the concentration of metal in soil, and offers the possibility to evaluate mobility of lead from soil to plant tissues and organs (Eq. 5.1) (Zhang et al. 2021).

$$\mathrm{TF} = C_{\mathrm{plant}}/C_{\mathrm{soil}} \tag{5.1}$$

where, C_{plant} and C_{soil} represent the concentration toxic metal in dry weight vegetal tissues and soil, respectively.

Some terms, bioconcentration factor and biomagnification, are applied in correlation with bioaccumulation. *Bioconcentration factor* (*BCF*) is used to evaluate the retention of lead in a living organism with the exclusion of dietary intake, only through the respiratory route from water in the case of aquatic organisms, or from the air in the case of terrestrial organisms. The bioconcentration process is characterized by the situation where the concentration of lead in living organisms exceeds its concentration in the abiotic environment (Eq. 5.2) (Chojnacka and Mikulewicz 2014).

$$\mathrm{BCF} = C_{\mathrm{organism}}/C_{\mathrm{water}} \tag{5.2}$$

Sometimes these two indices, *BCF* and *BAF*, are confused because they have a similar way of determination in stationary conditions, but the differences are major from a conceptual point of view (Savoca and Pace 2021), considering that *BCF* excludes the dietary exposure pathway.

Bioaccumulation factor (*BAF*) is an indicator applied usually to quantify the accumulation of toxic metals from the environment by living organisms due to uptake through all exposure routes, including transport across the body surface, respiration, and ingestion. BAF can be calculated as the ratio between the concentration of a specific metal or metalloid in biological tissues and the concentration of the metal in the environment (soil, sediments, and water) (Eq. 5.3) (Ali et al. 2019).

$$\mathrm{BAF} = C_{\mathrm{organism}}/C_{\text{abiotic environment}} \tag{5.3}$$

where, C_{organism} is the lead concentration in the organism tissue and $C_{\text{abiotic environment}}$ is the lead concentration in the abiotic medium.

If the *BCF* or *BAF* of aquatic species is greater than 2000 ($\log_{10}$ BAF or $\log_{10}$ BCF > 3.3), then the toxic compound is classified bioaccumulative, when the *BCF* or *BAF* is greater than 5000 ($\log_{10}$ BAF or $\log_{10}$ BCF > 3.7), the compound is considered highly bioaccumulative (Savoca and Pace 2021).

The assessing of metals accumulation in aquatics organisms applied the *biota-sediment accumulation factor* (*BSAF*), expresses as ratio between the contaminant concentration in biota and the concentration of the contaminant in the sediments (Savoca and Pace 2021), and described by Eq. (5.4).

$$\text{BSAF} = \frac{C_{\text{organisms tissue}}}{C_{\text{sediment}}} \tag{5.4}$$

where, $C_{\text{organism tissue}}$ is the organism tissue concentration, C_{sediment} is the sediment concentration of toxic (Melake et al. 2023).

According to Kobkeatthawin and co-workers (2021), the *BSAF* values are classified into three main groups: (1) deconcentrator when *BSAF* has values lower than 1; (2) microconcentrator if *BSAF* has values between 1 and 2, and (3) macroconcentrator when *BSAF* takes values higher than 2.

Biomagnification factor (*BMF*) describes the tendency of toxic compound to concentrate when it is transferred along the food web, generating a higher concentration into the next trophic level (Córdoba-Tovar et al. 2022). In this way, the species on the top of food chain, especially humans, present risks of accumulating higher concentration of chemicals by biomagnification in food chain. Studies conducted by Huang (2016), Córdoba-Tovar et al. 2022) revealed that in aquatic systems, biomagnification is mainly influenced by environmental, ecological, and biological parameters. The *BMF* can be determined with Eq. (5.5):

$$\text{BMF} = C_{\text{organism}}/C_{\text{prey/diet}} \tag{5.5}$$

where, C_{organism} and $C_{\text{prey/diet}}$ are the metal concentrations in organism and prey or food, respectively (Liu et al. 2019; Sun et al. 2020; Chormare and Kumar 2022). Considering the definitions of *BAF* and *BCF* discussed previously, *BMF* could also be determined according to Eq. (5.6) (Savoca and Pace 2021):

$$\text{BMF} = \text{BAF}_{\text{predator}}/\text{BAF}_{\text{prey}} \tag{5.6}$$

Moreover, estimating the *BMF* based only on the predator/prey comparison is insufficient considering the very complex predator–prey relationships (Sun et al. 2020). However, it is unknown whether species at all trophic levels are exposed to the hazards generated by specific level of pollutants, and is recommended to adjust the data by applying *trophic magnification factor* (*TMF*) (Madgett et al. 2021). *TMF* measures the biomagnification over the entire food chain or part of it, this factor is determined by field measurement, representing a weighted average of BMF per trophic levels, and can be determined from the slope of linear regression of $\log_{10} C_{\text{organism}}$ versus the position n of the target organism in the trophic web, and is calculated according to Eq. (5.7) (Zhang et al. 2021). According to Madgett et al. (2021), trophic magnification factor assumes that food represents the main route of exposure to metals and trophic level is the main driver of bioaccumulation in the food chain.

$$\log_{10}\text{TMF} = \left[\log_{10}(\text{Corganism } n) - \log_{10}(\text{Corganism } 1)\right]/(n-1) \tag{5.7}$$

where, $C_{\text{organism tissue}}$ and $C_{\text{organism food}}$ represent the lead concentrations in organism tissue and in food, respectively.

Values of *TMF* exceeding 1 reveal the *biomagnification* of that contaminant along the food chain (Conder et al. 2012; Savoca and Pace 2021; Chormare and Kumar 2022), while a value less than 1 indicates *trophic dilution* (Sun et al. 2020; 2019; Madgett et al. 2021). However, it is recommended that *TMF* estimates should be based on representative statistical analyses, in order to eliminate false positive and false negative errors in quantifying the bioaccumulation potential (Conder et al. 2012).

5.3.2 Trophic Transfer in the Terrestrial Systems

In terrestrial ecosystems, soil-to-plants transfer of lead is considered the main route for the entry of lead into the food chain, and represents an important pathway of exposure of humans (Naikoo et al. 2019). A relevant study assessing the trophic transfer of lead and other metal species, was carried out by Angelova et al. (2010), the analysis included the chain soil—rapeseed—rabbit, the results showed the transfer of metals in rapeseed and later in rabbits fed predominantly with these plants, the lead accumulation was primarily realized in the kidneys and liver. The purpose of the study was to compare the results for two target groups of animals fed with plants harvested at a distance of 0.5 km and respectively, 15 km from the source of metal pollution. Interesting is the fact that the Pb bioaccumulation factor found in rapeseed plants increased in the following order: stems, leaves, and flowers, indicating a high soil-to-plant transfer factor. The study shows that for the vertebrate animals, that consume the contaminated plants, the concentration of lead in the liver, kidneys, muscles, bones, and blood of rabbits increased in the following order: muscle, blood, liver, bones, and kidneys. The results show a direct correlation of the transfer of the toxic compounds to tissues and organs at superior trophic levels, in relation to the concentration in the soil. Thus, the pollution factors (PF), calculated as ratios of the lead levels in the two groups of rabbits, show values between 1.19 and 3.12, the most representative effect was observed for bones with a PF equal to 3.12, followed by kidneys with PF = 2.80, which is due to the detoxification function of this organ, known as the tissues of metal accumulation (Akele et al. 2022; Angelova et al. 2010).

Another representative way of transferring pollutants to humans is the ingestion of animal products (meat and dairy products) contaminated with lead, its accumulation in animals such as cows, sheep, goats, and other animals is a source of global concern, especially due to the consumption of dairy products mainly in children's nutrition. The analysis of the transfer of lead from the soil to the grass and later in the milk for human consumption represents a way to identify and quantify the potential risks to human health. Some studies performed near an industrial metallurgical site from Peru, indicated average concentrations of lead in soil, grass, and cow milk samples of 217.81, 20.09, and 0.58 mg/kg, respectively. The transfer factors calculated for the soil to grass/plants system recorded values of 0.095, and for the grass-milk system, it had values of 0.031. It should be emphasized that the level of pollutant concentration in milk exceeded 29 times the maximum allowed values established at the European level (Chirinos-Peinado et al. 2021), representing a serious public health concern. This statement is supported by the high concentration in Pb in the investigated area, thus it was proven by the fact that 85% of the children had more than 10 mg/dL (ranged from 6.17 to 34.53 mg/dL) of Pb in the blood hemoglobin (Astete et al. 2009), and over 55% had symptoms of lead intoxication.

Lead transfer has also been studied for other terrestrial food chains that do not include humans, such as soil–plant-aphid-ladybird (Naikoo et al. 2019). The results demonstrated a bioamplification phenomenon in the soil-to-plants system, with a transfer coefficient higher than 1, but at the second trophic level, in aphids, the biodilution phenomenon occurred, which was also maintained at the third trophic level, respectively, in ladybird. Similarly, transfer of lead in the soil-plant-mealybug-ladybird beetle trophic chain, showed a large transfer in plants, 2–4 times in the soil-to-roots system, but a decrease in Pb transfer coefficients was observed for roots-stems and stems-mealybug (Zhang et al. 2017). A study conducted by Dar et al. (2017) had evaluated the mustard-aphid-beetle food chain and the results show lead mobility in the third trophic level, even though it has usually been proven that lead is concentrated predominantly in the root-stem transfer.

However, it should be mentioned that limited number of studies are available on the trophic transfer of lead in terrestrial biota, and in biological systems with complex trophic web that include big mammals and predators (Zhang et al. 2021).

5.3.3 *Trophic Transfer in the Aquatic Systems*

The most representative pathway for human health exposure of lead as well as other metals is considered fish and seafood consumption (Ahmed et al. 2019; Lee et al. 2019; Sofoulaki et al. 2019); according to Griboff et al. (2018), accounting over 90% of the total exposure, followed by other routes such as skin contact and inhalation. Numerous studies have investigated bioaccumulation and trophic transfer of lead and other metals to different species of phytoplankton, zooplankton, crustacean, mollusks, fish in marine and freshwater ecosystems (Rubio-Franchini et al. 2016; Tingson et al. 2019; Lee et al. 2019; Sofoulaki et al. 2019; Gao et al. 2021; Madgett et al. 2021; Latif et al. 2022; Hu et al. 2023), the assessment of bioaccumulation in fish may be used as bioindicator for water quality evaluation (Mondal et al. 2018; Ahmed et al. 2019; Varol et al. 2022).

Madgett et al. (2021) showed that a large number of physiological and ecological parameters are involved in the bioaccumulation and transfer of heavy metals in the aquatic environment, such as age and dimension of the species, feeding habits, habitat, sex, exposure duration, type of affected tissue, etc. Lead speciation in marine aquatic systems is influenced by the presence of carbonates, chlorides, and natural ligands. The formation of inorganic complexes with lead depends on the pH of the water, the concentration of these complexes is proportional to the total concentration of lead in the sediments, generating potential hazards mainly for benthic organisms, but also for marine biota (Zuluaga Rodríguez et al. 2015).

There are three main pathways for lead uptake in fish via skin, gills, and dietary (Ali and Khan 2019; Sofoulaki et al. 2019; Latif et al. 2022). Skin and gills are organs which facilitate the interaction between soluble forms of lead and the fish body, the transfer of ions occurs at gills epithelium level by binding to the negative charged sites at the membrane level (Latif et al. 2022). The muscles tissues are the edible parts for human consumption and from this point of view, the muscles are the most representative organ in the assessment of lead accumulation in fish (Taweel et al. 2013). Furthermore, some researchers have shown that for marine environments, ideal indicators for the bioaccumulation analysis of Cu, Cr, Pb, Zn, Cd, and Ni are the tissues from the liver, gills, and muscles (Yin et al. 2018; Gao et al. 2021).

The exposure of fish to lead may induce toxic effects on membrane structure, generate the inflammation of the gut, cause oxidative stresses, neurotransmitter malfunction, changes in immunological responses, and other function due to its high affinity to red blood cells (Yin et al. 2018; Lee et al. 2019; Hu et al. 2023).

Studies on the bioaccumulation of lead (Łuszczek-Trojnaret al. 2013) in selected tissues of *Carassius gibelio* (Prussian carp) showed that lead concentrations accumulated in organs ranged from 2.0 to 7.4 mg/kg in the kidney, 3.0 to 5.2 mg/kg in the bone, and 4.5 mg/kg into hepatopancreas, by dietary exposure. A high rate of bioaccumulation was found at the beginning of the exposure period, the highest rates were recorded for muscles, hepatopancreas, intestines, and gills, suggesting a potential risk for human health by dietary exposure. Evaluation of the lead depuration revealed that the process depends on the type of organ, indicating a high elimination rate from soft tissues, while a very low depuration rate was recorded for scales and bones.

In addition, studies carried out under controlled conditions of lead exposure of the fish juveniles of some edible species *Tor putitora* (Mahseer) and *Ctenopharyngodon idella* (grass carp) have demonstrated that an exposure time of 60 days generated a predominant bioaccumulation in the following organs: gills > liver > intestine > swim bladder > muscle > skin, for both species. Moreover, the omnivorous species bioaccumulate higher concentrations of lead (Latif et al. 2022), as the result of the trophic transfer at superior trophic level. Ahmed et al. (2019) investigated costal ecosystem from Bangladesh in order to assess the bioaccumulation index in six species of commercially fishes, used for human consumption, study led to the conclusion that among all analyzed metals, lead is found in the highest concentration (13.88 mg/kg) and the bioaccumulation factor (BAFs) for Pb had a value of 913.66.

An extensive study carried out in a marine system, at different trophic levels, the food web included bivalves (4), gastropods (3), crustaceans (4), and fishes (6), showed that Pb in muscle tissues presents biodilution and Pb trophic magnification factors varied between 0.44 and 0.73 (Gao et al. 2021). An investigation of lead transfer along a trophic chain composed by four trophic levels: phytoplankton, zooplankton, shrimp, and fish; exhibited a concentration of Pb in fish 2–3 times higher than in control samples. Trophic level I accumulated metal with a bioconcentration factor from 930 to 3630, at the next trophic level in zooplankton, there was recorded a significant decrease of metal bioaccumulation, while at level III, from zooplankton to shrimp, BAF < 1, thus, at trophic level IV, in fish, Pb presented a BAF > 1.0. The high concentration of Pb (> 3 μg/g), distributed throughout the body of shrimps and fish, but predominantly in the muscles, leads to the conclusion that lead can generate risks for the health of consumers (Soto-Jiménez et al. 2011).

The biomagnification of lead in fish has been described in the study carried out in a freshwater ecosystem, Aras River, Iran, in the trophic chain represented by crustacean—fish, the amount of lead in zander muscle (*Sander lucioperca*) ranged between 0.51 and 0.93 μg/g dry weight and the corresponding concentrations in amphipods (*Gammarus* sp.) were 0.40–0.66 μg/g of dry weight. The results displayed a direct correlation of heavy metals between amphipods and zanders (Dehghani et al. 2022).

Hu et al. 2021 analyzed heavy metals in a food web consisting of the following trophic groups aquatic plants, crustaceans, mollusks, and fish, the results showed a tendency of Cd, Cr, Cu, and Pb to be efficiently biodiluted with the transition to a superior trophic level, presenting values of *TMF* lower than 1. However, the highest concentrations were recorded in mollusks, which are edible species. Moreover, Nasri et al. (2017) demonstrated the accumulation of lead by feeding in the species *Acanthodactylus boskianus* and its transfer from aquatic food webs to terrestrial food webs.

Generally, the transfer of HMs, and especially lead, through the food chain and the net accumulation in high concentrations in high level trophic species are a major issue in aquatic ecosystems and are particularly relevant to human food safety.

Moreover, understanding of trophic transfer of the metals in complex ecosystems is not totally known, the studies carried out use the bioassay method, which is simplified compared to natural complex systems (Soto-Jiménez et al. 2011). However, these studies can contribute to the development of models for human health risk assessment.

5.3.4 Bioindicators and Biomarkers

A number of microorganisms, plant, and animal species have been identified to have potential to be applied as biological indicators in order to evaluate or monitor pollution with certain heavy metals. *Bioindicator* is a term used to define biological species applied as sensitive tool in measuring or prediction of environmental stress, in monitoring synergetic and antagonistic effects of chemical pollutants on biota (Azzazy 2020; Parmar et al. 2016). Biomarkers and bioindicators are widely used as a tool to measure the response to chemical stress factors.

Species of mosses (*Hylocomium splendens*), lichens or vascular plants (*Phragmites australis, Cyperus rotundus, Eichhornia crassipes, Lygeum spartum, Atractylis serratuloides, Gymnocarpos decander*) have been considered bioindicators of environmental pollution with heavy metals and metalloids, including lead (Ali et al. 2008; Parmar et al. 2016; Jiang et al. 2018; Bayouli et al. 2021). Gupta et al. (2019) reported that lichen species were used as bioindicators for different isotopes of Pb. Mosses are the most used group as bioindicators and bioaccumulators of metal deposition in the environment, having a consistent application in the last decades (Jiang et al. 2018). The high sensitivity of plants to specific heavy metals reveals that a wide variety of plants can be bioindicators of heavy metal pollution in certain ecosystems.

Acoording to Pal et al. (2023), vascular plants (*Helianthus annum, Ficus religiosa*), lichen species (*Graphis scripta*), or invertebrates, insects (*Apis mellifera, Polistes*), and annelids (Hediste diversicolor), starfish, and birds (*Passer domesticus*) are the most common bioindicators for the detection of lead (Pb). Other species are indicators of quality; thus, frogs are considered bioindicators indicating changes in the environment, being sensitive species (Parmar et al. 2016). It is well known that fish have been considered indicator species of heavy metals in aquatic ecosystems, by determining some biochemical, haematological, or histological parameters of fish, the quality of the aquatic ecosystem can be estimated. Thus, it was found that fish *Cyprinus carpio* is a bioindicator species of Pb in water (Latif et al. 2022).

Environmental *biomarkers* include measurable parameters (chemical indicators, proteins, genes, etc.) useful to developing models regarding molecular toxicity mechanisms in different species of the investigated ecosystem (Azzazy 2020). Biological markers have the great advantage that presents the ability to illustrate the indirect biotic stress of pollutants, while other categories of analyses fail to achieve this goal (Zaghloul et al. 2020). Thus, has been proved that the presence of flavonoids and phenolic compounds reveal chemical stress in plants (Mongkhonsin et al. 2016; Jańczak-Pieniążek et al. 2022), also, the decreasing in the photosynthesis process with the decreasing in the chlorophyll pigments content, closing of the stomata and the accumulation of reactive oxidative species (ROS) (Krystofova et al. 2009; Azzazy 2020) show the heavy metal stress in plants. Phenolic compounds in plants are involved in scavenging ROS and chelating heavy metals (Michalak 2006; Mongkhonsin et al. 2016), a high number of phenolic acids with different capacities of action have been identified in plants (Chen et al. 2020).

In risk assessment, biomarkers can be useful tool for hazard identification and exposure assessment. The main benefits of using biomarkers in risk assessment summarize all exposure pathway at metal (Filipoiu et al. 2022). Usual *human biomarkers* applied for toxicological assessment are represented by blood, urine, nails, and hair. The most reliable biomarkers for human exposure to lead are blood lead (Pb-B), urine lead (Pb-U), and plasma lead (Pb-P). Pb-B is the most used biomarker and is determined in red cells, with the increase of the value of this marker, the values of Pb-U and Pb-P also increase and indicate a recent exposure situation (Sakai 2000).

The results of bioavailability of Pb tests showed that the half-life of Pb is 1 h in plasma, approximately 25–35 days in blood, 40 days in soft tissues, and 28 years in bones (Alimonti and Mattei 2008; Deshommes et al. 2012; Kumar et al. 2020). According to data presented by Kumar et al. (2020), an adult with an average weight of 70 kg accumulates an average of 120 mg of Pb, distributed as such 0.2 mg/L in blood, 5–50 mg/kg in bones, and 0,2–3 mg/kg in different tissues.

Some studies reported that the information provided by individual biomarkers are limited, therefore, the integrated use of biomarkers with bioindicators is recommended for a relevant and reliable assessment (Araújo et al. 2018).

5.4 Bioavailability of Lead

5.4.1 Bioavailability of Lead in Soil and Plants

Currently, the bioavailability of metals in the soil is a useful tool in human health risk assessment procedure (Yan et al. 2017). There is no comprehensive and fully accepted definition of this concept, however, bioavailability is recognized as a dynamic process (Mebane et al. 2020; Deshommes et al. 2012) that includes three stages: the first includes processes of a physicochemical nature called the *availability of the environment*, the second includes the process of uptake, represented by the *environment bioavailability,* and the third stage generates physiological changes or accumulation processes in organisms, called the *toxicological bioavailability* (Kim et al. 2015).

The bioaccumulation in plants is depending on the mobility of lead, and it is known that in natural conditions, lead is one of the least mobile metals in soil (Rolka and Wyszkowski 2021). The bioavailability of lead in soil depends on several factors such as soil reaction, organic matter content, clay minerals content, soil particle size, texture, capacity for cation exchange, organic colloids, aluminum and iron oxides, soil salinity, carbonates content, organic carbon content, total nitrogen, cation exchange capacity, aluminum compounds, the content of sulphate and phosphate, redox status, soil microorganisms activity, aeration, presence of amendments (Sharma and Dubey 2005; Keran et al. 2008; Ming et al. 2012; Kushwaha et al. 2018; Wieczorek et al. 2018; Kumar et al. 2020; Li et al. 2020; Rolka and Wyszkowski 2021; Vasile et al. 2021; Abedi et al. 2022; Mousavi et al. 2022; Wijayawardena et al. 2023).

Generally, Pb present in the soil matrix is strongly bonded to the organic matter or colloidal material or can exist in a precipitated form, which reduce the mobility of the compound and implicitly decrease the bioavailability for plant roots uptake and animals (Sharma and Dubey 2005).

However, some studies have shown that the bioavailability of Pb in soils is relatively high (Lamb et al. 2009; Oorts et al. 2021), it depends on the lead chemical form, thus soils freshly spiked with lead in form of salts indicated a higher bioavailability than long-term aged soils (Ming et al. 2012). The most stable forms of Pb in the environment are represented by the ionic forms, Pb (II), oxides and hydroxides and complexes with Pb oxyanions, under reducing conditions, has been established that Pb sulfides are the most stable forms (Vasile et al. 2021), while, Pb (II) is the most common and sensitive form of Pb (Mousavi et al. 2022). In the form of complexes, lead is immobilized at the soil matrix level, being unavailable for plant uptake and transport (Aslam et al. 2021).

The availability of lead in plants is influenced by the plant factors such as roots surface area, roots exudates, the presence of mycorrhizae, and rate of transpiration (Sharma and Dubey, 2005; Aslam et al. 2021). In addition, at rhizosphere level, the Pb tends to bind to carboxyl groups of the carbohydrates galacturonic acid and glucuronic acid in the cell wall, forming a protective barrier of the root system, which has the effect of limiting transport to the alloplasts (Sharma and Dubey 2005; Collin et al. 2022). It was found that lead moves mainly through the root apoplasts, crosses the cortex, and accumulates in the endoderm (Collin et al. 2022).

Absorption process in the root cells is governed by binding of Pb ions on the ion exchangeable sites located in the cell wall and the transport of extracellular precipitate, in the form of Pb carbonate that will accumulate in the cell wall (Sharma and Dubey 2005). Some authors reported that dicotyledonous plants can uptake higher amount of metal in their roots than monocotyledonous plants (Sharma and Dubey 2005).

The presence of other metal species in soil can influence the bioavailability of lead in soil, for example, study performed by Kushwaha et al. (2018) demonstrated a strong positive correlation between the concentration of Pb and Cd in the soil and their bioavailability. In addition, the relationships between metals are complex and further research should be performed for assessing the correlations of the metals bioaccessibility and for a comprehensive understanding these relationships (Wijayawardena et al. 2023).

5.4.2 Lead Bioaccessibility in Animals and Humans

Lead bioavailability in relation to animal species and humans refers to the fraction actually available and taken up by an organism, while bioaccessibility represents the fraction that interacts with dermal contact surface and is potentially available to be absorbed and adsorbed by the organism (McGeer et al. 2004; Wijayawardena et al. 2023). Absolute bioavailability (*ABA*) refers to the fraction of an ingested lead dose that crosses the gastrointestinal epithelium and is available for distribution to tissues and internal organs. Relative bioavailability (*RBA*) can be calculated as the ratio between the absolute bioavailability value of lead in soil and a value of a water-soluble reference form of lead. *ABA* and *RBA* are usually determined by applying animal models (EPA 2021).

In vivo and in vitro measurements were developed to determine the bioavailability of lead. In vivo models are considered to be expensive and complex, requiring a long time for application, while in vitro models have many advantages, are simply, highly reproducible, and easier to be managed, do not involve ethical elements, nevertheless may have a high degree of uncertainty. Uncertainties in various measurements create difficulties in developing predictability models (Deshommes et al. 2012; Yan et al. 2017).

Currently, to estimate the relative bioavailability (*RB*) of lead in soil, several animal models such as rats, mice, rabbits, monkeys, and pigs are used; the method is based on determining the Pb concentration in biological fluids (blood) or tissues (bones, liver, kidneys), and the measured parameters indicate the absorbed dose of metal (Deshommes et al. 2012). The models are indicative because, due to the physiological differences, they cannot supply reliable data regarding the bioavailability of Pb for humans (Kumar et al. 2020).

For in-vitro studies of a series, models for estimating single metal and metalloids bioaccessibility were developed and applied over time, such as Relative Bioavailability Leaching Procedure (RBALP), the Solubility Bioaccessibility Research Consortium assay (SBRC), Biotic ligand model (BML), Simplified Physiologically-Based Extraction Test (SBET), Physiologically-Based Extraction Test (PBET), In Vitro Gastrointestinal (IVG), In Vitro Digestion Model (RIVM), and the Unified BioAccessibility Research Group Europe (BARGE), other applied models are Bbiomet, PNEC-pro, and Biotic Ligand Model (BLM) (Kumar et al. 2020; Mebane et al. 2020; Wijayawardena et al. 2023. https://arche-consulting.be/tools/lead-chronic-biotic-ligand-model/).

McGree et al. (2004) claim that mechanistic models for the evaluation of bioavailability, bioaccumulation, and toxicity of metals are more reliable and justified in comparison to empirical methods, but require new approaches and studies in this direction. A Biotic Ligand Model (BLM) was developed for estimating the toxicity of lead in aquatic environment, this includes, in the analysis, the geochemical balance of the exposure environment and the relationship with the exposure conditions, as well as the characteristics of the exposed organism (*biotic ligand*). The BLM model combines the factors influencing lead speciation in water as well as various abiotic parameters (McGree et al. 2004; https://arche-consulting.be/tools/lead-chronic-biotic-ligand-model). Such BLM models have been developed and successfully applied for metals from freshwater—aluminum, cadmium, cobalt, copper, nickel, lead, zinc; also, for marine ecosystems—copper, nickel, and zinc.

5.5 Pathways of Human Exposure to Lead

Lead (Pb), a non-essential metal(loid) for living organisms, can be found in very low levels in the Earth crust (around 0.002%), therefore, background levels of Pb can be retrieved in soils based on geological sources. However, anthropogenic sources have a major role in the increasing of Pb concentrations in soil, mining activities being the most significant source of soil HMs contamination (Kumar et al. 2022a, b; Nag and Cummins 2022). Due to its properties (corrosion resistant, soft in texture, malleable, with good electrical conductivity) (NHMRC 2015; Efanny et al. 2019), lead has a large diversity of uses in manufacturing (glass industry, leather tanning, paint, metal plating, battery manufacturing). Environmental contamination with Pb may also occur during burning of coal, gas emission from vehicles, smelting of ores, oil combustion, agricultural practices (fertilizers, pesticides), as well as, from industrial and domestic wastes (Efanny et al. 2019; Kumar et al. 2022a, b) (Fig. 5.3).

General population may be exposed to lead via *ingestion* (oral consumption of Pb-contaminated water and food—cereals, vegetables, fruits, fish, eggs, meat), *inhalation* (direct breathing of Pb-laden dust, particulate matter, tobacco smoke), or *dermal exposure* (contaminated soil and dust) (Kumar et al. 2022a, b) (Fig. 5.3).

Lead, an toxic, hazardous, and persistent metal was classified as being probably carcinogenic to humans (Group 2A) by the International Agency for Research on Cancer (IARC), while World Health Organization (WHO) reported it among the ten chemicals of major public health threat (Cozma et al. 2019; EFSA 2012; Wang et al. 2019). Exposure to lead over a short period of time (*acute exposure*) may induce coma, shock, constipation, diarrhea, vomiting, seizures, or even death. Continuous exposure to lead across a longer period of time (*chronic exposure*) may decrease blood pressure and mental ability of brain, induce memory loss and anemia, damage kidney, liver, and brain and even may cause reproductive problems (to adults) (NHMRC 2015; Efanny et al. 2019; Aslam et al. 2021; Zyambo et al. 2022; Hoseini et al. 2023).

There are several papers reported on the evaluation of heavy metals concentration in different contaminated media (air, water, soil) and as a consequence, the presence of HMs in different food categories (cereals, fruits, eggs, fish, meat) and not only, is investigated. These works are very important to initiate epidemiological studies in order to estimate the health risks related to these pollutants. For example, Hoseini

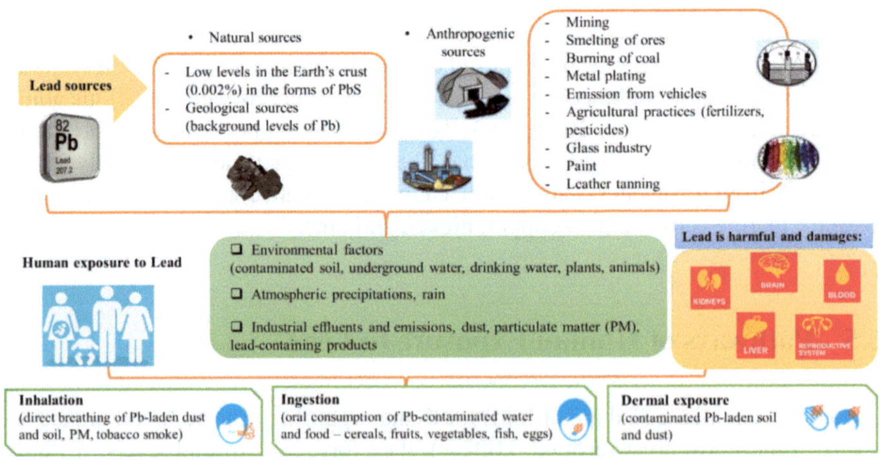

Fig. 5.3 Sources of lead in the environment and pathways to human exposure

et al. (2023) analyzed the level of Pb in hen eggs (considering a total of 42 eggs from 17 major brands), one of the main sources of nutritious food from Iran. The average Pb concentration in all samples was 7.16 ± 0.248 μg/kg, being however, lower than the maximum permitted levels established by the Joint Food and Agriculture Organization of the United Nations/World Health Organization (FAO/WHO) and the Institute of Standards and Industrial Research of Iran (ISIRI) (500 μg/kg and 50 μg/kg, respectively) (Hoseini et al. 2023). Di Bella et al. (2021) analyzed, among others, the concentration of trace metals (Cd, Pb, Hg) in the muscle tissue of farmed sea bass marketed in Sicily (Southern Italy). The mean concentrations of trace metals were as follows: 0.031422 μg/g, 0.109931 μg/g, 0.023099 μg/g wet weight for Cd, Pb, and Hg, respectively. However, the levels of Cd, Pb, and Hg were lower than the legal threshold of trace metals in fish muscle (mg/kg wet weight) established by European Commission/European Union and FAO (0.05/0.5 for Cd, 0.3/0.2 for Pb, and 0.5/0.5 for Hg). Baldi et al. (2021) studied the bioaccumulation and translocation of lead in seven herbaceous plants (barley, castor bean, common bean, Indian mustard, sorghum, spinach, and tomato) grown in urban and peri-urban soil (from Montepaldi region, Italy) polluted with lead in different concentrations (300, 650, 1000 mg Pb/kg). Even none of the species were not classified as lead hyperaccumulators, the Pb concentration in the edible parts of the plants exceeded the safe limit set by FAO/WHO. For example, barley and sorghum accumulated a Pb concentration of 3.4–30.3 and respectively, 8–27.2 times higher than the maximum allowable safe limit (ML) imposed by the Codex Alimentarius. Thus, numerous studies carried out in various geographical regions, on basic food categories, have shown that the presence of lead was detected in different concentrations, many samples exceeding the maximum allowable safe limit (ML), data that are summarized in Table 5.1. Also, it is concerning that in infant foods, the lead was identified in high level, in all the analyzed data exceeds the MCL (Table 5.1).

It is well known that lead has a poor bioavailability for plants due to its extreme insolubility compared to other metals, and as a consequence, only 5% of Pb is transported into aerial components and 95% remains accumulated in roots (Aslam et al. 2021; Roșca et al. 2021).

Ahmed et al. (2020) conducted a study to evaluate the potential health risk of workers exposed to HMs, from one of the largest plastic companies (from Dhaka region, Bangladesh). In this regard, blood samples of the workers were collected based on ages and smoking status, considering also the measurement of indoor industrial dust samples. The results indicated a tendency of the toxicity of the HMs in human blood indicated as below: Pb > Zn > Ni > Cd, while the highest concentration level ranges were registered for Pb (14.50–48.00 μg/L). Regarding the concentration of HMs collected from dust samples, higher concentrations were obtained for Pb (47.24 mg/kg) in the pellet manufacturing section and for Cd (4.20 mg/kg) in the plastic waste recycling section (Ahmed et al. 2020). In a study performed in Iran, from 200 subjects (workers exposed to lead), with a mean age of 45.8 ± 11.8 years, from which 92% were male, the mean blood lead level (BLL) was 27.77 ± 39.45 μg/dL. The final results suggested that people with chronic lead exposure with BLL higher than 10 μg/dL are at risk of renal, liver, and hematologic disorders (Nakhaee et al.

Table 5.1 The content of lead in different food categories

Product	Mean concentration (mg kg^{-1})	Max concentration (mg kg^{-1})	Sample no	Country/region	EU standard (2021/1317) (mg kg^{-1})	FAO (Codex Alimentarius 2019) (mg kg^{-1})	References
Cereals and pulse							
Wheat	16.98	20.61	9	Punjab, India	0.20	0.2	Sharma et al. (2018)
Rice	17.13	20.30	13	Punjab, India	0.20	0.2	Sharma et al. (2018)
Maize	18.28	21.30	5	Punjab, India	0.20	0.2	Sharma et al. (2018)
Rice	0.058	–	6	Iran, Isfahan	0.20	0.2	Sanaei et al. (2021)
Rice	0.15	0.6	38	China, Guangxi Province	0.20	0.2	Zhang et al. (2022)
Rice	0.011 ± 0.015	0.069	162	USA	0.20	0.2	Norton et al. (2014)
Rice	0.185 ± 0.004	2.749	122	China (mine impacted)	0.20	0.2	Norton et al. (2014)
Rice	0.046 ± 0.064	0.333	88	China	0.20	0.2	Norton et al. (2014)
Rice	0.012 ± 0.027	0.130	24	France	0.20	0.2	Norton et al. (2014)
Rice	0.023 ± 0.04	0.266	64	India	0.20	0.2	Norton et al. (2014)
Rice	0.024 ± 0.026	0.333	43	Ghana	0.20	0.2	Norton et al. (2014)

(continued)

Table 5.1 (continued)

Product	Mean concentration (mg kg^{-1})	Max concentration (mg kg^{-1})	Sample no	Country/region	EU standard (2021/1317) (mg kg^{-1})	FAO (Codex Alimentarius 2019) (mg kg^{-1})	References
Brown rice (in μg/kg)	7.4 μg/kg	34 μg/kg	28	USA, Louisiana	0.20	0.2	TatahMentan et al. (2020)
White rice	5.6 μg/kg	32 μg/kg	11	USA, Louisiana	0.20	0.2	TatahMentan et al. (2020)
White rice	14 μg/kg	96 μg/kg	11	USA (origin Italy, India, Thailand)	0.20	0.2	TatahMentan et al. (2020)
Grains (lentils, barleys, beans, oats, wheat and pea)	9.7 μg/kg	80 μg/kg	–	USA, Louisiana	0.20	0.1	TatahMentan et al. (2020)
Black gram	0.70	–	–	Bangladesh	0.20	0.1	Kumar et al. (2022a, b)
Chickpea	0.90	–	–	Bangladesh	0.20	0.1	Kumar et al. (2022a, b)
Lentil	0.53	–	–	Bangladesh	0.20	0.1	Kumar et al. (2022a, b)
Grass peas	1.08 [1]	–	–	Bangladesh	0.20	0.1	Kumar et al. (2022a, b)
Mung bean	0.51	–	–	Bangladesh	0.20	0.1	Kumar et al. (2022a, b)
Beans	0.054	–	6	Iran, Isfahan	0.20	0.1	Sanaei et al. (2021)
Root and tuber vegetables							
Beetroot (dried)	0.115	0.348	5	Poland	0.1	0.1	Rusin et al. (2021)
Carrot (dried)	0.206	0.348	7	Poland	0.1	0.1	Rusin et al. (2021)

(continued)

Table 5.1 (continued)

Product	Mean concentration (mg kg^{-1})	Max concentration (mg kg^{-1})	Sample no	Country/region	EU standard (2021/1317) (mg kg^{-1})	FAO (Codex Alimentarius 2019) (mg kg^{-1})	References
Celery (dried)	0.259	0.259	4	Poland	0.1	0.1	Rusin et al. (2021)
Tomatoes (dried)	0.081	0.133	5	Poland	0.1	0.1	Rusin et al. (2021)
Ginger	0.11	–	–	India, Solapur district	0.1	0.1	Mawari et al. (2022)
Garlic	0.774	–	–	India, Solapur district	0.1	0.1	Mawari et al. (2022)
Onions	0.046	–	–	India, Solapur district	0.1	0.1	Mawari et al. (2022)
Carrot	0.087	–	–	India, Solapur district	0.1	0.1	Mawari et al. (2022)
Sugarcane	1.741	–	–	India, Solapur district	0.1	0.1	Mawari et al. (2022)
Carrot	0.09	2.11	100 g	Romania, Banat County	0.1	0.1	Manea et al. (2020)
Parsley root	0.66	2.45	100 g	Romania, Banat County	0.1	0.1	Manea et al. (2020)
Onion	0.054	–	6 types	Iran, Isfahan	0.1	0.1	Sanaei et al. (2021)
Potato	0.087	–	6 types	Iran, Isfahan	0.1	0.1	Sanaei et al. (2021)
Potato	0.45	1.06	100 g	Romania, Banat County	0.1	0.1	Manea et al. (2020)
Onion	0.13	0.5	100 g	Romania, Banat County	0.1	0.1	Manea et al. (2020)

(continued)

Table 5.1 (continued)

Product	Mean concentration ($mg\ kg^{-1}$)	Max concentration ($mg\ kg^{-1}$)	Sample no	Country/region	EU standard (2021/1317) ($mg\ kg^{-1}$)	FAO (Codex Alimentarius 2019) ($mg\ kg^{-1}$)	References
Potato	0.244	0.855	–	Slovak Republic	0.1	0.1	Musilova et al. (2015)
Potato	0.542	–	–	India, Solapur district	0.1	0.1	Mawari et al. (2022)
Potato	0.064	–	45	Romania, Galați County	0.1	0.1	Bora et al. (2022)
Parsley	0.109	–	–	Romania, Copsa Mica	0.1	0.1	Muntean et al. (2010)
Potatoes	5.272	–	–	Romania, Copsa Mica	0.1	0.1	Muntean et al. (2010)
Fruiting vegetables							
Long Cucumber	0.076	–	80	Romania, Galați County	0.05	0.1	Bora et al. (2022)
Cucumber	0.019	0.72	100 g	Romania, Banat County	0.05	0.1	Manea et al. (2020)
Tomato	0.045	0.093	45	Romania, Galați County	0.05	0.1	Bora et al. (2022)
Tomato		0.85	100 g	Romania, Banat County	0.05	0.1	Manea et al. (2020)
Tomato	0.016			Poland	0.05	0.1	Rusin et al. (2021)
Tomato	2.63 ± 0.11	–	1 kg	Ethiopia, Koka area	0.05	0.1	Bayissa and Gebeyehu, (2021)

(continued)

Table 5.1 (continued)

Product	Mean concentration (mg kg^{-1})	Max concentration (mg kg^{-1})	Sample no	Country/region	EU standard (2021/1317) (mg kg^{-1})	FAO (Codex Alimentarius 2019) (mg kg^{-1})	References
Green beans	0.19	0.35	100 g	Romania, Banat County	0.05	0.1	Manea et al. (2020)
White Beans	43.60	59.45	45	Romania, Galați County	0.05	0.1	Bora et al. (2022)
Zucchini	0.028	32.75	45	Romania, Galați County	0.05	0.1	Bora et al. (2022)
Sugarcane	1.741	–	–	India	0.05	0.1	Mawari et al. (2022)
Sweetcorn	0.16	–	–	Brasil, São Paulo State	0.1	0.05	Guerra et al. (2012)
Corn	0.141	–	–	Romania, Copsa Mica	0.1	0.05	Muntean et al. (2010)
Vegetable leaf							
Cabbage	7.26 ± 0.05	–	1 kg	Ethiopia, Koka area	0.1	0.1	Bayissa and Gebeyehu, (2021)
Sugar beet leaf	90.23 ppm	149.50	5	India, Bangalore District	0.1	0.3	Ramesh and Murthy (2012)
Coriander	64.47 ppm	75.50	5	India, Bangalore District	0.1	0.3	Ramesh and Murthy (2012)
Lettuce	0.21	0.62	10	Romania, Banat County	0.3	0.1	Manea et al. (2020)
Lettuce	0.073	0.094	–	Romania, Galați County	0.3	0.1	Bora et al. (2022)

(continued)

Table 5.1 (continued)

Product	Mean concentration (mg kg^{-1})	Max concentration (mg kg^{-1})	Sample no	Country/region	EU standard (2021/1317) (mg kg^{-1})	FAO (Codex Alimentarius 2019) (mg kg^{-1})	References
Parsley leaf	0.5	1.97	100 g	Romania, Banat Area	0.1	0.3	Manea et al. (2020)
Chard	0.58	–	–	Brasil, São Paulo State	0.1	0.3	Guerra et al. (2012)
Watercress	0.86	–	–	Brasil, São Paulo State	0.1	0.1	Guerra et al. (2012)
Iceberg Lettuce	0.41	–	–	Brasil, São Paulo State	0.3	0.1	Guerra et al. (2012)
Coriander	1.24	–	–	Brasil, São Paulo State	0.1	0.1	Guerra et al. (2012)
Cabbage	1.66	–	–	Brasil, São Paulo State	0.1	0.1	Guerra et al. (2012)
Broccoli Brassica oleracea L	0.99	–	–	Brasil, São Paulo State	0.1	0.1	Guerra et al. (2012)
Cauliflower *Brassica oleracea* var *botrytis*	0.36	–	–	Brasil, São Paulo State	0.1	0.1	Guerra et al. (2012)
Spinach *Tetragonia expansa*	1.05	–	–	Brasil, São Paulo State	0.3	0.30	Guerra et al. (2012)
Endive *Cichorium endivia* L	0.49	–	–	Brasil, São Paulo State	0.3	0.30	Guerra et al. (2012)
Purple Cabbage *Brassica oleracea* L	0.63	–	–	Brasil, São Paulo State	0.1	0.1	Guerra et al. (2012)

(continued)

Table 5.1 (continued)

Product	Mean concentration (mg kg^{-1})	Max concentration (mg kg^{-1})	Sample no	Country/region	EU standard (2021/1317) (mg kg^{-1})	FAO (Codex Alimentarius 2019) (mg kg^{-1})	References
White Cabbage *Brassica oleracea* L. var. capitata	0.60	–	–	Brasil, São Paulo State	0.1	0.1	Guerra et al. (2012)
Parsley *Petroselinum crispum* (Mill.) Nym	1.02	–	–	Brasil, São Paulo State	0.1	0.1	Guerra et al. (2012)
Celery *Apium graveolens* L	0.47	–	–	Brasil, São Paulo State	0.1	0.1	Guerra et al. (2012)
Rocket *Eruca sativa* L	0.76	–	–	Brasil, São Paulo State	0.3	0.3	Guerra et al. (2012)
Crisphead Lettuce *Lactuca sativa* L	0.48	–	–	Brasil, São Paulo State	0.3	0.3	Guerra et al. (2012)
Smooth Lettuce *Lactuca sativa* L	0.44	–	–	Brasil, São Paulo State	0.3	0.3	Guerra et al. (2012)
Vegetables	0.148	–	6	Iran, Isfahan	0.1	0.3	Sanaei et al. (2021)
Tea	0.002		6 types	Iran, Isfahan	-	-	Sanaei et al. (2021)
Fruits							
Apple	0.199	–	–	Romania, Copsa Mica	0.1	0.1	Muntean et al. (2010)
Apple	0.13	0.56	100 g	Romania, Banat County	0.1	0.1	Manea et al. (2020)
Apple	0.009	0.024	57	Poland	0.1	0.1	Rusin et al. (2021)
Pear	0.008	0.017	12	Poland	0.1	0.1	Rusin et al. (2021)

(continued)

Table 5.1 (continued)

Product	Mean concentration (mg kg^{-1})	Max concentration (mg kg^{-1})	Sample no	Country/region	EU standard (2021/1317) (mg kg^{-1})	FAO (Codex Alimentarius 2019) (mg kg^{-1})	References
Peach	0.16	0.77	100 g	Romania, Banat County	0.1	0.1	Manea et al. (2020)
Plum	-	0.49	100 g	Romania, Banat County	0.1	0.1	Manea et al. (2020)
Grape	-	0.71	100 g	Romania, Banat County	0.1	0.1	Manea et al. (2020)
Cranberry	0.004	0.005	4	Poland	0.2	0.2	Rusin et al. (2021)
Grapes	0.005	0.009	15	Poland	0.1	0.1	Rusin et al. (2021)
Grapes	0.024	–	45	Romania, Galați County	0.1	0.1	Bora et al. (2022)
Raspberry	0.012	0.033	7	Poland	0.2	0.1	Rusin et al. (2021)
Strawberry	0.009	0.027	49	Poland	0.2	0.1	Rusin et al. (2021)
Strawberry	0.01078	0.0109	–	Romania, Galați County	0.2	0.1	
Eggs	0.007	–	–	Iran	-	-	Hoseini et al. (2023)
Eggs	0.51			India	-	-	Giri and Singh (2019)
Eggs	0.001	0.146	39	Italy, Campania region	-	-	Esposito et al. (2016)
Eggs	0.10	1.798	201	USA, Boston, Massachusetts area	-	-	Leibler et al. (2018)

(continued)

Table 5.1 (continued)

Product	Mean concentration (mg kg^{-1})	Max concentration (mg kg^{-1})	Sample no	Country/region	EU standard (2021/1317) (mg kg^{-1})	FAO (Codex Alimentarius 2019) (mg kg^{-1})	References
Fish *R. rutilus*	0.0399 ± 0.0200	–	muscle tissues, 19	Poland, Dratów	0.30	0.3	Chałabis-Mazurek et al. (2021)
Fish *E. Lucius*	0.0360 ± 0.0010	–	muscle tissues, 19	Poland, Dratów	0.3	0.3	Chałabis-Mazurek et al. (2021)
Fish *P. fluviatilis*	0.0533 ± 0.0325	–	muscle tissues, 19	Poland, Dratów	0.3	0.3	Chałabis-Mazurek et al. (2021)
Fish *R. rutilus*	0.0672 ± 0.1472	–	muscle tissues	Poland, Lake Syczyńskie	0.3	0.3	Chałabis-Mazurek et al. (2021)
Fish *E. Lucius*	0.0305 ± 0.0064		muscle tissues	Poland, Lake Syczyńskie	0.3	0.3	Chałabis-Mazurek et al. (2021)
Fish *P. fluviatilis*	0.1057 ± 0.0787	–	muscle tissues	Poland, Lake Syczyńskie	0.3	0.3	Chałabis-Mazurek et al. (2021)
Fish	0.018	0.308	–	Bangladesh	0.3	0.3	Majumder et al. (2021)
Salami var-1 (beef, pork and pork bacon)	0.0087	0.0116	180	Slovakia, Western Slovakia region	0.1	0.1	Lukáčová et al. (2014)
Salami var -2 (beef, pork and pork bacon)	0.007	0.0096	180	Slovakia, Western Slovakia region	0.1	0.1	Lukáčová et al. (2014)

(continued)

Table 5.1 (continued)

Product	Mean concentration (mg kg^{-1})	Max concentration (mg kg^{-1})	Sample no	Country/region	EU standard (2021/1317) (mg kg^{-1})	FAO (Codex Alimentarius 2019) (mg kg^{-1})	References
Milk	0.129	0.850	3	India, East Singhbhum (mining areas)	0.02	0.02	Giri and Singh (2019)
Milk	0.089	0.536	3	India, West Singhbhum (mining areas)	0.02	0.02	Giri and Singh (2019)
Raw milk	0.047	0.25	118	Iran, Shahroud Southwest Iran	0.02	0.02	Reza et al. (2018)
Raw milk	0.00175 ± 0.00373	0.0386	997	China, (10 main milk producing areas)	0.02	0.02	Zhou et al. (2019)
Milk	0.018	–	10	Romania, Baia Mare	0.02	0.02	Miclean et al. (2019)
Milk	0.230	0.476	42	Brasil, Vale do Paraiba region	0.02	0.02	Soares et al. (2010)
Milk	0.58	0.60	20	Peru, (metallurgical complex)	0.02	0.02	Chirinos-Peinado et al. (2021)
Milk	0.234 ± 0.079	–	40	Poland, Upper Silesia	0.02	0.02	Sujka et al. (2019)
Raw milk	0.045	630	75	Egypt, West Delta	0.02	0.02	Amer et al. (2021)
Milk powder–infant formula	0.001	0.007	22	Romania	0.02	0.01	Muntean et al. (2013)

(continued)

Table 5.1 (continued)

Product	Mean concentration (mg kg^{-1})	Max concentration (mg kg^{-1})	Sample no	Country/region	EU standard (2021/1317) (mg kg^{-1})	FAO (Codex Alimentarius 2019) (mg kg^{-1})	References
Milk-based Infant formula	0.007	0.0249	28	Turkey, Ankara	0.01	0.01	Sipahi et al. (2014)
Milk infant formula	0.134	0.336	3	Egypt, Alexandria	0.01	0.01	Kotb et al. (2017)
Cereals, Infant formula	0.003	0.01	15	Romania	0.02	0.01	Muntean et al. (2013)
Cereals with rice, Infant formula	0.0057	0.097	4	Romania	0.02	0.01	Muntean et al. (2013)
Infant formula, cereal-based (a single type or any combination of wheat, rice, corn and oats)	0.0696	0.017	23	Turkey, Ankara	0.02	0.01	Sipahi et al. (2014)
Infant formula, Mixed (combination of cereals, milk, fruit, and vegetables)	0.0062	0.0122	12	Turkey, Ankara	0.02	0.01	Sipahi et al. (2014)
Milk powder–infant formula	0.1692 ± 0.0324	–	8	Albania	0,02	0.01	Dhamo and Shabani (2014)

2018). Occupational exposure of adults from Benin City, Nigeria (mechanics, auto electricians, petrol attendants etc.) to environmental lead with increased blood lead level associated with liver dysfunction was also reported by Onyeneke and Omokaro (2016). These results were to be expected because it is well-documented that BLL > 10 μg/dL may pose hazardous effects on various human organs (especially in case of infants, children, and pregnant woman) (Boskabady et al. 2018; NHMRC 2015), while for workers in occupational exposure, a BLL > 30 μg/dL is considered unsafe (Boskabady et al. 2018). Jusko et al. (2008) noted that children's intellectual ability, including intelligence quotient (IQ), at 6 years of age is affected by blood lead concentrations substantially below 10 μg/dL. Lanphear et al. (2005) also concluded that environmental lead exposure in children with maximum BLL < 7.5 μg/dL is linked to intellectual deficits.

Average-specific guideline values (e.g. for drinking water quality and air quality guidelines, maximum limits in food) developed by international organizations (e.g. WHO, FAO/WHO) are now available for several chemicals (WHO 2010). For example, the permissible limit of Pb in drinking water was set at 10 μg/L in the European Union (WHO 2008; SCHER 2011). Directive 86/278/EEC for Protection of the Environment (European Communities Council 1986) limits the level of Pb in the agricultural soil between 50 and 300 mg/kg (Nag and Cummins 2022).

5.6 Lead Human Health Risk Assessment

Human health risk assessment (HHRA) is a broad evaluation approach based on predictive tools, that connects environmental pollution and human health, by characterization and quantification of any possible adverse effects on human health associated with their exposure to contaminated media (Jia et al. 2015; Yang et al. 2015). HHRA is the first component of risk analysis process followed by risk management and risk communication (WHO 2021). Risk assessment involves understanding the probability of an adverse effect occurring within a specified time period taking into account the amount of chemicals present in the environment (e.g. soil, water, air) or in food products (vegetables, fruits); routes of exposure; exposure time; toxicity of the chemical substance; category of population exposed/potential receptors (Yang et al. 2015; Cozma et al. 2018; Minut et al. 2020).

HHRA includes five general steps (https://www.epa.gov/risk/human-health-risk-assessment; Hlihor et al. 2018; WHO 2021): (1) *problem formulation* (defining the scope and the objective of the assessment); (2) *hazard identification* (stressor identification including also the examination of available scientific data for a particular chemical); (3) *evaluation of the dose–response relationship* (relationship between exposure and effects, considering reference dose (*RfD*) - the concentration of a chemical for which adverse effects on human health are known to occur); (4) *exposure assessment* (data about the population exposed, chemical sources, magnitude, frequency, duration, route of exposure); (5) *risk characterization* (risk estimation and risk description, providing important qualitative or quantitative information for decision-making stakeholders).

The HHRA of metals is usually evaluated considering average daily dose (*ADD*), hazard quotient (*HQ*), hazard index (*HI*), and cancer risk (*CR*), according to USEPA (United States Environmental Protection Agency) model. To estimate metals daily intake, *ADD* for non-carcinogenic and carcinogenic is calculated based on the following equation (Eq. 5.8) (USEPA 2004; Hlihor et al. 2017, 2018; Taiwo et al. 2019):

$$\mathrm{ADD} = \frac{\mathrm{C} \times \mathrm{IR} \times \mathrm{EF} \times \mathrm{ED}}{\mathrm{BW} \times \mathrm{AT}} \tag{5.8}$$

where: ADD = average daily dose of metal (mg/kg/day); C = concentration of metal detected in the sample (mg/kg); IR = ingestion rate (g/day); ED = exposure duration (years) = 70 years (average lifetime for carcinogens) and 5 years (for non-carcinogens); EF = exposure frequency (day/year) = 365 day/year; AT = averaging time of life expectancy for adults considered in the study = $EF \times ED$ (days); BW = body weight of the different age category (children, adolescents, adults) (kg). *IR*, *EF*, *ED*, *AT*, and *BW* values are standard values according to USEPA. More information can be found on USEPA which developed multiple guidance materials on human health risk assessment (USEPA 1992, 2004, 2011).

Depending on exposure pathways, the average daily dose of metal through ingestion, inhalation, or dermal contact may be calculated. For each case, there are different features that should be considered. Regarding dietary exposure, food ingestion rate varies along with the type of food, source of food, and age category. In case of dermal exposure, skin adherence factor, dermal absorption factor, and exposed skin area are taken into account. More details are presented in the literature (Peng et al. 2017).

Further, the risk (for non-carcinogenic substances) can be determined by the hazard quotient (*HQ*) which is the ratio of the average daily dose (*aADD*—for acute exposure, *cADD*—for chronic exposure) to the corresponding reference dose (*ARfD*—acute reference dose, in mg/kg/day or Acceptable Daily Intake (*ADI*), for chronic exposure, in mg/kg/day).

For *cancer risk* (*CR*), the probability of a person to develop cancer, over a lifetime, as a result of exposure to a substance that is a potential carcinogen—a cancer slope factor, *SF*, in $(\text{mg/kg/day})^{-1}$, is considered, which can be extrapolated from the Integrated Risk Information System (IRIS) Database (https://www.epa.gov/iris) (Hlihor et al. 2018; USEPA 1998) (Fig. 5.4). *ARfD* and *ADI* values for different chemicals are available on IRIS database (chemical search task); Agency for Toxic Substances and Disease Registry (ATSDR) (https://www.atsdr.cdc.gov/toxprofiledocs/index.html); Risk Assessment Information System (RAIS) (http://rais.ornl.gov/tools/profile.php); eChemPortal (https://www.oecd.org/env/ehs/risk-assessment/echemportalglobalportaltoinformationonchemicalsubstances.htm), or INCHEM (http://www.inchem.org/).

Human dietary exposure to trace metals generally originates from the food consumption of the population (Di Bella et al. 2021). In this regard, some important

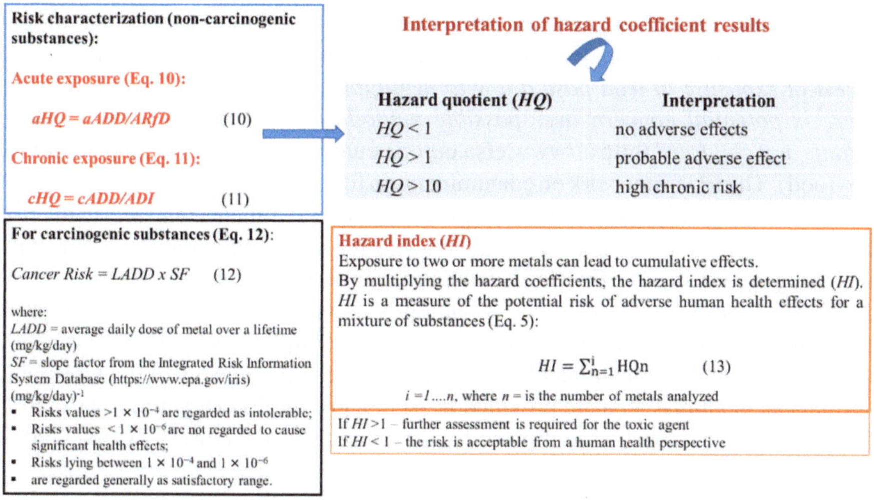

Fig. 5.4 Risk characterization and interpretation

exposure assessment tools and guidelines by food can be found on USEPA database (https://www.epa.gov/expobox/exposure-assessment-tools-media-food) (e.g. International Programme on Chemical Safety (IPCS); USEPA. (2007)—Framework for Metals Risk Assessment; World Health Organization (WHO)—Guidelines for the Study of Dietary Intakes of Chemical Contaminants, etc.).

United States EPA ExpoBox (https://www.epa.gov/expobox/exposure-factors-interactive-resource-scenarios-tool-expofirst) is an important tool providing a compilation of exposure assessment that links guidance documents, databases, templates, reference materials designed and released by EPA's Office of Research and Development in 2013 (https://www.epa.gov/expobox/basic-information-about-epa-expobox).

ConsExpo Web (https://www.consexpoweb.nl/) is a tool released in 2016 by National Institute for Public Health and the Environment (RIVM) of the Netherlands (WHO 2021), especially applied for assessing the exposure to chemical substances from daily consumer products (industrial chemical and biocides). Also, since 2002, European Food Safety Authority (EFSA) developed risk assessments studies for over 5000 substances in food and feed (Dorne et al. 2021). ***OpenFoodTox*** is a chemical hazard database designed by EFSA, which comprises open source data for the chemical compounds characterization (https://www.efsa.europa.eu/en/data-report/chemical-hazards-database-openfoodtox), including pesticides, flavorings, nutritional sources, feed additives, contaminants—persistent organic pollutants, marine mycotoxin, melamine, heavy metal ions, and metalloids etc. (Dorne et al. 2021).

European Commission or Member States have previously sent requests to EFSA to develop risk assessments on several metals, including lead, cadmium, arsenic, chromium, mercury, nickel, and uranium. In 2010, the experts from EFSA published a scientific opinion on possible health risks associated to the presence of lead in food; cereals, vegetables, and tap water were considered the major routes to lead dietary exposure for the majority of Europeans. The experts concluded that: "*current levels of exposure to lead pose a low to negligible health risk for most adults but there is potential concern over possible neurodevelopmental effects in foetuses, infants and children*" (https://www.efsa.europa.eu/en/topics/topic/metals-contaminants-food). The EU framework on contaminants in food comprises: Regulation 315/93/EEC containing the principles of EU legislation on contaminants in food; Regulation EC 1881/2006 establishing the maximum levels for some contaminants in foodstuff, including nitrate, mycotoxins, lead, cadmium, mercury, inorganic tin; Regulation EC 333/2007 comprising the methods of sampling and analysis for the official control of the maximum levels of these metals (CR, 2021) (https://www.efsa.europa.eu/en/topics/topic/metals-contaminants-food). Further, the maximum levels of undesirable substances in animal feed are provided in Annex I of the EU Directive 2002/32/EC (available on https://eur-lex.europa.eu/legal-content/EN/ALL/?uri=CELEX:32002L0032).

EFSA delivered several information on the levels of trace metals elements (e.g. Cd, Pb) detected in a series of foods (e.g. fish, eggs, fruits, vegetables, drinking water, alcoholic, and non-alcoholic beverages etc.), from the European market and further, estimated human exposure to these chemicals by using individual data available on the European food consumption database (Di Bella et al. 2021). In this regard, EFSA questioned the value of lead provisional tolerable weekly intake (PTWI) of 25 μg/kg b.w., established by European Commission's Scientific Committee for Food (SCF), as being completely non-safe for the population. Moreover, the Joint FAO/WHO Expert Committee on Food Additives (JECFA) estimated, for this threshold, a decrease up to 3 IQ points in children and an increase in systolic blood pressure of almost 3 mmHg for adults (Wang et al. 2019). Even though this threshold was withdrawn, there is no evidence of other new limit (EFSA 2012; Wang et al. 2019; Koch et al. 2022). On the contrary, in 2010, EFSA Panel on Contaminants in the Food Chain (CONTAM) reported a scale of values "*for the 95% lower confidence limit of the benchmark dose of 1% extra risk* ($BMDL_{01}$) *for each endpoint*" (Koch et al. 2022). These BMDL (benchmark dose lower confidence limit, %) values describe the interconnection of dietary intake, metal content, and the risk of intoxication associated with health problems occurrence (EFSA 2012; Koch et al. 2022):

- For children, a dietary intake of 0.50 μg/kg b.w. per day ($BMDL_{01}$) was associated with developing neurotoxicity effects;
- For adults, a dietary intake of 1.50 μg/kg b.w. ($BMDL_{01}$) and 0.63 μg/kg b.w. ($BMDL_{10}$) was associated with cardiovascular problems and chronic kidney diseases, respectively.

EFSA recommended the margin of exposure (MOE) approach to evaluate the health risks from dietary exposure to lead (Wang et al. 2019). The MOE can be defined "*as the ratios of the observable effect level* (e.g. *benchmark dose lower bound* (*BMDL*)) *on the dose–response curve to the critical effect and the exposure level of the population*" (Wang et al. 2019) and can be calculated with the following formula (Eq. 5.9):

$$MOE = BMDL/EXP \tag{5.9}$$

where EXP is the daily dietary lead intake (μg/kg b.w./day) being similar with ADD (Eq. 5.10):

$$EXP = \frac{C \times IR}{BW} \tag{5.10}$$

The risk is interpreted as follows: the values of MOE < 1 indicates a high health risk, while MOE > 1 indicates an acceptably low risk (Wang et al. 2019).

Koch et al. (2022) assessed the exposure of young Polish adults (man and female) to Cd, Pb, Hg, and Ni by dietary intake. From different food categories, water and beverages were the main dietary sources of Pb (31% of the total daily intake). Evaluating the risk related to dietary exposure to Pb resulted in estimated dietary intake of 1.12 μg/kg b.w. for women and 1.02 μg/kg b.w. for men, suggesting a potential risk of developing nephrotoxicity in both genders (Koch et al. 2022). Wang et al. (2019) applied MOE approach to evaluate health risks of dietary lead exposure of the residents (different age groups age group of 3–6, 7–17, 18–59, and 60 years) of Guangzhou, China. Rice, leafy vegetables, and wheat flour were the main food sources of lead exposure. Considering mean exposure levels, the MOE values for all age groups tested were higher than 1, unless preschool children for whom MOE < 1. In general, the risk was considered low for Guangzhou population, but however, high for young children. Taiwo et al. (2019) evaluated the health risk of Zn, Cr, Cd, Ni, Pb in staple foods (beans, maize, rice etc.) by applying USEPA method. Comparative to other metals, Pb was below the detection limit in all samples tested. The content of metals in food sample decreased in the following order: Zn > Cr > Cd > Ni > Pb.

Winiarska-Mieczan et al. (2023) conducted a health risk assessment of Polish adults exposed to cadmium and lead concentration in drinking instant coffee and coffee substitutes. The estimated safety consumption of these beverages by Polish adults was performed based on the following parameters: BMDL (%), chronic daily intake (CDI), hazard quotient (HQ), and hazard index (HI). Analyzing the content of metals from tested beverages, the authors observed that the instant coffee contained higher Pb compared to other substances tested (0.089 mg Pb per 1 kg), and moreover, the level of Pb was higher compared to Cd. Based on their results, the authors concluded that consumption of these beverages can be considered safe for the adult population. However, taking into account the ability of HMs to accumulate in the tissue of living organisms and their long half-life (e.g. more than 10 years in bones for Pb), no safe limits of HMs intake may exist (Winiarska-Mieczan et al. 2023). Li et al. (2017) have taken into account, in their research, the bioaccessibility of Pb in soil when evaluated human health risks associated to this metal. They found that bioaccessible contents of Pb were substantially lower compared to total contents. Regarding the risk exposure pathway for adults, oral ingestion (76.79%) and inhalation (49.56%) were the predominant routes. Human health risk of population from Daye followed USEPA method and considered the exposure to both total content and bioaccessible content of Pb. No carcinogenic risk of Pb was obtained, however, HI and HQ coefficients for bioaccessible content were lower compared to total content. This means that traditional health risk method may overestimate the actual risk (Li et al. 2017).

5.7 Conclusions

Lead pollution in agroecosystems represents a global environmental issue, due to its high toxicity, it is classified as one of the most toxic heavy metals in the environment. Pb and its compounds persist in soil for long periods of time, bioaccumulate in agricultural plants through various pathways—soil, water, and air followed by subsequent trophic transfer to other trophic levels. Human exposure to lead occurs by different pathways such as inhalation, oral ingestion, and dermal absorption. The literature analysis shows that lead has a poor bioavailability for plants due to its extreme insolubility, and as a consequence, only 5% of Pb is transported into aerial components and a major part 95% remains accumulated in roots, thus limiting the transport in the food chain. However, it has been proved that in some food chains, the phenomenon of lead biomagnification occurs, while in other food chains, biodilution has been observed. Furthermore, it has been found that lead generates considerable effects on human health, numerous clinical studies show high levels of BLL for children and adults, and the occurrence of various diseases.

The exposure of lead is a significant public health issue and requires a coherent and global approach. Thus, in order to reduce the level of exposure to lead-contaminated food, a series of standards regarding the quality of food in terms of maximum allowed limit, were developed and applied at the European Union and globally level. These are useful tools applied as references, nevertheless, literature studies show that in some cases the level of lead in market food exceeds the allowed maximum limits.

Human health risk assessment involves assessing the probability of an adverse effect occurring taking into account the dose of chemicals present in the environment or in food products, routes of exposure, exposure time, toxicity of the chemical substance, and the exposed population. In this regard, the paper presents and discusses some important tools and guidelines for the assessment of exposure and risks generated by the presence of lead in food, which were developed by USEPA and EFSA and successfully demonstrated the applicability. Public health measures should focus on awareness of lead exposure, policies to prevent and reduce lead exposure by reducing the applications, and use of the metal and its compounds, and by minimizing emissions containing lead.

References

Abedi T, Gavanji S, Mojiri A (2022) Lead and zinc uptake and toxicity in maize and their management. Plants 11(15):1922. https://doi.org/10.3390/plants11151922

Agency for Toxic Substances and Disease Registry (2007) ATSDR, toxicological profile for lead. U.S. Department of Health and Human Services, Public Health Service, Division of Toxicology and Environmental Medicine/Applied Toxicology Branch

Ahmed ASS, Sultana S, Habib A, Ullah H, Musa N, Hossain MB, Rahman MM, Sarker MSI (2019) Bioaccumulation of heavy metals in some commercially important fishes from a tropical river estuary suggests higher potential health risk in children than adults. PLoS ONE 14(10):e0219336. https://doi.org/10.1371/journal.pone.0219336.31622361

Ahmed MS, Yesmin M, Jeba F, Hoque MJ, Jamee AR, Salam A (2020) Risk assessment and evaluation of heavy metals concentrations in blood samples of plastic industry workers in Dhaka, Bangladesh. Toxicol Rep 7. https://doi.org/10.1016/j.toxrep.2020.10.003

Ahmed S, Tuj-Zohra F, Mahdi MM, Nurnabi M, Alam MZ, Choudhury TR (2022) Health risk assessment for heavy metal accumulation in leafy vegetables grown on tannery effluent contaminated soil. Toxicol Rep 9:346–355. https://doi.org/10.1016/j.toxrep.2022.03.009

Akele ML, Desalegn SK, Asfaw TB, Assefa AG, Alemu AK, de Oliveira RR (2022) Heavy metal contents in bovine tissues (kidney, liver and muscle) from Central Gondar Zone, Ethiopia. Heliyon 8(12):e12416. https://doi.org/10.1016/j.heliyon.2022.e12416

Al-Saleh I, Abduljabbar M (2017) Heavy metals (lead, cadmium, methylmercury, arsenic) in commonly imported rice grains (Oryza sativa) sold in Saudi Arabia and their potential health risk. Int J Hyg Environ Health 220:168–1178. https://doi.org/10.1016/j.ijheh.2017.07.007

Ali H, Khan E (2019) Trophic transfer, bioaccumulation, and biomagnification of non-essential hazardous heavy metals and metalloids in food chains/webs-concepts and implications for wildlife and human health. Hum Ecol Risk Assess 25(6):1353–1376. https://doi.org/10.1080/10807039.2018.1469398

Ali MA, Mohamed HF, Amer WM (2008) Biophysical measurements of lead in some bioindicator plants. Romanian J Biophys 18(1):57–66

Ali N, Sardar K, Said M, Salma K, Sadia A, Sadaf S, Toqeer A, Miklas S (2015) Toxicity and bioaccumulation of heavy metals in spinach (*Spinacia oleracea*) grown in a controlled environment. Int J Environ Res Public Health 12(7):7400–7416. https://doi.org/10.3390/ijerph120707400

Ali H, Khan E, Ilahi I (2019) Environmental chemistry and ecotoxicology of hazardous heavy metals: environmental persistence, toxicity, and bioaccumulation. J Chem 6730305. https://doi.org/10.1155/2019/6730305

Alimonti A, Mattei D (2008) Biomarkers for human biomonitoring. In: WIT transactions on state of the art in science and engineering, vol 30. WIT Press

Amer AAE, El-Makarem HSA, El-Maghraby MA, Abou-Alella SA (2021) Lead, cadmium, and aluminum in raw bovine milk: residue level, estimated intake, and fate during artisanal dairy manufacture. J Adv Vet Anim Res 8(3):454–464. https://doi.org/10.5455/javar.2021.h534

Angelova VR, Ivanova RV, Todorov JM, Ivanov KI (2010) Lead, cadmium, zinc, and copper bioavailability in the soil-plant-animal system in a polluted area. Sci World J 10:273–285. https://doi.org/10.1100/tsw.2010.33

Anguilano L, Onwukwe U, Dekhli A, Venditti S, Aryani D, Reynolds A (2022) Hyperaccumulation of lead using *Agrostis tenuis*. Environ Syst Res 11:30. https://doi.org/10.1186/s40068-022-00279-z

Araújo FG, Morado CN, Parente TTE, Paumgartten FJR, Gomes ID (2018) Biomarkers and bioindicators of the environmental condition using a fish species (*Pimelodus maculatus* Lacepède, 1803) in a tropical reservoir in Southeastern Brazil. Braz J Biol 78(2). https://doi.org/10.1590/1519-6984.167209

Ashraf U, Sheka Kanu A, Mo Z, Hussain S, Anjum SA, Khan I, Abbas RN, Tang X (2015) Lead toxicity in rice: effects, mechanisms, and mitigation strategies-a mini review. Environ Sci Pollut Res Int 22(23):18318–21832. https://doi.org/10.1007/s11356-015-5463-x

Aslam M, Aslam A, Sheraz M, Ali B, Ulhassan Z, Najeeb U, Zhou W, Gill RA (2021) Lead toxicity in cereals: mechanistic insight into toxicity, mode of action, and management. Front Plant Sci 11. https://doi.org/10.3389/fpls.2020.587785

Astete J, Cáceres W, Gastañaga M, Lucero M, Sabastizagal I, Oblitas T, Pari J, Rodriguez F (2009) Lead intoxication and other health problems in children population who live near mine tailing. Rev Peru Med Exp Salud Publica 26(1):15–19. https://doi.org/10.17843/rpmesp.2009.261.1327
Azzazy MF (2020) Plant bioindicators of pollution in Sadat City, Western Nile Delta. Egypt. Plos One. 15(3):e0226315
Balali-Mood M, Naseri K, Tahergorabi Z, Khazdair MR, Sadeghi M (2021) Toxic mechanisms of five heavy metals: mercury, lead, chromium, cadmium, and arsenic. Front Pharmacol 12:643972
Baldi A, Cecchi S, Grassi C, Camillo AZ, Orlandini S, Napoli M (2021) Lead bioaccumulation and translocation in herbaceous plants grown in urban and peri-urban soil and the potential human health risk. Agronomy 11(12):2444. https://doi.org/10.3390/agronomy11122444
Bayissa LD, Gebeyehu HR (2021) Vegetables contamination by heavy metals and associated health risk to the population in Koka area of central Ethiopia. PLoS ONE. https://doi.org/10.1371/journal.pone.0254236
Bayouli IT, Bayouli HT, Dell'Oca A, Meers E, Sun J (2021) Ecological indicators and bioindicator plant species for biomonitoring industrial pollution: eco-based environmental assessment. Ecol Indic 125:107508. https://doi.org/10.1016/j.ecolind.2021.107508
Bharagava RN, Mishra S (2018) Hexavalent chromium reduction potential of *Cellulosimicrobium* sp. isolated from common effluent treatment plant of tannery industries. Ecotoxicol Environ Saf 147:102–109
Bora FD, Bunea A, Pop SR, Baniță SI, Duşa DŞ, Chira A, Bunea CI (2022) Quantification and reduction in heavy metal residues in some fruits and vegetables: a case study Galați County, Romania. Horticulturae 8:1034. https://doi.org/10.3390/horticulturae8111034
Borgå K, Kidd KA, Muir DC, Berglund O, Conder JM, Gobas FA, Kucklick J, Malm O, Powell DE (2011) Trophic magnification factors: considerations of ecology, ecosystems, and study design. Integr Environ Assess Manag 8(1):64–84
Borgå K (2013) Ecotoxicology: bioaccumulation. In: Reference module in earth systems and environmental science. https://doi.org/10.1016/B978-0-12-409548-9.00765-X
Borgulat J, Łukasik W, Borgulat A, Nadgórska-Socha A, Kandziora-Ciupa M (2021) Influence of lead on the activity of soil microorganisms in two Beskidy landscape parks. Environ Monit Assess 193:839. https://doi.org/10.1007/s10661-021-09503-2
Boskabady M, Marefati N, Farkhondeh T, Shakeri F, Farshbaf A, Boskabady MH (2018) The effect of environmental lead exposure on human health and the contribution of inflammatory mechanisms, a review. Environ Int 120:404–420. https://doi.org/10.1016/j.envint.2018.08.013
Briffa J, Sinagra E, Blundell R (2020) Heavy metal pollution in the environment and their toxicological effects on humans. Heliyon 6:e04691. https://doi.org/10.1016/j.heliyon.2020.e04691
CR (2021) Commission regulation (EU) 2021/1317 of 9 August 2021 amending Regulation (EC) No 1881/2006 as regards maximum levels of lead in certain foodstuffs. O. J. L 286/1
Cardwell RD, DeForest DK, Brix KV, Adams WJ (2013) Do Cd, Cu, Ni, Pb, and Zn biomagnify in aquatic ecosystems? Rev Environ Contam T 226. https://doi.org/10.1007/978-1-4614-6898-1_4
Chałabis-Mazurek A, Rechulicz J, Pyz-Łukasik R (2021) A food-safety risk assessment of mercury, lead and cadmium in fish recreationally caught from Three Lakes in Poland. Animals 11:3507. https://doi.org/10.3390/ani11123507
Chen S, Lin R, Lu H, Wang Q, Yang J, Liu J, Yan C (2020) Effects of phenolic acids on free radical scavenging and heavy metal bioavailability in *Kandelia obovata* under cadmium and zinc stress. Chemosphere 249:126341. https://doi.org/10.1016/j.chemosphere.2020.126341
Chen CY, Folt CL (2000) Bioaccumulation and diminution of arsenic and lead in a fresh water food web. Environ Sci Technol 34(18). https://doi.org/10.1021/es991070c
Chirinos-Peinado D, Castro-Bedriñana J, García-Olarte E, Quispe-Ramos R, Gordillo-Espinal S (2021) Transfer of lead from soil to pasture grass and milk near a metallurgical complex in the Peruvian Andes. Transl Anim Sci 5:1–9. https://doi.org/10.1093/tas/txab003

Chiwetalu U, Egwuagu O, Ugwu KC (2022) Lead accumulation in maize grains, leaves and other parts of the plant: a possible route for lead poisoning. FJS 6(5):16–23. https://doi.org/10.33003/fjs-2022-0605-1083

Chojnacka K, Mikulewicz M (2014) Bioaccumulation. In: Wexler P (eds) Encyclopedia of toxicology (Third Edition). Academic Press, pp 456–460. https://doi.org/10.1016/B978-0-12-386454-3.01039-3

Chojnacka K, Mikulewicz M (2022) Bioaccumulation. In: Reference module in biomedical sciences, Elsevier. https://doi.org/10.1016/B978-0-12-824315-2.00351-1

Chormare R, Kumar MA (2022) Environmental health and risk assessment metrics with special mention to biotransfer, bioaccumulation and biomagnification of environmental pollutants. Chemosphere 134836. https://doi.org/10.1016/j.chemosphere.2022.134836

Codex Alimentarius (2019) Commission report of the eighth session of the codex committee on contaminants in foods. In: Proceedings of the joint FAO/WHO food standards programme Codex Alimentarius Commission 37th session; Geneva, Switzerland

Collin S, Baskar A, Geevarghese DM, Ali MNVS, Bahubali P, Choudhary R, Lvov V, Tovar GI, Senatov F, Koppala S, Swamiappan S (2022) Bioaccumulation of lead (Pb) and its effects in plants: a review. J Hazard Mater Letters 3:100064. https://doi.org/10.1016/j.hazl.2022.100064

Conder JM, Gobas FAPC, Borgå K, Muir DCG, Powell DE (2012) Use of trophic magnification factors and related measures to characterize bioaccumulation potential of chemicals. Integr Environ Assess Manag 8(1):85–97. https://doi.org/10.1002/ieam.216

Córdoba-Tovar L, Marrugo-Negrete J, Barón PR, Díez S (2022) Drivers of biomagnification of Hg, As and Se in aquatic food webs: a review. Environ Res 204:112226. https://doi.org/10.1016/j.envres.2021.112226

Cozma P, Gavrilescu M, Roşca M, Apostol LC, Hlihor RM, Gavrilescu M (2018) Evaluation of human health risks associated with pesticide dietary intake—an overview on quantitative uncertainty analysis. Environ Eng Manag J 17:2263–2274

Cozma P, Asiminicesei D-M, Hlihor R-M, Rosca M, Gavrilescu M (2019) Influence of cadmium and lead ions toxicity on lettuce seeds germination: a preliminary analysis. In: Proceedings of the 7th IEEE international conference on e-health and bioengineering conference (EHB), Iasi, Romania, pp 1–4. https://doi.org/10.1109/EHB47216.2019.8969968

Dar MI, Green ID, Naikoo MI, Khan FA, Ansari AA, Lone MI (2017) Assessment of biotransfer and bioaccumulation of cadmium, lead, and zinc from fly ash amended soil in mustard-aphid-beetle food chain. Sci Total Environ 584–585:1221–1229. https://doi.org/10.1016/j.scitotenv.2017.01.186

Dehghani A, Aminjan AR, Dehghani A (2022) Trophic transfer, bioaccumulation, and health risk assessment of heavy metals in Aras River: case study-Amphipoda–zander–human. Environ Sci Pollut Res 29:30764–30773

Deshommes E, Tardif R, Edwards M, Sauvé S, Prévost M (2012) Experimental determination of the oral bioavailability and bioaccessibility of lead particles. Chem Cent J 6(1):138. https://doi.org/10.1186/1752-153X-6-138

Dhamo K, Shabani L (2014) Testing infant milk formulae for lead and cadmium. IRJES 3(3):51–53

Di Bella C, Calagna A, Cammilleri G, Schembri P, Lo Monaco D, Ciprì V, Battaglia L, Barbera G, Ferrantelli V, Sadok S, Tliba I, Lo Dico GM, Allegro R (2021) Risk assessment of cadmium, lead, and mercury on human health in relation to the consumption of farmed sea bass in Italy: a meta-analytical approach. Front Mar Sci 8:616488. https://doi.org/10.3389/fmars.2021.616488

Dorne JLCM, Richardson J, Livaniou A, Carnesecchi E, Ceriani L, Baldin R, Kovarich S, Pavan M, Saouter E, Biganzoli F, Pasinato L, Zare Jeddi M, Robinson TP, Kass GEN, Liem AKD, Toropov AA, Toropova AP, Yang C, Tarkhov A, Georgiadis N, Di Nicola MR, Mostrag A, Verhagen H, Roncaglioni A, Benfenati E, Bassan A (2021) EFSA's OpenFoodTox: an open source toxicological database on chemicals in food and feed and its future developments. Environ Int 146. https://doi.org/10.1016/j.envint.2020.106293

Efanny M, Andarwulan N, Yuliana ND (2019) Dietary exposure assessment and risk characterization of lead based on lead contaminant research (online) in Indonesia and Indonesian Individual Food

Consumption Survey (IFCS). In: The 3rd EIW IOP conference series: earth and environmental science, vol 278, p 012021. https://doi.org/10.1088/1755-1315/278/1/012021
EFSA (2012) Lead dietary exposure in the European population. EFSA J 10:2831. https://doi.org/10.2903/j.efsa.2012.2831
El-Hassanin AS, Samak MR, Abdel-Rahman GN, Yahia H-S, Saleh EM (2020) Risk assessment of human exposure to lead and cadmium in maize grains cultivated in soils irrigated either with low-quality water or freshwater. Toxicol Rep 7:10–15. https://doi.org/10.1016/j.toxrep.2019.11.018
EPA (2021) Fact sheet: lead RBA and IVBA. Relative bioavailability and in vitro bioaccessibility of lead in soil. Environmental Protection Agency of USA, January 4, 2021, Online at: https://semspub.epa.gov/work/HQ/100002717.pdf
Esposito M, Cavallo S, Chiaravalle E (2016) Trace elements in free-range hen eggs in the Campania region (Italy) analyzed by inductively coupled plasma mass spectrometry (ICP-MS). Environ Monit Assess 188:32
European Communities Council (1986) Council Directive 86/278/EEC of 12 June 1986 on the protection of the environment, and in particular of the soil, when sewage sludge is used in agriculture. Official J L 181:0006–0012
Filipoiu DC, Bungau SG, Endres L, Negru PA, Bungau AF, Pasca B, Radu A-F, Tarce AG, Bogdan MA, Behl T, Nechifor AC, Hassan SSU, Tit DM (2022) Characterization of the toxicological impact of heavy metals on human health in conjunction with modern analytical methods. Toxics 10:716. https://doi.org/10.3390/toxics10120716
Gao Y, Wang R, Li Y, Ding X, Jiang Y, Feng J, Zhu L (2021) Trophic transfer of heavy metals in the marine food web based on tissue residuals. Sci Total Environ 772:145064. https://doi.org/10.1016/j.scitotenv.2021.145064
Ghani A, Shah AU, Akhtar U (2010) Effect of lead toxicity on growth, chlorophyll and lead (Pb^+). Pak J Nutri 9:887–891. https://doi.org/10.3923/pjn.2010.887.891
Giri S, Singh AK (2019) Heavy metals in eggs and chicken and the associated human health risk assessment in the mining areas of Singhbhum copper belt, India. Arch Environ Occup Health 74(4):161–170
Griboff J, Horacek M, Wunderlin DA, Monferran MV (2018) Bioaccumulation and trophic transfer of metals, As and Se through a freshwater food web affected by antrophic pollution in Córdoba, Argentina. Ecotoxicol Environ Saf 148:275–284. https://doi.org/10.1016/j.ecoenv.2017.10.028
Gu Q, Yu T, Yang Z, Ji J, Hou Q, Wang L, Wei X, Zhang Q (2019) Prediction and risk assessment of five heavy metals in maize and peanut: a case study of Guangxi, China. Environ Toxicol Pharmacol 70:103199. https://doi.org/10.1016/j.etap.2019.103199
Gu X, Wang Z, Wang J, Ouyang W, Wang B, Xin M, Lian M, Lu S, Lin C, He M, Liu X (2022) Sources, trophodynamics, contamination and risk assessment of toxic metals in a coastal ecosystem by using a receptor model and Monte Carlo simulation. J Hazard Mater 424(Part B). https://doi.org/10.1016/j.jhazmat.2021.127482
Guerra F, Trevizam AR, Muraoka T, Marcante NC, Canniatti-Brazaca SG (2012) Heavy metals in vegetables and potential risk for human health. Sci Agric 69(1):54–60. https://doi.org/10.1590/S0103-90162012000100008
Guo G, Lei M, Wang Y, Song B, Yang J (2018) Accumulation of As, Cd, and Pb in sixteen wheat cultivars grown in contaminated soils and associated health risk assessment. Int J Environ Res Public Health 15(11):2601. https://doi.org/10.3390/ijerph15112601
Gupta N, Gedam VV, Moghe C, Labhasetwar P (2019) Comparative assessment of batch and column leaching studies for heavy metals release from coal fly ash bricks and clay bricks. Environ Technol Innov 16:100461. https://doi.org/10.1016/j.eti.2019.100461
Hernández-Pitalúa D, Hernández-Orduña MG, Martínez-Escalante GA, Lagunes-Gómez I (2022) Influence of lead (Pb) and its relationship with the pH of water on the growth of creole maize (*Zea mays* L.). Agriculture 12:749. https://doi.org/10.3390/agriculture12060749
Hlihor RM, Pogăcean MO, Cozma P, Rosca M, Gavrilescu M (2017) Dissipation behavior of pesticides applied in multiple treatments in apples. Sci Papers Ser Hortic 60:73–82

Hlihor RM, Gavrilescu M, Pogacean MO, Gavrilescu M, Cozma P, Simion IM, Rosca M (2018) Modelling the dynamics of pesticides in plant products and estimating risks to human health, (in Romanian). Politehnium Publishing House, Iasi

Hoseini H, Abedi AS, Mohammadi-Nasrabadi F, Salmani Y, Esfarjani F (2023) Risk assessment of lead and cadmium concentrations in hen's eggs using Monte Carlo simulations. Food Sci Nutr 2023:1–12. https://doi.org/10.1002/fsn3.3268

Hu C, Shui B, Yang X, Wang L, Dong J, Zhang X (2021) Trophic transfer of heavy metals through aquatic food web in a seagrass ecosystem of Swan Lagoon, China. Sci Total Environ 762(25):143139. https://doi.org/10.1016/j.scitotenv.2020.143139

Hu Y, Lin S, Tang J, Li Y, Wang X, Jiang Y, Zhang H, Wang B (2023) Effects of microplastics and lead exposure on gut oxidative stress and intestinal inflammation in common carp (*Cyprinus carpio* L.). Environ. Pollut. 327(15), 121528. https://doi.org/10.1016/j.envpol.2023.121528

Huang JH (2016) Arsenic trophodynamics along the food chains/webs of different ecosystems: a review. Chem Ecol 32:803–828. https://doi.org/10.1080/02757540.2016.1201079

Huang X, Zhao B, Wu Y, Tan M, Shen L, Feng G (2022) The lead and cadmium content in rice and risk to human health in China: a systematic review and meta-analysis. PLoS ONE 17(12):e0278686. https://doi.org/10.1371/journal.pone.0278686

Hussain T, Murtaza G, Wang X, Zia MN, Aziz H, Ali S, Murtaza B, Fiaz S (2021) Bioassimilation of lead and zinc in rabbits fed on spinach grown on contaminated soil. Ecotoxicol Environ Saf 224:112638. https://doi.org/10.1016/j.ecoenv.2021.112638

Jańczak-Pieniążek M, Cichoński J, Michalik P, Chrzanowski G (2022) Effect of heavy metal stress on phenolic compounds accumulation in winter wheat plants. Molecules 28(1):241. https://doi.org/10.3390/molecules28010241

Jia J, Li X, Wu P, Liu Y, Han C, Zhou L, Yang L (2015) Human health risk assessment and safety threshold of harmful trace elements in the soil environment of the Wulantuga open-cast coal mine. Minerals 5:837–848. https://doi.org/10.3390/min5040528

Jiang Z, Zhang H, Qin R, Zou J, Wang J, Shi Q, Jiang W, Liu D (2014) Effects of lead on the morphology and structure of the nucleolus in the root tip meristematic cells of *Allium cepa* L. Int J Mol Sci 15(8):13406–13423. https://doi.org/10.3390/ijms150813406

Jiang Y, Fan M, Hu R, Zhao J, Wu Y (2018) Mosses are better than leaves of vascular plants in monitoring atmospheric heavy metal pollution in urban areas. Int J Environ Res Public Health 15(6):1105. https://doi.org/10.3390/ijerph15061105

Jusko TA, Henderson CR, Lanphear BP, Cory-Slechta DA, Parsons PJ, Canfield RL (2008) Blood lead concentrations < 10 microg/dL and child intelligence at 6 years of age. Environ Health Perspect 116:243–248. https://doi.org/10.1289/ehp.10424

Kanwal A, Farhan M, Sharif F, Hayyat MU, Shahzad L, Ghafoor GZ (2020) Effect of industrial wastewater on wheat germination, growth, yield, nutrients and bioaccumulation of lead. Sci Rep 10:11361. https://doi.org/10.1038/s41598-020-68208-7

Keran H, Odobašić A, Ćatić S, Salkić M, Bratovčić A, Šestan I (2008) The influence of different soil conditions to the lead bioavailability to plants. In: 12th international research/expert conference trends in the development of machinery and associated technology. TMT 2008, Istanbul, Turkey, 26–30 August (2008)

Khan ZI, Ahmad K, Yasmeen S, Akram NA, Ashraf M, Mehmood N (2017) Potential health risk assessment of potato (*Solanum tuberosum* L.) grown on metal contaminated soils in the central zone of Punjab, Pakistan. Chemosphere 66:157–162

Kim RY, Yoon JK, Kim TS, Yang JE, Owens G, Kim KR (2015) Bioavailability of heavy metals in soils: definitions and practical implementation-a critical review. Environ Geochem Health 37(6):1041–1061. https://doi.org/10.1007/s10653-015-9695-y

Kobkeatthawin T, Sirivithayapakorn S, Nitiratsuwan T, Muenhor D, Loh P-S, Pradit S (2021) Accumulation of trace metal in sediment and soft tissue of *Strombus canarium* in a tropical remote island of Thailand. J Mar Sci Eng 9:991. https://doi.org/10.3390/jmse9090991

Koch W, Czop M, Iłowiecka K, Nawrocka A, Wiącek D (2022) Dietary intake of toxic heavy metals with major groups of food products-results of analytical determinations. Nutrients 14:1626. https://doi.org/10.3390/nu14081626

Kotb AA, Farahat MF, El Daree HB, Agamy NFM (2017) Lead and cadmium contents of infant milk formulas sold in Alexandria, Egypt Marwa. Can J Clin Nutr 5(2):116–128

Krystofova O, Shestivska V, Galiova M, Novotny K, Kaiser J, Zehnalek J, Babula P, Opatrilova R, Adam V, Kizek R (2009) Sunflower plants as bioindicators of environmental pollution with lead(II) ions. Sensors 9:5040–5058. https://doi.org/10.3390/s90705040

Kumar A, Kumar AMMSCP, Chaturvedi AK, Shabnam AA, Subrahmanyam G, Mondal R, Gupta DK, Malyan SK, Kumar S, Khan SA, Yadav KK (2020) Lead toxicity: health hazards, influence on food chain, and sustainable remediation approaches. Int J Environ Res Public Health 17(7):2179. https://doi.org/10.3390/ijerph17072179

Kumar S, Islam R, Akash PB, Khan MHR, Proshad R, Karmoker J, Mac Farlane GR (2022b) Lead (Pb) contamination in agricultural products and human health risk assessment in Bangladesh. Water Air Soil Pollut 233:257. https://doi.org/10.1007/s11270-022-05711-9

Kumar S, Islam R, Akash PB, Khan MH, Proshad R, Karmoker J, MacFarlane GR (2022) Lead (Pb) contamination in agricultural products and human health risk assessment in Bangladesh. Water Air Soil Pollut 233:257. https://doi.org/10.1007/s11270-022-05711-9

Kushwaha A, Hans N, Kumar S, Rani R (2018) A critical review on speciation, mobilization and toxicity of lead in soil-microbe-plant system and bioremediation strategies. Ecotoxicol Environ Saf 147:1035–1045. https://doi.org/10.1016/j.ecoenv.2017.09.049

Labudda M, Dziurka K, Fidler J, Gietler M, Rybarczyk-Płońska A, Nykiel M, Prabucka B, Morkunas I, Muszyńska E (2022) The alleviation of metal stress nuisance for plants-a review of promising solutions in the face of environmental challenges. Plants 11(19):2544. https://doi.org/10.3390/plants11192544

Lamb DT, Ming H, Megharaj M, Naidu R (2009) Heavy metal (Cu, Zn, Cd and Pb) partitioning and bioaccessibility in uncontaminated and long-term contaminated soils. J Hazard Mater 171:1150–1158

Lanphear B P, Hornung R, Khoury J, Yolton K, Baghurst P, Bellinger D C, Canfield R L, Dietrich KN, Bornschein R, Greene T, Rothenberg SJ, Needleman HL, Schnaas L, Wasserman G, Graziano J, Roberts R (2005) Low-level environmental lead exposure and children's intellectual function: an international pooled analysis, Environ Health Perspect. 113:894–899. https://doi.org/10.1289/ehp.7688

Latif M, Zahoor M, Muhammad A, Naz S, Kamran AW, Ullah R, Shah AB, Almeer R, Sayedg A (2022) Bioaccumulation of lead in different organs of *Ctenopharyngodon idella* (grass fish) and *Tor putitora* (Mahseer) fish. Braz J Biol 84:e260355. https://doi.org/10.1590/1519-6984.260355

Lee JW, Choi H, Hwang UK, Kang JC, Kang YJ, Kim KI, Kim JH (2019) Toxic effects of lead exposure on bioaccumulation, oxidative stress, neurotoxicity, and immune responses in fish: a review. Environ Toxicol Pharmacol 68:101–108

Leibler JH, Basra K, Ireland T, McDonagh A, Ressijac C, Heiger-Bernays W, Vorhees D, Rosenbaum M (2018) Lead exposure to children from consumption of backyard chicken eggs. Environ Res 167:445–452. https://doi.org/10.1016/j.envres.2018.08.013

Li X, Huang S, McBride MB (2020) Rhizosphere effect on Pb solubility and phytoavailability in Pb-contaminated soils. Environ Pollut 268:115840. https://doi.org/10.1016/j.envpol.2020.115840

Li Q, Li F, Xiao MS, Cai Y, Xiong L, Huang JB, Fu JT (2017) Bioaccessibility and human health risk assessment of lead in soil from Daye City. In: IOP conferences series: earth and environmental science, vol 108, p 042116. https://doi.org/10.1088/1755-1315/108/4/042116

Liu L, Hu LL, Tang JJ, Li YF, Zhang Q, Chen X (2012) Food safety assessment of planting patterns of four vegetable-type crops grown in soil contaminated by electronic waste activities. J Environ Manag 93(1):22–30

Liu J, Cao L, Dou S (2019) Trophic transfer, biomagnification and risk assessments of four common heavy metals in the food web of Laizhou Bay, the Bohai Sea, Sci Total Environ. 670: 508–522. https://doi.org/10.1016/j.scitotenv.2019.03.140

Liu J, Ma X, Wang M, Sun X (2013) Genotypic differences among rice cultivars in lead accumulation and translocation and the relation with grain Pb levels. Ecotoxicol Environ Saf 90:35–40

Lu Y, Yao H, Shan D, Jiang Y, Zhang S, Yang J (2015) Heavy metal residues in soil and accumulation in maize at long-term wastewater irrigation area in Tongliao, China. J Chem 628280. https://doi.org/10.1155/2015/628280

Lukáčová A, Golian J, Massányi P, Formicki G (2014) Lead concentration in meat and meat products of different origin. Potravinarstvo 8(1):43–47. https://doi.org/10.5219/334

Łuszczek-Trojnar E, Drąg-Kozak E, Popek W (2013) Lead accumulation and elimination in tissues of Prussian carp, Carassius gibelio (Bloch, 1782), after long-term dietary exposure, and depuration periods. Environ Sci Pollut Res Int 20(5):3122–3132. https://doi.org/10.1007/s11356-012-1210-8

Madgett AS, Yates K, Webster L, McKenzie C, Moffat CF (2021) The concentration and biomagnification of trace metals and metalloids across four trophic levels in a marine food web. Mar Pollut Bull 173(Part A):112929. https://doi.org/10.1016/j.marpolbul.2021.112929

Majumder AK, Al Nayeem A, Islam M, Akter MM, Carter WS (2021) Critical review of lead pollution in Bangladesh. J Health Pollut. 11(31):210902. https://doi.org/10.5696/2156-9614-11.31.210902

Manea DN, Ienciu AA, Ştef R, Şmuleac IL, Gergen II, Nica DV (2020) Health risk assessment of dietary heavy metals intake from fruits and vegetables grown in selected old mining areas—a case study: The Banat Area of Southern Carpathians. Int J Environ Res Public Health 17:5172. https://doi.org/10.3390/ijerph17145172

Mawari G, Kumar N, Sarkar S, Kumar Daga M, Meghachandra Singh M, Joshi TK, Khan NA (2022) Heavy metal accumulation in fruits and vegetables and human health risk assessment: findings from Maharashtra, India. Environ Health Insights 16:1. https://doi.org/10.1177/11786302221119151

McGeer J, Henningsen G, Lanno R, Fisher N, Sappington K, Drexler J, Beringer M (2004) Issue paper on the bioavailability and bioaccumulation of metals. U.S. Environmental Protection Agency. Risk Assessment Forum

Mebane CA, Chowdhury MJ, De Schamphelaere KAC, Lofts S, Paquin PR, Santore RC, Wood CM (2020) Metal bioavailability models: current status, lessons learned, considerations for regulatory use, and the path forward. Environ Toxicol Chem 39:60–84

Melake BA, Endalew SM, Alamirew TS, Temesegen LM (2023) Bioaccumulation and biota-sediment accumulation factor of metals and metalloids in edible fish: a systematic review in Ethiopian Surface. Waters Environ Health Insights 17:11786302231159348

Metanat K, Ghasemi-Fasaei R, Ronaghi A, Yasrebi J (2019) Lead phytostabilization and cationic micronutrient uptake by maize as influenced by Pb levels and application of low molecular weight organic acids. Commun Soil Sci Plant Anal 50:1–10

Mganga N, Manoko MLK, Rulangaranga ZK (2011) Classification of plants according to their heavy metal content around north Mara gold mine, Tanzania: implication for phytoremediation. Tanzania J Sci 37(1):109–119

Michalak A (2006) Phenolic compounds and their antioxidant activity in plants growing under heavy metal stress. Pol J Environ Stud 15(4):523–530

Ming H, He W, Lamb DT, Megharaj M, Naidu R (2012) Bioavailability of lead in contaminated soil depends on the nature of bioreceptor. Ecotoxicol Environ Saf 78:344–350. https://doi.org/10.1016/j.ecoenv.2011.11.045

Minut M, Rosca M, Hlihor R-M, Cozma P, Gavrilescu M (2020) Modelling of health risk associated with the intake of pesticides from Romanian fruits and vegetables. Sustainability 12(23):10035. https://doi.org/10.3390/su122310035

Mitra S, Chakraborty AJ, Tareq AM, Emran TB, Nainu F, Khusro A, Idris AM, Khandaker MU, Osman H, Alhumaydhi FA, Simal-Gandara J (2022) Impact of heavy metals on the environment and human health: novel therapeutic insights to counter the toxicity. J King Saud Univ Sci 34(3):101865. https://doi.org/10.1016/j.jksus.2022.101865

Mondal K, Ghosh S, Haque S (2018) A review on contamination, bioaccumulation and toxic effect of cadmium, mercury and lead on freshwater fishes. Int J Zool Stud 3(2):153–159

Mongkhonsin B, Nakbanpote W, Hokura A, Nuengchamnong N, Maneechai S (2016) Phenolic compounds responding to zinc and/or cadmium treatments in *Gynura pseudochina* (L.) DC. Extracts and biomass. Plant Physiol Biochem 109:549–560

Mousavi SM, Brodie G, Payghamzadeh K, Raiesi T, Strivastava AK (2022) Lead bioavailability in the environment: its exposure and effects. J Adv Environ Health Res 10(1):1–14. https://doi.org/10.32598/JAEHR.10.1.1256

Muntean E, Muntean N, Mihăiescu T (2010) Cadmium and lead soil pollution in Copsa Mica area in relation with the food chain. Res J Agricult Sci 42(3):731–734

Muntean E, Muntean N, Creta C, Dud M (2013) Occurrence of lead and cadmium in some baby foods and cereal products. ProEnvir. 6:587–590

Musilova J, Bystricka J, Vollmannova A, Janotova B, Volnova B, Hegedusova A (2015) Factors affecting heavy metals accumulation in potato tubers. Env Prot Nat Res 26(3):54–59. https://doi.org/10.1515/OSZN-2015-0018

Nag R, Cummins E (2022) Human health risk assessment of lead (Pb) through the environmental-food pathway. Sci Total Environ 810:151168. https://doi.org/10.1016/j.scitotenv.2021.151168

Nag R, O'Rourke SM, Cummins E (2022) Risk factors and assessment strategies for the evaluation of human or environmental risk from metal(loid)s—a focus on Ireland. Sci Total Environ 802. https://doi.org/10.1016/j.scitotenv.2021.149839

Naikoo MI, Dar MI, Khan FA, Raghib F, Rajakaruna N (2019) Trophic transfer and bioaccumulation of lead along soil-plant-aphid-ladybird food chain. Environ Sci Pollut Res. https://doi.org/10.1007/s11356-019-05624-x

Najmi A, Albratty M, Al-Rajab AJ, Alhazmi HA, Javed SA, Ahsan W, Rehman ZU, Hassani R, Alqahtani SS (2023) Heavy metal contamination in leafy vegetables grown in Jazan Region of Saudi Arabia: assessment of possible human health hazards. Int J Environ Res Public Health 20(4):2984. https://doi.org/10.3390/ijerph20042984

Nakhaee S, Amirabadizadeh A, Brent J, Mehrpour M (2018) Impact of chronic lead exposure on liver and kidney function and haematologic parameters. Basic Clin Pharmacol Toxicol 124(5):621–628. https://doi.org/10.1111/bcpt.13179

Nasri I, Hammouda A, Hamza F, Zrig A, Selmi S (2017) Heavy metal accumulation in lizards living near a phosphate treatment plant: possible transfer of contaminants from aquatic to terrestrial food webs. Environ Sci Pollut Res 24:12009–12014. https://doi.org/10.1007/s11356-015-5390-x

Ngole VM (2011) Using soil heavy metal enrichment and mobility factors to determine potential uptake by vegetables. Plant Soil Environ 57(2):75–80

NHMRC (2015) National Health and Medical Research Council, NHMRC information paper: evidence on the effects of lead on human health. National Health and Medical Research Council, Canberra. Online at: www.nhmrc.gov.au/guidelines/publications/eh58

Norton GJ, Williams PN, Adomako EE, Price AH, Zhu Y, Zhao FJ, McGrath S, Deacon CM, Villada A, Sommella A, Lu Y, Ming LP, De Silva MCS, Brammer H, Dasgupta T, Islam MR, Meharg AA (2014) Lead in rice: analysis of baseline lead levels in market and field collected rice grains. Sci Total Env 485:428–434

Onyeneke EC, Omokaro EU (2016) Effect of occupational exposure to lead on liver function parameter. Int J Pharm Pharm Sci 6(1):15–19. https://doi.org/10.5829/idosi.ijpms.2016.6.1.1120

Oorts K, Smolders E, Lanno R, Chowdhury MJ (2021) Bioavailability and ecotoxicity of lead in soil: implications for setting ecological soil quality standards. Environ Toxicol. https://doi.org/10.1002/etc.5051

Pal S, Sequeira JS, Joshi DM, Darshni S, Jaiswal A (2023) Bioindicators in heavy metal detection. Int J Eng Res Technol 10(02):206–222

Parmar TK, Rawtani D, Agrawal YK (2016) Bioindicators: the natural indicator of environmental pollution. Front Life Sci 9(2):110–118

Peng X, Shi GL, Liu GR, Xu J, Tian YZ, Zhang YF, Feng YC, Russel AG (2017) Source apportionment and heavy metal health risk (HMHR) quantification from sources in a Southern city in China, using an ME2-HMHR model. Environ Pollut 221:335–342. https://doi.org/10.1016/j.envpol.2016.11.083

Pourrut B, Shahid M, Dumat C, Winterton P, Pinelli E (2011) Lead uptake, toxicity, and detoxification in plants. In: Whitacre D (ed) Reviews of environmental contamination and toxicology. Reviews of environmental contamination and toxicology, vol 213. Springer, New York

Proc K, Bulak P, Kaczor M, Bieganowski AA (2021) New approach to quantifying bioaccumulation of elements in biological processes. Biology 10:345. https://doi.org/10.3390/biology10040345

Rahman Farooqi ZU, Hussain MM, Ayub MA, Qadir AA, Ilic P (2022) Potentially toxic elements and phytoremediation: opportunities and challenges. In: Bhat RA, Policarpo Tonelli FM, Dar GA, Hakeem K (eds) Phytoremediation. Academic Press, pp 19–36

Rai PK, Lee SS, Zhang M, Tsang YF, Kim KH (2019) Heavy metals in food crops: health risks, fate, mechanisms, and management. Environ Int 125:365–385. https://doi.org/10.1016/j.envint.2019.01.067

Raletsena MV, Munyai R, Mongalo NI (2023) Accumulation of heavy metals, yield and nutritional composition of potato (*Solanum tuberosum* L.) cultivars irrigated with fly-ash-treated acid mine drainage. Sustainability 15(2):1327. https://doi.org/10.3390/su15021327

Ramesh HL, Moorthy YVN (2012) Assessment of heavy metal contamination in green leafy vegetables grown in Bangalore urban district of Karnataka. Adv Life Sci Technol 6:40–51

Rees N, Fuller R (2020) The toxic truth: children's exposure to lead pollution undermines a generation of future potential. A Report of UNICEF and Pure Earth,2nd edn. New York

Reza N, González-Montaña JR, Martínez-Pastor F, Hosseini H, Shahrouzian A, Khabazkhoob M, Malayeri FA, Bandani HM, Paknejad M, Foroughi-Nia B, Moghaddam AF (2018) Lead and cadmium levels in raw bovine milk and dietary risk assessment in areas near petroleum extraction industries. Sci Total Environ 1(635):308–314. https://doi.org/10.1016/j.scitotenv.2018.04.138

Rolka E, Wyszkowski M (2021) Availability of trace elements in soil with simulated cadmium, lead and zinc pollution. Minerals 11(8):879. https://doi.org/10.3390/min11080879

Rosca M, Cozma P, Minut M, Hlihor R-M, Bețianu C, Diaconu M, Gavrilescu M (2021). New evidence of model crop *Brassica napus* L. in soil clean-up: comparison of tolerance and accumulation of lead and cadmium. Plants 10:2051. https://doi.org/10.3390/plants10102051

Rubio-Franchini I, López-Hernández M, Ramos-Espinosa MG, Rico-Martínez R (2016) Bioaccumulation of metals arsenic, cadmium, and lead in zooplankton and fishes from the Tula River Watershed, Mexico. Water Air Soil Pollut 227(5). https://doi.org/10.1007/s11270-015-2702-1

Rubio-Franchini I, Rico-Martínez R (2011) Evidence of lead biomagnification in invertebrate predators from laboratory and field experiments. Environ Pollut 159(7):1831–1835. https://doi.org/10.1016/j.envpol.2011.03.021

Rusin M, Domagalska J, Rogala D, Razzaghi M, Szymala I (2021) Concentration of cadmium and lead in vegetables and fruits. Sci Rep 11:11913. https://doi.org/10.1038/s41598-021-91554-z

Saadaoui W, Gamboa-Rosales H, Sifuentes-Gallardo C, Durán-Muñoz H, Abrougui K, Mohammadi A, Tarchoun N (2022) Effects of lead, copper and cadmium on bioaccumulation and translocation factors and biosynthesis of photosynthetic pigments in *Vicia faba* L. (Broad Beans) at different stages of growth. Appl Sci 12(18):8941. https://doi.org/10.3390/app12188941

Sakai T (2000) Biomarkers of lead exposure. Ind Health 38(2):127–142

Sanaei F, Amin MM, Alavijeh ZP, Esfahani RA, Sadeghi M, Bandarrig NS, Atehizadeh A, Taheri E, Rezakazemi M (2021) Health risk assessment of potentially toxic elements intake via food crops consumption: Monte Carlo simulation-based probabilistic and heavy metal pollution index. Environ Sci Pollut Res 28:1479–1490. https://doi.org/10.1007/s11356-020-10450-7

Savoca D, Pace A (2021) Bioaccumulation, biodistribution, toxicology and biomonitoring of organofluorine compounds in aquatic organisms. Int J Mol Sci 22:6276. https://doi.org/10.3390/ijms22126276

SCHER (Scientific Committee on Health and Environmental Risks) (2011) Opinion on Lead Standard in Drinking Water, 11 January 2011, Online at: https://ec.europa.eu/health/scientific_committees/environmental_risks/docs/scher_o_128.pdf
Schützendübel A, Polle A (2002) Plant responses to abiotic stresses: heavy metal-induced oxidative stress and protection by mycorrhization. J Exp Bot 53(372):1351–1365. https://doi.org/10.1093/jexbot/53.372.1351
Shahid M, Pinelli E, Pourrut B, Silvestre J, Dumat C (2011) Lead-induced genotoxicity to *Vicia faba* L. roots in relation with metal cell uptake and initial speciation. Ecotoxicol Environ Saf 74:78–84
Shahid M, Dumat C, Khalid S, Schreck E, Xiong T, Niazi NK (2017) Foliar heavy metal uptake, toxicity and detoxification in plants: a comparison of foliar and root metal uptake. J Hazard Mater 325:36–58
Sharma S, Nagpal AK, Kaur I (2018) Heavy metal contamination in soil, food crops and associated health risks for residents of Ropar wetland, Punjab, India and Its Environs. Food Chem 255:15–22. https://doi.org/10.1016/j.foodchem.2018.02.037
Sharma P, Dubey SR (2005) Lead toxicity in plants. Braz J Plant Physiol 17
Sipahi H, Eken A, Aydın A, Şahin G, Baydar T (2014) Safety assessment of essential and toxic metals in infant formulas. Turk J Pediatr 56:385–391
Soares VA, Kus MMM, Peixoto ALC, Carrocci JS, Salazar RFS, Filho HJI (2010) Determination of nutritional and toxic elements in pasteurized bovine milk from Vale do Paraiba region (Brazil). Food Control 21(1):45–49. https://doi.org/10.1016/j.foodcont.2009.03.010
Sofoulaki K, Kalantzi I, Machias A, Pergantis SA, Tsapakis M (2019) Metals in sardine and anchovy from Greek coastal areas: public health risk and nutritional benefits assessment. Food Chem Toxicol 23:113–124. https://doi.org/10.1016/j.fct.2018.10.053
Soliman MM, Hesselberg T, Mohamed AA, Renault D (2022) Trophic transfer of heavy metals along a pollution gradient in a terrestrial agro-industrial food web. Geoderma 413:115748. https://doi.org/10.1016/j.geoderma.2022.115748
Soto-Jiménez MF, Arellano-Fiore C, Rocha-Velarde R, Jara-Marini ME, Ruelas-Inzunza J, Páez-Osuna F (2011) Trophic transfer of lead through a model marine four-level food chain: tetraselmis suecica, Artemia franciscana, Litopenaeus vannamei, and Haemulon scudderi. Arch Environ Contam Toxicol 61:280–291. https://doi.org/10.1007/s00244-010-9620-4
Souahi H, Chebout A, Akrout K, Massaoud N, Gacem R (2021) Physiological responses to lead exposure in wheat, barley and oat. Environ Challenges 4:100079. https://doi.org/10.1016/j.envc.2021.100079
Sujka M, Pankiewicz U, Kowalski R, Mazurek A, Ślepecka K, Góral M (2019) Determination of the content of Pb, Cd, Cu, Zn in dairy products from various regions of Poland. Open Chem 17(1):694–702. https://doi.org/10.1515/chem-2019-0072
Sulaiman FR, Ibrahim NH, Syed Ismail SN (2020) Heavy metal (As, Cd, and Pb) concentration in selected leafy vegetables from Jengka, Malaysia, and potential health risks. SN Appl Sci 2:1430. https://doi.org/10.1007/s42452-020-03231-x
Suman J, Uhlik O, Viktorova J, Macek T (2018) Phytoextraction of heavy metals: a promising tool for clean-up of polluted environment? Front Plant Sci 9. https://doi.org/10.3389/fpls.2018.01476
Sun T, Wu H, Wang X, Ji C, Shan X, Li F (2020) Evaluation on the biomagnification or biodilution of trace metals in global marine food webs by meta-analysis. Environ Pollut 264:113856. https://doi.org/10.1016/j.envpol.2019.113856
Swaringen BF, Gawlik E, Kamenov GD, McTigue NE, Cornwell DA, Bonzongo JCJ (2022) Children's exposure to environmental lead: a review of potential sources, blood levels, and methods used to reduce exposure. Environ Res 204:112025. https://doi.org/10.1016/j.envres.2021.112025
Taiwo AM, Oyeleye OF, Majekodunmi BJ, Anuobi VE, Afolabi SA, Idowu OE, Ojekunle ZO, Taiwo OT (2019) Evaluating the health risk of metals (Zn, Cr, Cd, Ni, Pb) in staple foods from

Lagos and Ogun States, Southwestern Nigeria. Environ Monit Assess 191:167. https://doi.org/10.1007/s10661-019-7307-5

Tang B, Xu H, Song F, Ge H, Yue S (2022) Effects of heavy metals on microorganisms and enzymes in soils of lead–zinc tailing ponds. Environ Res 207:112174

TatahMentan M, Nyachoti S, Scott L, Phan N, Okwori FO, Felemban N, Godebo TR (2020) Toxic and essential elements in rice and other grains from the United States and other countries. Int J Environ Res Public Health 17:8128. https://doi.org/10.3390/ijerph17218128

Taweel A, Shuhaimi-Othman M, Ahmad AK (2013) Assessment of heavy metals in tilapia fish (*Oreochromis niloticus*) from the Langat river and Engineering lake in Bangi, Malaysia, and evaluation of the health risk from tilapia consumption. Ecotoxicol Environ Saf 93:45–51. https://doi.org/10.1016/j.ecoenv.2013.03.031.PMid:23642778

Tchounwou PB, Yedjou CG, Patlolla AK, Sutton DJ (2012) Heavy metal toxicity and the environment. Exp. Suppl. 101:133–164

Thakur S, Singh L, Zularisam AW, Sakinah M, Din MFM (2017) Lead induced oxidative stress and alteration in the activities of antioxidative enzymes in rice shoots. Biol Plant 61:595–598

Tingson KN, Zafaralla MT, Macandog DM, Rañola RF (2019) Lead biomagnification in a food web of the open waters along Sta. Rosa Subwatershed, Philippines. J Environ Sci 21(1):110–119. https://doi.org/10.47125/jesam/2018_1/11

Uddin MM, Zakeel MCM, Zavahir JS, Marikar FMMT, Jahan I (2021) Heavy metal accumulation in rice and aquatic plants used as human food: a general review. Toxics 9(12):360. https://doi.org/10.3390/toxics9120360

USEPA (1992) Guidelines for exposure assessment. U.S. Environmental Protection Agency, Risk Assessment Forum, Washington, DC, EPA/600/Z-92/001

USEPA (1998) Human health risk assessment protocol for hazardous waste combustion facilities, vol 1. Peer Review Draft, Online at: https://nepis.epa.gov/Exe/ZyPDF.cgi/10000IGP.PDF?Dockey=10000IGP.PDF

USEPA (2004) Example exposure scenarios. National Center for Environmental Assessment, Washington, EPA/600/R-03/036

USEPA (2011) Exposure factors handbook 2011 edition (Final), (EPA/600/R-09/052F). Washington

Usman K, Abu-Dieyeh MH, Zouari N, Al-Ghouti MA (2020) Lead (Pb) bioaccumulation and antioxidative responses in Tetraena qataranse. Sci Rep 10:17070. https://doi.org/10.1038/s41598-020-73621-z

van den Brink Nico W, Arblaster JA, Bowman SR, Conder JM, Elliott JE, Johnson MS, Muir DCG, Natal-da-Luz T, Rattner BA, Sample BE, Shore RF (2015) Use of terrestrial field studies in the derivation of bioaccumulation potential of chemicals. Integr Environ Assess Manag 12(1):135–145

Varol M, Kaçar E, Sünbül MR, Md Towfiqul Islam AR (2022) Levels of metals and elements in tissues of fish species in the Kızılırmak River (Turkey) and assessment of health risks and nutritional benefits. Environ Res 214:113791. https://doi.org/10.1016/j.envres.2022.113791

Vasile GG, Tenea AG, Dinu C, Iordache AMM, Gheorghe S, Mureseanu M, Pascu LF (2021) Bioavailability, accumulation and distribution of toxic metals (As, Cd, Ni and Pb) and their impact on *Sinapis alba* plant nutrient metabolism. Int J Environ Res Public Health 18(24):12947. https://doi.org/10.3390/ijerph182412947

Wang M, Liang B, Zhang W, Chen K, Zhang Y, Zhou H, Cheng Y, Liu H, Zhong X, Li Y, Liu Y (2019) Dietary lead exposure and associated health risks in Guangzhou, China. Int J Environ Res Public Health 16:1417. https://doi.org/10.3390/ijerph16081417

Wang Z-W, Nan Z-R, Wang S-L, Zhao Z-J (2011), Accumulation and distribution of cadmium and lead in wheat (Triticum aestivum L.) grown in contaminated soils from the oasis, north-west China. J Sci Food Agric. 91(2):377–84. https://doi.org/10.1002/jsfa.4196

Wani AL, Ara A, Usmani JA (2015) Lead toxicity: a review. Interdiscip Toxicol 8(2):55–64

Wei LL, Zhou Q, Xie CX, Wang J, Li J (2016) Bioaccumulation and biomagnification of heavy metals in three gorges reservoir and effect of biological factors. Huanjing Kexue 37:325–334

WHO (2008) Guidelines for drinking-water quality, 3rd edn. Incorporating first and second addenda. Geneva, World Health Organization. http://www.who.int/water_sanitation_health/dwq/gdwq3rev/en/index.html
WHO (2010) WHO human health risk assessment toolkit: chemical hazards (IPCS harmonization project document; no.8)
WHO (2021) WHO human health risk assessment toolkit: chemical hazards, 2nd edn. (IPCS harmonization project document; no.8)
Wieczorek J, Baran A, Urbański K, Mazurek R, Klimowicz-Pawlas A (2018) Assessment of the pollution and ecological risk on lead and cadmium in soils. Environ Geochem Health 40:2325–2342
Wijayawardena MAA, Yan K, Liu Y, Naidu R (2023) Can the mouse model successfully predict mixed metal(loid)s bioavailability in humans from contaminated soils? Chemosphere. 311, Part 2, 137113. https://doi.org/10.1016/j.chemosphere.2022.137113
Wilson J, Dixon SL, Wisinski C, Speidel C, Breysse J, Jacobson M, Crisci S, Jacobs DE (2022) Pathways and sources of lead exposure: Michigan Children's Lead Determination (the MI CHILD study). Environ Res 215, Part 2, 114204, https://doi.org/10.1016/j.envres.2022.114204
Winiarska-Mieczan A, Jachimowicz K, Kislova S, Kwiecień M, Zasadna Z, Yanovych D (2023) Cadmium and lead concentration in drinking instant coffee, instant coffee drinks and coffee substitutes: Safety and health risk assessment. Biol Trace Elem Res 201:425–434. https://doi.org/10.1007/s12011-022-03129-2
Wu X, Cai Q, Xu Q, Zhou Z, Shi J (2020) Wheat (*Triticum aestivum* L.) grains uptake of lead (Pb), transfer factors and prediction models for various types of soils from China. Ecotoxicol Environ Saf 206, 111387. https://doi.org/10.1016/j.ecoenv.2020.111387
Yan K, Dong Z, Ayanka Wijayawardena MA, Liu Y, Naidu R, Semple K (2017) Measurement of soil lead bioavailability and influence of soil types and properties: a review. Chemosphere 184:27–42. https://doi.org/10.1016/j.chemosphere.2017.05.143
Yang ZH, Lien PJ, Huang WS, Surampalli RY, Kao CM (2015) Development of the risk assessment and management strategies for TPH-contaminated sites using TPH fraction methods. J Hazard Toxic Radioact Waste 21(1). https://doi.org/10.1061/(ASCE)HZ.2153-5515.0000290
Yin Y, Zhang P, Yue X, Du X, Li W, Yin Y, Yi C, Li C (2018) Effect of sub-chronic exposure to lead (Pb) and *Bacillus subtilis* on *Carassius auratus gibelio*: bioaccumulation, antioxidant responses and immune responses. Ecotoxicol Environ Saf 161:755–762
Zaghloul A, Saber M, Gadow S, Awad F (2020) Biological indicators for pollution detection in terrestrial and aquatic ecosystems. Bull Natl Res Cent 44:127. https://doi.org/10.1186/s42269-020-00385-x
Zhang W, Chen L, Liu K, Chen L, Lin K, Guo J, Liu L, Cui C, Yan Z (2014) Lead accumulations and toxic effects in earthworms (*Eisenia fetida*) in the presence of decabromodiphenyl ether. Environ Sci Pollut Res Int 21(5):3484–3490
Zhang C, Wang X, Ashraf U, Qiu B, Ali S (2017) Transfer of lead (Pb) in the soil-plant-mealybug-ladybird beetle food chain, a comparison between two host plants. Ecotoxicol Environ Saf 143:289–295. https://doi.org/10.1016/j.ecoenv.2017.05.032
Zhang H, Zhao Y, Wang Z, Liu Y (2021) Distribution characteristics, bioaccumulation and trophic transfer of heavy metals in the food web of grassland ecosystems. Chemosphere 278:130407. https://doi.org/10.1016/j.chemosphere.2021.130407
Zhang C, Zou X, Yang H, Liang J, Zhu T (2022) Bioaccumulation and risk assessment of potentially toxic elements in soil-rice system in karst area. Southwest China Front Environ Sci 10:866427. https://doi.org/10.3389/fenvs.2022.866427
Zhou X, Xueyin Q, Zheng N, Su C, Wang J, Soyeurt H (2019) Large scale study of the within and between spatial variability of lead, arsenic, and cadmium contamination of cow milk in China. Sci Total Environ 650:3054–3061. https://doi.org/10.1016/j.scitotenv.2018.09.094
Zulfiqar U, Farooq M, Hussain S, Maqsood M, Hussain M, Ishfaq M, Ahmad M, Anjum MZ (2019) Lead toxicity in plants: impacts and remediation. J Environ Manage 15(250):109557. https://doi.org/10.1016/j.jenvman.2019.109557

Zuluaga Rodríguez J, Gallego Ríos SE, Ramírez Botero CM (2015) Content of Hg, Cd, Pb and As in fish species: a review. Vitae 22(2). https://doi.org/10.17533/udea.vitae.v22n2a09
Zyambo G, Yabe J, Muzandu K, M'kandawire E, Choongo K, Kataba A, Chawinga K, Liazambi A, Nakayama SM, Nakata H, Ishizuka M (2022) Human health risk assessment from lead exposure through consumption of raw cow milk from free-range cattle reared in the vicinity of a lead–zinc mine in Kabwe. Int J Environ Res Public Health 19:4757. https://doi.org/10.3390/ijerph19084757

Web Sites

https://www.epa.gov/risk/human-health-risk-assessment
https://www.epa.gov/iris
https://www.epa.gov/expobox/exposure-assessment-tools-media-food
https://www.atsdr.cdc.gov/toxprofiledocs/index.html
http://rais.ornl.gov/tools/profile.php
https://www.epa.gov/expobox/exposure-factors-interactive-resource-scenarios-tool-expofirst
https://www.epa.gov/expobox/basic-information-about-epa-expobox
https://www.consexpoweb.nl/
https://www.oecd.org/env/ehs/risk-assessment/echemportalglobalportaltoinformationonchemicalsubstances.htm
http://www.inchem.org/
https://www.efsa.europa.eu/en/data-report/chemical-hazards-database-openfoodtox
https://www.efsa.europa.eu/en/topics/topic/metals-contaminants-food
https://arche-consulting.be/tools/lead-chronic-biotic-ligand-model

Chapter 6
Cellular and Neurological Effects of Lead (Pb) Toxicity

Shubham Gudadhe, Sushma Kumari Singh, and Jawaid Ahsan

Abstract Lead (Pb^{2+}), a naturally occurring common heavy metal found in the earth's crust, replaces other cations in living creatures, disturbing many biological processes such as metal transport, energy metabolism, apoptosis, and cell signalling. Additionally, it has a significant influence on the central nervous system, specifically on the developing brain. It has severe neurotoxic effects on youngsters. Lead can act as a calcium ion replacement, crossing the blood–brain barrier and causing damage in brain areas, resulting in neurological problems. It possesses genotoxic characteristics and disrupts cellular activity. Neurotoxicity is a major problem, especially in the developing central nervous system, where it can cause long-term cognitive, motor, and behavioural deficits. Paediatric lead poisoning is more common, and early detection requires a high level of precision. The molecular processes and cellular effects of lead toxicity are discussed in this chapter. The pathophysiology, aetiology, and epidemiology of lead exposure are also reviewed in this chapter. It also investigates the neuropsychological issues linked with Intelligence Quotient (IQ), memory, executive functioning, attention, processing speed, language, visuospatial skills, motor skills, and effects on mood. The chapter also discusses lead-induced oxidative stress and its consequences. It will provide an in-depth understanding of the neuropsychological effects of lead toxicity at different levels, which will be helpful for its better management and finding remedies for the related toxic effects.

Keywords Lead · Cognition · Neurodevelopment · Oxidative stress · Signalling pathway · Toxicity

---

S. Gudadhe · S. K. Singh · J. Ahsan (✉)
Department of Biotechnology, Central University of South Bihar (CUSB), Panchanpur, Gaya, Bihar 824236, India
e-mail: jahsan@cub.ac.in

N. Kumar and A. K. Jha (eds.), *Lead Toxicity Mitigation: Sustainable Nexus Approaches*, Environmental Contamination Remediation and Management,
https://doi.org/10.1007/978-3-031-46146-0_6

6.1 Introduction

Lead (Pb) is considered to be a significant and naturally occurring toxic metal among the various heavy metals present in the Earth's crust. Lead, which has an atomic number of 82 and is derived from the Latin word Plumbum, is a prevalent toxic substance found throughout various locations (Patra et al. 2011). In ancient times, lead was used for several purposes (Maiti et al. 2017). The presence of lead can be identified in both living organisms and non-living surroundings. The increase in anthropogenic activities and vehicle emissions is primarily accountable for the increase in lead concentration within the human body through inhalation, ingestion, and dermal contact. In particular, the liver, spleen, and kidney have been recognised as significant target areas for lead poisoning. Lead in the form of a toxin generates a variety of biochemical, physiological, and behavioural dysfunctions (Bandyopadhyay et al. 2014). Lead is one of the most toxic heavy chemicals to people for humans and has been for thousands of years. Lead makes us sick when it gets into our bodies through food, air, and water because it reacts with biological molecules that contain sulphur, oxygen, or nitrogen (Maiti et al. 2017). Lead poisoning is usually found when the amount of lead in the blood rises. But short-term exposure to lead can cause problems like neurobehavioral and brain damage, memory problems, high blood pressure, and damage to the kidneys. The parts and systems of the body that are most likely to be affected by high levels of lead are the blood, kidney, reproductive, and central nervous systems (Assi et al. 2016). Jalali et al. state that when the amount of malondialdehyde (MDA) increases, the activities of erythrocyte superoxide dismutase (SOD) and glutathione peroxidase (GPx) increase along with the total number of erythrocytes (Jalali et al. 2017). Rats exposed to lead had a low number of cells, lymphocytes, and neutrophils, leading to microcytic anaemia. Chelation treatment is generally recommended for low levels of lead poisoning that have caused brain damage (encephalopathy). But researchers are still looking at treatments that use less medicine but last longer. An important part of treating chronic diseases is determining how much lead is in the body and what happens when people are exposed to low levels of lead in the surroundings (Singh et al. 2017). Heavy metal lead (Pb) is a common pollution in the environment, and it has been said to cause poisoning in many people (Karri et al. 2016). The detrimental impact of Pb-induced oxidative stress on the Central Nervous System (CNS) is widely acknowledged. Exposure of rodents to Pb has been found to be associated with persistent alterations in brain-derived neuronal factor (BDNF), β-amyloid (Aβ) aggregation, and oxidative damage. These findings pose significant environmental and public health challenges due to their close association with impaired intelligence and growth (Feng et al. 2015; Li et al. 2018).

A research study has shown that developmental exposure to Pb results in an over accumulation of Pb in the hippocampus, which is associated with a decline in cognitive abilities that is directly proportional to the dose of Pb (Wei et al. 2022). It is worth noting that exposure to environmental insults during developmental stages, specifically prepuberty and adolescence, has a substantial influence on neural plasticity and subsequent behaviour in adulthood (Encinas et al. 2006; Sanders et al. 2015). Studies have shown that exposure to Pb during early stages of life in animals such as rodents and primates can lead to cognitive impairment and a subsequent increase in amyloid biomarkers that are relevant to Alzheimer's disease in later stages of life (Bihaqi et al. 2014a; Liu et al. 2014). The presence of increased apoptotic markers has been reported in conjunction with the aforementioned condition. The issue of childhood lead poisoning persists (Chandramouli et al. 2009).

6.2 Sources of Lead Exposure

Lead is a naturally occurring heavy metal that is very poisonous. Lead can be found everywhere in nature, but most of it comes from human actions such as mining, making things, and burning fossil fuels. There are three distinct forms of lead, namely metallic lead, inorganic lead, and lead compounds, also known as lead salts, as well as organic lead that contains carbon. Lead in the environment rarely occurs in its elemental state but rather in its + 2 oxidation state (Pb^{2+}) in various ores throughout the earth. Lead has been found in at least 1272 of the 1684 National Priority List (NPL) sites identified by the United States (U.S.) Environmental Protection Agency (EPA) (Gerberding and Falk 2005). Lead is one of the most durable heavy metals in nature. Groundwater, soil, dust from metal ores, brass plumbing fixtures, several industrial activities, folk remedies, burning petroleum, making lead battery, paint industries, and mining processes, contaminating food, and certain herbal products made with lead are all sources of lead in the environment (Fig. 6.1). People are always getting lead from things such as contaminated air, water, earth, house dust, and food, as well as by breathing it in. Lead paints and lead chips are the main and most common ways for children to get too much lead (Patra et al. 2011). Lead has various applications, such as in leaded petrol, paints, ceramics, ammunition, water pipes, solders, hair dye, cosmetics, farm equipment, aeroplanes, shielding for X-ray machines and in the production of corrosion and acid resistant materials utilised in the construction sector (Sanders et al. 2009). Various sources of lead poisoning include the production of ammunition, ceramic glazes, circuit boards, caulking, sheet lead, solder, certain brass and bronze plumbing, radiation shields, intravenous pumps, foetal monitors, as well as specific surgical and military equipment, such as jet turbine engines and military tracking systems, among others (Fig. 6.1). Employees are at an increased risk of being exposed to lead at different construction locations (Levin and Goldberg 2000; Mitra et al. 2017). When taking part in hobbies or activities that increase exposure, kids can be exposed to lead-based paint that is peeling or flaking or weathered powdered paint. Particularly at risk are kids with pica, which is the compulsive, habitual ingestion of non-food substances (Mitra et al. 2017). The severity of the toxic reaction depends on a number of things, such as the dose, the age of the person exposed, the stage of a woman's life (children, breastfeeding, menopause), the person's job, the length of time they were exposed, their health and lifestyle, and their nutritional status.

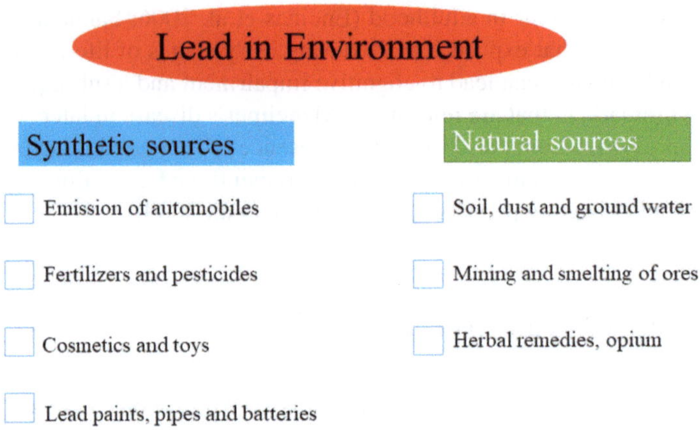

Fig. 6.1 Lead in the environment originates from both synthetic and natural sources

6.3 Lead Exposure in Humans

Exposure to lead (Pb) is still a major public health issue around the world. Pb is a toxic metal that can be found in the environment because of things like lead mining, battery recycling, and the use of lead petrol. Children and pregnant women are especially vulnerable to the effects of Pb exposure. The quantification of the exposure of Pb and its body burden in human studies is primarily accomplished by measuring the measurement of metal concentration in both blood and bone. There is a lack of consensus regarding the exposure levels required to elicit the initial symptoms of neurotoxicity in individuals who are occupationally exposed. However, the majority experts concur that overt neurotoxic effects can manifest at blood Pb levels of 60 μg/dL whole blood. Consequently, it is recommended that workers maintain a maximum concentration of approximately 40 μg/dL (CDC 2018).

But other studies found a link between exposure to lead and changes in thinking in workers whose blood lead levels were between 20 and 40 g/dL (Barth et al. 2002; Lucchini et al. 2012; Murata et al. 2009). The World Health Organisation says that adults who live in communities should keep the amount of lead in their blood below 10 g/dL. But there does not seem to be a safe amount of exposure to Pb, and levels of 1–3 g/dL have been linked to subtle neurotoxic effects (Kosnett et al. 2007). The concentration of Pb in bone is believed to be a measure of total exposure. It is measured mostly by K-shell X-ray fluorescence spectroscopy in the tibia and patella, which are cortical and trabecular bone, respectively. The half-life of Pb in bone is reported to be different depending on where it is in the body and on factors like age, previous exposure, and other situations that affect bone turnover (Farooqui et al. 2017).

According to a study done in China, children's mean BLL was 4.71 g/dL, with 41.4% of those having BLLs higher than 5 g/dL (Li et al. 2020).

6.4 Neuropsychological Effects of Lead Toxicity

Lead exposure has a wide range of adverse effects on cognitive functioning. Prenatal exposure, as assessed by lead levels in umbilical cord blood, has been linked to Cord blood, was associated with worse scores on the Bayley Scales of Infant Development in the sensorimotor and visuomotor subscales (Koller et al. 2004; McMichael et al. 1988). Numerous cross-sectional and longitudinal studies on children have demonstrated that lead exposure reduces children's overall cognitive functioning, but the majority of these studies examine global measures of intellectual functioning rather than domain-specific effects. Chronic exposure to lead is more detrimental to cognitive function in adults than acute exposure (Bellinger 2004; Koller et al. 2004; Lidsky and Schneider 2003; Needleman 2004). Studies on domain-specific cognitive affects are listed below.

6.4.1 *Intelligence*

Most of the time, when children are exposed to lead, their intelligence scores go down. Reviewing paediatric cross-sectional studies on brain problems caused by exposure to lead, it was found that IQ dropped by three points when blood lead levels went from 5 to 20 g/dL and dropped by 5.3 points when blood lead levels went from 5 to 50 g/dL (Winneke et al. 1996). When lead levels in the blood went from 10 to 20 g/dL, there was a pretty consistent link between a drop and a three-point drop (Pocock et al. 1994; Winneke et al. 1996). Based on these results, it seems that exposing someone to lead lowers their intelligence in a way that depends on how much lead they are exposed to. Even though it has not been seen as often in adults as it has in kids, some adults have shown signs of having less intelligence. The Task Group on the Effects of Inorganic Lead of the World Health Organisation's Programme for Chemical Safety (Joint FAO/WHO Expert Committee on Food Additives 2002). After conducting a comprehensive analysis of the existing literature, it was determined that human intellectual functioning may be negatively affected by blood levels below 25 μg/dL. Furthermore, it was found that for every 10 μg/dL increase in blood lead levels, there is a predicted decrease in IQ of 1–5 points. The findings suggest that there is a correlation between occupational lead exposure and decreased cognitive and intelligence scores in adults, with the effect being dependent on the dosage (Khalil et al. 2009). When researchers first looked at the effects of lead on the brain, they focused on how it affected the brain as a whole. However, more recent research shows that it is important to examine how lead affects the brain in different areas.

6.4.2 Memory

Several studies have indicated a decrease in learning and memory performance among adults who have been exposed to lead in their occupation. The findings indicate that lead exposure has a more pronounced negative impact on cognitive function in the elderly population, as evidenced by reduced scores in learning and memory tasks, among other cognitive impairments. Specifically, individuals 55 years and above appear to be more vulnerable to the deleterious effects of exposure to lead. Although older adults are particularly vulnerable, research has also observed reduced memory performance in individuals under 55 years of age who have been exposed to elevated levels of lead. Subjects exhibited a decline in their ability to recall verbal and visual information after exposure to lead (Khalil et al. 2009; Stewart and Schwartz 2007). There has been constant evidence of lower visuospatial memory scores, which suggests that lead exposure affects spatial skills and the ability to remember what you see. Lead exposure on the job is also linked to lower visual memory scores, especially a delay in remembering a complex figure (Schwartz et al. 2000). Lead exposure has also been associated with lower verbal memory scores, which affects instant recall, delayed recall, and identification. Chronic contact seems to not only affect both vocal and nonverbal memories, but also to cause them to get worse over time. In this group, the results on both verbal and nonverbal memory tests kept going down over time. This means that long-term contact may cause gradual loss of memory over many years (Mason et al. 2014).

6.4.3 Processing Speed

Lead poisoning has been shown to slow processing speed, and the results suggest that the link is dose-dependent. People exposed to high amounts of lead took longer to make decisions and respond. For example, significant slowing down of decision-making speed and wider gaps in a detection/reaction time task have been found to be caused by contact (Winneke et al. 1996). These results also revealed slight deficiencies in classification speed and precision during a category search task. Only individuals with blood lead concentrations of 40 g/dL or higher exhibited these deficits. The dose-dependence of neurobehavioral deficits was confirmed by a follow-up study with the same participants and testing battery. However, the primary finding of both studies was a delayed sensory-motor reaction time, which may have artificially hampered overall processing speed (Stollery et al. 1991).

6.4.4 *Executive Functioning and Attention*

Several investigations have demonstrated that occupational exposure to lead decreases executive functioning. Impaired executive functioning abilities in switching and inhibition tasks (Trails Making Test B and Stroop Task, respectively) were also observed in a group with a maximum lead exposure of 20 g/g (tibia bone lead measurement). Lower executive functioning scores were also discovered in earlier studies employing comparable assessments and scores (Schwartz et al. 2000, 2005).

6.5 Cellular Effects of Lead Neurotoxicity

In recent decades, new information about how lead affects cells and how it works has helped us to learn more about its neurotoxicity. Using cellular models of learning and memory, researchers have investigated how lead might cause brain problems. A new study shows that exposure to lead is bad for the Central Nervous System (CNS), that environmental factors make people more sensitive to lead, and that being exposed to lead as a child can cause neurodegeneration as an adult.

As the CNS is the main target of lead poisoning, the brain is the most studied when it comes to lead poisoning. Lead neurotoxicity occurs when the CNS is exposed to enough lead to change how it normally works and cause damage to the CNS. Lead's direct neurological effects include apoptosis (programmed cell death), excitotoxicity, which affects neurotransmitter storage and release and changes neurotransmitter receptors, mitochondria, second messengers, cerebrovascular endothelial cells, and both astroglia and oligodendroglia. Loss of memory, vision, cognitive and behavioural problems, and brain damage/mental retardation are some of the symptoms that can show up right away or later (Sanders et al. 2009). Although most of the early studies focused on the neurocognitive effects of lead, more recent research has shown that higher exposures are linked to morbidities such as antisocial behaviour, delinquency, and violence. To explain the mechanism of lead toxicity on the CNS, several theories have been put forth (Hwang 2007).

6.6 Effect of Lead on Signalling Pathways

The first publication pertaining to lead-mediated oxidative stress was released in 1965. The present study revealed that certain metals have the ability to increase the rate of oxidation of crucial fatty acids. The efficacy of lead as a material during that period was reportedly inadequate. Subsequent to a considerable period of time, it was noted that lead was responsible for the escalation in lipid peroxidation, as determined by the analysis of Malondialdehyde (MDA). The lead-induced lipid peroxidation in rat brain was also documented by a number of researchers. A positive correlation was found between elevated lead concentration and increased lipid peroxidation, similar effect was observed in hepatic tissues as well (Shafiq-ur-Rehman 2003). Lead-induced oxidative stress is primarily attributed to cellular membrane and DNA, as well as inhibition of key enzymes such as catalase, GPx, SOD, and G6PD, and non-enzymatic antioxidant molecules such as thiols (GSH) in mammalian organisms (Flora et al. 2008; Valko et al. 2005).

Several studies have suggested that metal-induced toxicity involves a multifactorial mechanism, as illustrated in Fig. 6.3. Multifactorial mechanisms may be linked to various biological processes such as oxidative stress, enzyme inhibition, DNA damage, alterations in gene expression, and phenomena such as adventitious mimicry. The mechanism of metal-induced generation of free radicals, particularly Reactive Oxygen Species (ROS).

The precise mechanisms underlying lead-induced oxidative stress remain unclear, likely due to the limited capacity of lead to undergo rapid valence changes. Lead exhibits a propensity for covalent bonding with sulphydryl groups because of its electron-sharing affinities. The interaction between lead and GSH is crucial for the manifestation of its toxic effects (Hultberg et al. 2001).

In the context of a signaling pathway, lead acts as a calcium mimic and binds to the calmodulin protein (a Ca^{2+} 134 binding protein) that has been implicated in the induction of lead toxicity. The findings indicate that lead binding exhibits a higher relative affinity compared to calcium (Kirberger et al. 2013), as illustrated in Fig. 6.2. Various mechanisms for lead-mediated oxidative stress have been suggested.

6.7 Lead-Induced Neurotoxicity and Its Mechanisms of Action

One of the most vulnerable parts of the body to lead is the nervous system. In general, it damages the nervous system, but it affects children's brains a lot more. Neurotoxicity is also linked to the production of too many free radicals, which can change how the brain works. Lead quickly penetrates the Blood–Brain Barrier (BBB) and replaces calcium ions, disrupting intracellular calcium regulation in brain cells. Long-term lead poisoning in children can cause comas, seizures, and changes in their mental state. Several clinical studies have been conducted on the link between lead poisoning and the way the brain develops and works (Brochin et al. 2008). Blood lead levels are negatively correlated with neurological development and function. Lead-poisoned children exhibited abnormal behaviour such as melancholy, aggression, destruction, social withdrawal, and atypical body movements (Hou et al. 2013; Mărginean et al. 2016).

Neurological differences are mostly caused by the way ions work. When lead replaces calcium ions, it becomes able to cross the BBB at a good rate (Fig. 6.3). After crossing the BBB, lead builds up in astroglial cells with lead-binding proteins.

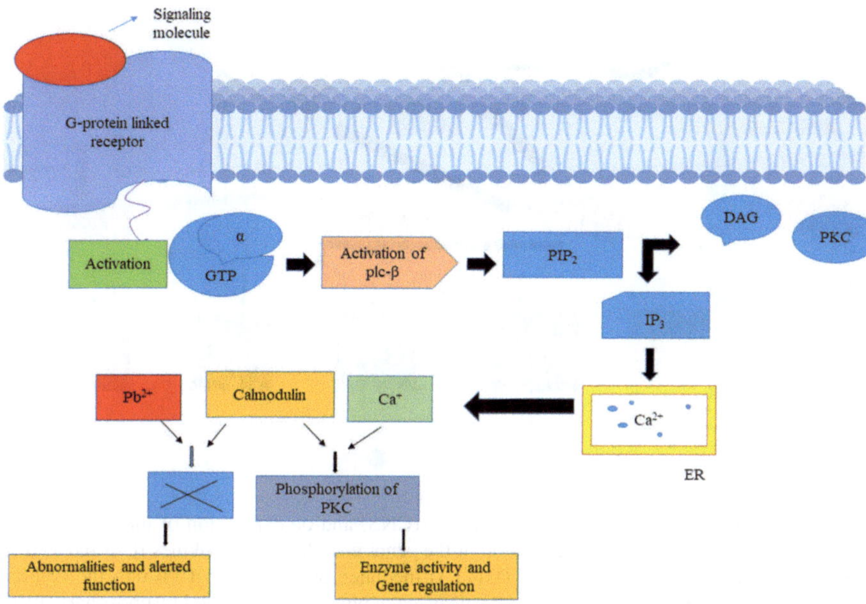

Fig. 6.2 Lead interference with calcium (Ca)-dependent inositol trisphosphate or inositol 1,4,5-trisphosphate pathway (ER = endoplasmic reticulum)

Lead is more dangerous for growing nervous systems because they do not have enough mature astroglial cells. Immature astroglial cells do not have any proteins that bind to lead. Lead can easily harm undeveloped astroglial cells and interfere with the development of myelin sheaths (Wang et al. 2011). Lead is also moved by Divalent Metal Transporter 1 (DMT1), a protein with 12 transmembrane domains that is found in capillary cells. DMT-1's job is to move essential metals, but it also moves toxic metals that look like important minerals (Moos et al. 2006). Protein Kinase C (PKC) is an enzyme that plays a crucial role in many physiological processes, including cell proliferation and brain development, and can be stimulated by subnanomolar concentrations of lead ions.

6.8 Lead Affects Movement of Calcium

Lead changes the brain and behaviour in complicated ways that are hard to understand. Still, work on cells and molecules has led to a better understanding of how lead affects how the brain works. The effects of lead on biological processes that rely on calcium are especially important. Calcium is an important ion for neural function, such as cell growth and development, the release of neurotransmitters, and biochemical reactions inside the cell.

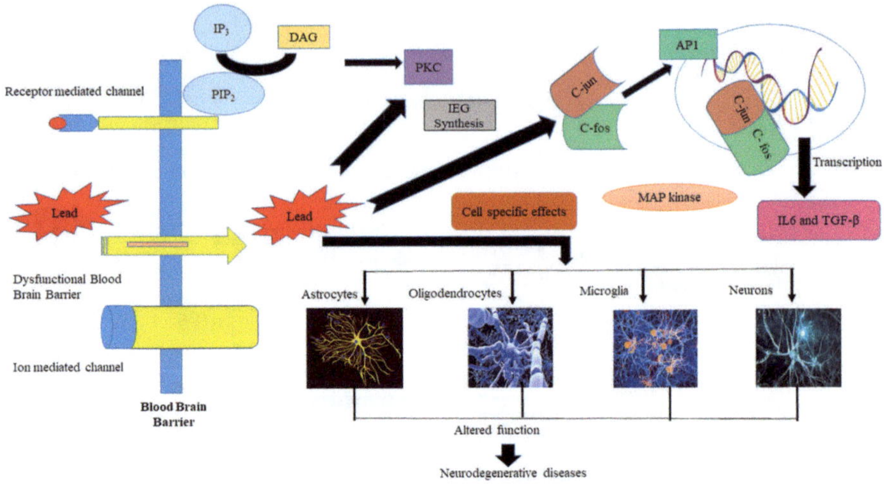

Fig. 6.3 Effect of lead on central nervous system (CNS) and on expression of interleukin-6 and TGF-β1. Lead exposure alters the expression of the genes encoding the cytokines IL-6 and TGF-1. Gene expression of cytokines IL-6 and TGF-1 is mediated by the entry of Pb into the cell and mobilisation of calcium ions, followed by the cleavage of phosphatidylinositol bisphosphate (PIP2) into inositol triphosphate (IP3) and diacylglycerol (DAG). DAG activates PKC, which increases the expression of IEG jun and fos genes. The mitogen-activated protein kinase (MAPK) pathway is essential for dimerization aphosphorylation of the c-jun and c-fos proteins, which resulted in the formation of the nuclear transcription factor AP-1 (activator protein 1) and increased expression of IL-6 and 214 TGF-1. Directly, lead increases the activation of PKC, c-fos, and c-jun protein expression. Lead's powerful effect on CNS cells results in neurodegeneration

Lead and calcium are divalent cations that share similarities in terms of their ionic charge and size. The capacity of lead to imitate or hinder calcium-mediated impacts is fundamental to its biological and behavioural consequences. A less regulated ligand in the human body in comparison to calcium.

A heavy metal that lacks regulation. Lead has the ability to bind to the same sites as calcium and can enter the cell via calcium channels. This results in the displacement, inhibition, substitution, and/or activation of calcium-dependent processes (Bridges and Zalups 2005; Habermann et al. 1983; Kerper and Hinkle 1997).

The widespread occurrence of calcium in cellular signalling and the crucial significance of the spatial and temporal arrangement of calcium signals in cellular operation imply that interference with calcium-dependent mechanisms can result in significant cellular outcomes. This notion is supported by various studies (Berridge et al. 2003; Bootman 2012; Bootman et al. 2001, 2002; Bridges and Zalups 2005). The impact of lead on the calcium dynamics of neurons provides insight into numerous extensive alterations in brain activity and conduct.

6.9 Effect of Lead on NMDA Receptor

Lead is an antagonist of the N-Methyl-D-aspartate receptor (NMDA-R) that operates in a non-competitive manner.

The N-methyl-D-aspartate receptors (NMDA-Rs) are a type of ionotropic receptor that is stimulated by the neurotransmitter glutamate. These receptors play a crucial role in various physiological processes, such as neural development, neuronal plasticity, learning and memory, and long-term potentiation, which is a physiological manifestation of learning (Cory-Slechta et al. 1997; Gilbert and Lasley 2007; Hubbs-Tait et al. 2005; Nihei and Guilarte 2001).

When glutamate binds to NMDA-Rs, calcium flows in through a ligand-gated ion channel. This can cause an excitatory post-synaptic potential and has a big effect on how neurons work by starting second messenger pathways that depend on calcium. The blocking of postsynaptic NMDA-Rs by lead results in the inhibition of activity-dependent calcium influx. This can subsequently interfere with NMDA receptor-dependent developmental processes, neural plasticity, learning and memory, as well as Long-Term Potentiation (LTP). The induction of LTP is hindered by chronic and developmental exposure to lead across a broad spectrum of concentrations, resulting in a higher threshold. This phenomenon is linked to compromised learning and memory (Lasley et al. 2001; Lasley and Gilbert 2000, 2002; Luo et al. 2011; Nihei and Guilarte 2001). Blocking NMDA receptors or other effects of lead on calcium-dependent processes may have something to do with how well LTP and learning work.

Apoptosis is another thing that happens when NMDA receptors are blocked. This is a type of cell death that is planned and caused by a well-known biological process (Anastasio et al. 2009; Hansen et al. 2004; Léveillé et al. 2010; Lyall et al. 2009; Yuede et al. 2010). During brain growth, apoptosis is usually used to get rid of unwanted links and 'sculpt' the brain. Pathological apoptosis, on the other hand, can happen in some situations. Low amounts of lead during development have also been shown to cause apoptosis and mess up brain development in both human and zebrafish models by blocking NMDA receptors (Dou and Zhang 2011; Dribben et al. 2011; Liu et al. 2010).

Due to the important role NMDA receptors play in many neuro and behavioural processes and the fact that lead can block NMDA receptors, knowing how lead affects the brain and behaviour depends on these receptors.

6.10 Effect of Lead on Calmodulin

Lead also targets calmodulin (CaM), or 'calcium-modulated protein', a significant intracellular calcium-activated protein (Heizmann and Hunzlker 1991). Calmodulin is involved in calcium signalling, neurotransmitter receptors, ion channels, and neural plasticity (McCue et al. 2010). Calmodulin possesses four distinct binding sites that are naturally bound by calcium ions. Calmodulin exhibits functional activity upon complete binding of calcium to all four of its sites (Costa 1998).

According to several studies (Fullmer et al. 1985; Habermann et al. 1983; Sandhir and Gill 1994; Shirran and Barran 2009), at levels that are relevant to physiological processes, calmodulin exhibits a higher binding affinity towards lead compared to calcium, thereby leading to the activation of the protein. Upon the occurrence of this event, calmodulin undergoes activation in a manner that is not consistent with normal physiological processes. The signalling of calmodulin undergoes a state of tonic activation and becomes independent of external stimuli. The extensive involvement of calmodulin in calcium signalling implies that uncontrolled activation of calmodulin can result in various outcomes, including but not limited to the disruption of signal transduction that is dependent on calmodulin and interference with calmodulin-mediated learning and memory (Rocha and Trujillo 2019).

6.11 Effect of Lead on Protein Kinase C

Protein Kinase C (PKC) is an intracellular signalling enzyme that is dependent on calcium and phospholipids and is involved in diverse cellular functions (Markovac and Goldstein 1988). Protein Kinase C (PKC) catalyses the phosphorylation of proteins through the transfer of phosphate groups from Adenosine Triphosphate (ATP). The regulation of cellular growth and differentiation is reliant on the phosphorylation of transport proteins via PKC. The Protein Kinase C (PKC) has been found to be involved in cytoskeletal function and signal transduction (Pears 1995). Additionally, PKC has been observed to have a significant impact on learning and memory, as noted (Van der Zee et al. 1992; Xu et al. 2014).

Lead replaces calcium in the activation of PKC at a clinically meaningful picomolar dose, raising intracellular calcium, and obstructing neurotransmitter release (Goldstein 1993). According to Bouton et al. (2001), lead mimics calcium at the synaptotagmin site and competes for it with higher affinity than calcium. Extended exposure to lead results in elevated PKC activity, which in turn triggers a compensatory reduction in activity, potentially through downregulation or decreased effectiveness of calcium activity.

The significance of PKC in calcium-mediated long-term potentiation (LTP) has been established. Studies have shown that PKC inhibitors, such as polymyxin B, impede the initiation and preservation of calcium-induced LTP (Cheng et al. 1994). The negative impact of lead on cognitive abilities such as learning and memory is believed to be caused, at least partially, by interference with typical PKC operation. Furthermore, the influence of lead on PKC activity has consequences for various cellular processes such as cell division, neural communication, neural plasticity, and cytoskeletal organisation (Bressler et al. 1999) Additionally, it affects cellular proliferation and differentiation (Markovac and Goldstein 1988).

6.12 Lead as Neurotransmitter Releaser

Typically, the depolarization of neurons results in the activation of voltage-gated calcium channels, thereby facilitating the entry of calcium ions into the presynaptic terminal. Upon calcium influx, a series of enzymes are activated, thereby facilitating the fusion of the synaptic vesicle with the cellular membrane and subsequent liberation of neurotransmitters. Lead has been found to have a converging impact on neurotransmitter release by binding to voltage-gated calcium channels and subsequently decreasing the influx of calcium. Furthermore, it has been observed that lead engages in competition with calcium for the binding sites of various proteins that play a role in the release of neurotransmitters, such as calmodulin, CaM kinase II (CaMKII), and synaptotagmin (Bouton et al. 2001; Kern et al. 2000; Westerink et al. 2002).

Collectively, these measures lead to a decrease in the discharge of neurotransmitters at the presynaptic terminal. The inhibition of neuronal release of glutamate and GABA can be observed at nanomolar concentrations of lead. The perturbation of regular neurotransmitter release can result in diverse outcomes for the brain and conduct, contingent on the particular neurotransmitter and its placement within the brain (Braga et al. 1999).

6.13 Lead and Neurodegenerative Diseases

Recent studies offer compelling evidence that lead exposure has detrimental impacts on the CNS in both adult and paediatric populations. Lead-induced damage within the brain can result in various neurological disorders, including but not limited to brain damage, mental retardation, behavioural problems, nerve damage, and potential development of Alzheimer's disease, Parkinson's disease, and schizophrenia. The prefrontal cerebral cortex, hippocampus, and cerebellum are particularly vulnerable to such damage. These findings suggest the need for further investigation into the potential long-term effects of lead exposure on the brain (Sanders et al. 2009).

6.13.1 Alzheimer's Disease (AD)

Numerous research studies have examined the impact of lead exposure on cognitive abilities and IQ in children. However, investigations into developmental lead exposure in non-human primates and rodents have revealed associations with the onset of Alzheimer's disease during the later stages of life. Alzheimer's disease is widely recognised as the prevailing neurodegenerative disorder. The condition is distinguished by cognitive decline and dementia, accompanied by brain pathology consisting of proteinaceous plaques composed of Amyloid beta (Aβ). The globus pallidus, dentate gyrus, temporal cortex, and temporal white matter of postmortem human brains affected by Alzheimer's disease have exhibited significantly elevated levels of lead in comparison to control healthy brains of the same age group, as per the findings of Haraguchi et al. (2001). The exposure to Pb has been found to raise the mRNA of Amyloid Precursor Protein (APP) and the aggregation of Aβ in rats, leading to amyloidogenesis and the deposition of senile plaques. Additionally, in nonhuman primates who were exposed to lead during infancy, there was an upregulation of APPs (Bihaqi et al. 2014a, b; Wu et al. 2008). Exposed mice to lead across the course of different life spans and discovered that there is a window of sensitivity to lead toxicity in the developing brain. Cognitive impairment only occurred in mice exposed to Pb as newborns, not as adults (Bihaqi et al. 2014a). According to Bihaqi et al. (2014a) and Masoud et al. (2016), the exposure of mice to lead during their early life stages results in increased expression of tau protein associated with Alzheimer's disease and changes in epigenetic markers linked to the development of the same disease (Bihaqi et al. 2014a; Masoud et al. 2016).

The relationship between lead exposure during infancy and AD is being explained by an emerging theory that suggests an epigenetic basis for the increased production of proteins relevant to AD and cognitive decline. Exposures experienced during the foetal or early developmental stages have the potential to induce epigenetic modifications in the brain, thereby resulting in gene reprogramming. According to Schneider et al. (2013), a study was conducted on rats that were exposed to Pb either in utero or in postnatally. The results indicated a reduction in the expression of DNA methyl transferase in the hippocampus of female rats that were exposed to Pb. This suggests that there may be a decrease in DNA methylation, which could lead to the expression of genes that are typically suppressed (Schneider et al. 2013).

The investigation conducted by Schneider involved the examination of gene expression pertaining to DNA methyl transferases, which was carried out at postnatal day (Schneider et al. 2013). On the other hand, a study was conducted by Dosunmu wherein infant mice exposed to Pb were subjected to genome-wide expression and methylation profiling until postnatal day 700. The results showed that a specific group of genes, which are typically expressed in aged mice, were repressed. The aforementioned genes were found to be implicated in the immune response, metal binding, and metabolism. Suppression of their expression resulting from developmental exposure to Pb has been observed to impede the brain's capacity to counteract stressors associated with ageing (Dosunmu et al. 2012).

6.13.2 Parkinson's Disease (PD)

According to research findings, lead has been observed to decrease the production of catecholamine as well as synaptic neurotransmission. The decrease in GABA (gamma-aminobutyric acid) could be a common factor in all human neurodegenerative disorders caused by unusual levels of calcium inside cells (Błaszczyk 2016). The occurrence of oxidative stress resulting from chronic lead intoxication has been verified through the observation of elevated levels of lipid peroxide in the brain and liver of rats. Exposure to lead has been found to diminish dopaminergic neurotransmission through mechanisms such as mitochondrial dysfunction, oxidative stress, and heightened gliofilament expression in astrocytes (Patra et al. 2011).

Lead toxicity poses a greater risk to children through dietary exposure and can result in adverse effects on the nervous system and pica behaviour, as documented in literature (Zeng et al. 2016). The study conducted by Loikkanen et al. provides evidence that lead has an impact on cellular processes through the regulation of calcium and calcium-binding proteins. Additionally, the study suggests that lead affects the release and reuptake of various neurotransmitters. The aforementioned study indicates that it inhibits the acetylcholine and dopamine releases that are dependent on Ca^{2+} and activity (Loikkanen et al. 1998).

The hippocampus region of the brain is subject to tau hyperphosphorylation and α-synuclein accumulation, which are the primary factors that trigger apoptosis and autophagy. This phenomenon has been extensively studied and documented (Zhang et al. 2012). The study conducted by Rogers et al. revealed that the APP is a significant contributor to lead toxicity via iron regulatory pathways, as observed in human dopaminergic SH-SY5Y neuroblastoma cells.

The involvement of PKC in dopamine transport function and the induction of oxidative stress through PKC activation by lead, leading to neurotoxicity, has been reported (Rogers et al. 2016).

Lead is easily able to cross the BBB and binds to sulfhydryl groups, which changes anti-oxidant enzymes and raises the amount of lipid peroxidation. In the same way, lead poisoning can happen when –ALAD (Delta-aminolevulinic acid dehydratase) is stopped from working and too much of its substrate, –ALA, builds up. –ALA quickly oxidises to make free radicals and release ferrous ions, which start the process of lipid peroxidation (Ashafaq et al. 2016).

6.14 Conclusion

The neurotoxic effects of lead exposure and its considerable effects on human neuropsychology are discussed here. Pb toxicity can cause the central nervous system to suffer from a variety of negative consequences, altering cognitive functioning. Lead works by a mechanism that interferes with calcium dynamics, which are critical to many cellular activities. Lead obstructs calcium's ability to regulate itself, which impairs synaptic transmission and neuronal activity. In addition, mounting data point to a possible connection between lead exposure and the emergence of neurodegenerative disorders. An increased risk of neurodegenerative diseases including Alzheimer's and Parkinson's has been linked to chronic lead exposure. Complex processes, including oxidative stress, inflammation, protein aggregation, and mitochondrial dysfunction, underlie these correlations. It is crucial for public health to comprehend lead's neurotoxic effects and how they affect neuropsychology in humans. Reduced lead exposure is essential for preventing neurodegenerative illnesses and long-term cognitive deficits in sensitive populations like children. To reduce the neurotoxic effects of lead on human neuropsychology, more study is required to understand the underlying mechanisms and create efficient preventive, early detection, and intervention measures.

References

Anastasio NC, Xia Y, O'Connor ZR, Johnson KM (2009) Differential role of N-methyl-D-aspartate receptor subunits 2A and 2B in mediating phencyclidine-induced perinatal neuronal apoptosis and behavioral deficits. Neuroscience 163(4):1181–1191

Ashafaq M, Tabassum H, Vishnoi S, Salman M, Raisuddin S, Parvez S (2016) Tannic acid alleviates lead acetate-induced neurochemical perturbations in rat brain. Neurosci Lett 617:94–100. https://doi.org/10.1016/j.neulet.2016.02.001

Assi MA, Hezmee MNM, Sabri MYM, Rajion MA (2016) The detrimental effects of lead on human and animal health. Vet World 9(6):660

Bandyopadhyay D, Ghosh D, Chattopadhyay A, Firdaus SB, Ghosh AK, Paul S, Bhowmik D, Mishra S, Dalui K (2014) Lead induced oxidative stress: a health issue of global concern. J Pharm Res 8(9):1198–1207

Barth A, Schaffer A, Osterode W, Winker R, Konnaris C, Valic E, Wolf C, Rüdiger H (2002) Reduced cognitive abilities in lead-exposed men. Int Arch Occup Environ Health 75(6):394–398. https://doi.org/10.1007/s00420-002-0329-1

Bellinger DC (2004) Lead. Pediatrics 113(Supplement_3):1016–1022. https://doi.org/10.1542/peds.113.S3.1016

Berridge MJ, Bootman MD, Roderick HL (2003) Calcium signalling: dynamics, homeostasis and remodelling. Nat Rev Mol Cell Biol 4(7):517–529

Bihaqi SW, Bahmani A, Adem A, Zawia NH (2014a) Infantile postnatal exposure to lead (Pb) enhances tau expression in the cerebral cortex of aged mice: relevance to AD. Neurotoxicology 44:114–120. https://doi.org/10.1016/j.neuro.2014.06.008

Błaszczyk JW (2016) Parkinson's disease and neurodegeneration: GABA-collapse hypothesis. Front Neurosci. https://doi.org/10.3389/fnins.2016.00269

Bihaqi SW, Bahmani A, Subaiea GM, Zawia NH (2014b) Infantile exposure to lead and late-age cognitive decline: relevance to AD. Alzheimer's Dement 10(2):187–195

Bootman MD (2012) Calcium signaling. Cold Spring Harb Perspect Biol 4(7):a011171

Bootman MD, Lipp P, Berridge MJ (2001) The organisation and functions of local Ca^{2+} signals. J Cell Sci 114(12):2213–2222

Bootman MD, Berridge MJ, Roderick HL (2002) Calcium signalling: more messengers, more channels, more complexity. Curr Biol 12(16):R563–R565

Bouton CM, Frelin LP, Forde CE, Godwin HA, Pevsner J (2001) Synaptotagmin I is a molecular target for lead. J Neurochem 76(6):1724–1735

Braga MF, Pereira EF, Albuquerque EX (1999) Nanomolar concentrations of lead inhibit glutamatergic and GABAergic transmission in hippocampal neurons. Brain Res 826(1):22–34

Bressler J, Kim K, Chakraborti T, Goldstein G (1999) Molecular mechanisms of lead neurotoxicity. Neurochem Res 24:595–600

Bridges CC, Zalups RK (2005) Molecular and ionic mimicry and the transport of toxic metals. Toxicol Appl Pharmacol 204(3):274–308

Brochin R, Leone S, Phillips D, Shepard N, Zisa D, Angerio A (2008) The cellular effect of lead poisoning and its clinical picture. GUJHS 5(2):1–8

CDC (Centers for disease control and prevention) (2018) The national institute for occupational safety and health (NIOSH). Adult blood lead epidemiology and surveillance (ABLES). https://www.cdc.gov/niosh/topics/ables/description

Chandramouli K, Steer CD, Ellis M, Emond AM (2009) Effects of early childhood lead exposure on academic performance and behaviour of school age children. Arch Dis Child 94(11):844–848. https://doi.org/10.1136/adc.2008.149955

Cheng G, Rong X-W, Feng T-P (1994) Block of induction and maintenance of calcium-induced LTP by inhibition of protein kinase C in postsynaptic neuron in hippocampal CA1 region. Brain Res 646(2):230–234

Cory-Slechta DA, McCoy L, Richfield EK (1997) Time course and regional basis of Pb-induced changes in MK-801 binding: reversal by chronic treatment with the dopamine agonist apomorphine but not the D1 agonist SKF-82958. J Neurochem 68(5):2012–2023

Costa LG (1998) Signal transduction in environmental neurotoxicity. Annu Rev Pharmacol Toxicol 38(1):21–43

Dosunmu R, Alashwal H, Zawia NH (2012) Genome-wide expression and methylation profiling in the aged rodent brain due to early-life Pb exposure and its relevance to aging. Mech Ageing Dev 133(6):435–443

Dou C, Zhang J (2011) Effects of lead on neurogenesis during zebrafish embryonic brain development. J Hazard Mater 194:277–282

Dribben WH, Creeley CE, Farber N (2011) Low-level lead exposure triggers neuronal apoptosis in the developing mouse brain. Neurotoxicol Teratol 33(4):473–480

Encinas JM, Vaahtokari A, Enikolopov G (2006) Fluoxetine targets early progenitor cells in the adult brain. Proc Natl Acad Sci 103(21):8233–8238

Farooqui Z, Bakulski KM, Power MC, Weisskopf MG, Sparrow D, Spiro A, Vokonas PS, Nie LH, Hu H, Park SK (2017) Associations of cumulative Pb exposure and longitudinal changes in Mini-Mental Status Exam scores, global cognition and domains of cognition: the VA Normative Aging Study. Environ Res 152:102–108. https://doi.org/10.1016/j.envres.2016.10.007

Feng X, Chen A, Zhang Y, Wang J, Shao L, Wei L (2015) Central nervous system toxicity of metallic nanoparticles. Int J Nanomed 10:4321

Flora SJS, Mittal M, Mehta A (2008) Heavy metal induced oxidative stress & its possible reversal by chelation therapy. Indian J Med Res 128(4):501–523

Fullmer CS, Edelstein S, Wasserman RH (1985) Lead-binding properties of intestinal calcium-binding proteins. J Biol Chem 260(11):6816–6819

Gerberding JL, Falk H (2005) Agency for Toxic Substances and Disease Registry justification of appropriation estimates for Appropriations Committees fiscal year 2006

Gilbert ME, Lasley SM (2007) Developmental lead (Pb) exposure reduces the ability of the NMDA antagonist MK-801 to suppress long-term potentiation (LTP) in the rat dentate gyrus, in vivo. Neurotoxicol Teratol 29(3):385–393

Goldstein GW (1993) Evidence that lead acts as a calcium substitute in second messenger metabolism. Neurotoxicology 14(2–3):97–101

Habermann E, Crowell K, Janicki P (1983) Lead and other metals can substitute for Ca^{2+} in calmodulin. Arch Toxicol 54:61–70

Hansen HH, Briem T, Dzietko M, Sifringer M, Voss A, Rzeski W, Zdzisinska B, Thor F, Heumann R, Stepulak A (2004) Mechanisms leading to disseminated apoptosis following NMDA receptor blockade in the developing rat brain. Neurobiol Dis 16(2):440–453

Haraguchi T, Ishizu H, Kawai K, Tanabe Y, Uehira K, Takehisa Y, Terada S, Tsuchiya K, Ikeda K, Kuroda S (2001) Diffuse neurofibrillary tangles with calcification (a form of dementia): X-ray spectrometric evidence of lead accumulation in calcified regions. Neuroreport 12(6):1257–1260. https://doi.org/10.1097/00001756-200105080-00040

Heizmann CW, Hunzlker W (1991) Intracellular calcium-binding proteins: more sites than insights. Trends Biochem Sci 16:98–103

Hou S, Yuan L, Jin P, Ding B, Qin N, Li L, Liu X, Wu Z, Zhao G, Deng Y (2013) A clinical study of the effects of lead poisoning on the intelligence and neurobehavioral abilities of children. Theor Biol Med Model 10(1):1–9

Hubbs-Tait L, Nation JR, Krebs NF, Bellinger DC (2005) Neurotoxicants, micronutrients, and social environments: individual and combined effects on children's development. Psychol Sci Public Interest 6(3):57–121

Hultberg B, Andersson A, Isaksson A (2001) Interaction of metals and thiols in cell damage and glutathione distribution: potentiation of mercury toxicity by dithiothreitol. Toxicology 156(2–3):93–100

Hwang L (2007) Environmental stressors and violence: lead and polychlorinated biphenyls. Rev Environ Health 22(4):313–328

Jalali SM, Najafzadeh H, Mousavi SM (2017) Comparative effect of silymarin and D-penicillamine on lead induced hemotoxicity and oxidative stress in rat. Iran J Toxicol 11(3):11–18

Joint FAO/WHO Expert Committee on Food Additives (2002) Safety evaluation of certain food additives and contaminants. World Health Organization

Karri V, Schuhmacher M, Kumar V (2016) Heavy metals (Pb, Cd, As and MeHg) as risk factors for cognitive dysfunction: a general review of metal mixture mechanism in brain. Environ Toxicol Pharmacol 48:203–213

Kern M, Wisniewski M, Cabell L, Audesirk G (2000) Inorganic lead and calcium interact positively in activation of calmodulin. Neurotoxicology 21(3):353–363

Kerper LE, Hinkle PM (1997) Cellular uptake of lead is activated by depletion of intracellular calcium stores. J Biol Chem 272(13):8346–8352

Khalil N, Morrow LA, Needleman H, Talbott EO, Wilson JW, Cauley JA (2009) Association of cumulative lead and neurocognitive function in an occupational cohort. Neuropsychology 23(1):10

Kirberger M, Wong HC, Jiang J, Yang JJ (2013) Metal toxicity and opportunistic binding of Pb^{2+} in proteins. J Inorg Biochem 125:40–49

Koller K, Brown T, Spurgeon A, Levy L (2004) Recent developments in low-level lead exposure and intellectual impairment in children. Environ Health Perspect 112(9):987–994

Kosnett MJ, Wedeen RP, Rothenberg SJ, Hipkins KL, Materna BL, Schwartz BS, Hu H, Woolf A (2007) Recommendations for medical management of adult lead exposure. Environ Health Perspect 115(3):463–471. https://doi.org/10.1289/ehp.9784

Lasley SM, Gilbert ME (2000) Glutamatergic components underlying lead-induced impairments in hippocampal synaptic plasticity. Neurotoxicology 21(6):1057–1068

Lasley SM, Gilbert ME (2002) Rat hippocampal glutamate and GABA release exhibit biphasic effects as a function of chronic lead exposure level. Toxicol Sci 66(1):139–147

Lasley SM, Green MC, Gilbert ME (2001) Rat hippocampal NMDA receptor binding as a function of chronic lead exposure level. Neurotoxicol Teratol 23(2):185–189

Léveillé F, Papadia S, Fricker M, Bell KF, Soriano FX, Martel M-A, Puddifoot C, Habel M, Wyllie DJ, Ikonomidou C (2010) Suppression of the intrinsic apoptosis pathway by synaptic activity. J Neurosci 30(7):2623–2635

Levin SM, Goldberg M (2000) Clinical evaluation and management of lead-exposed construction workers. Am J Ind Med 37(1):23–43

Li H, Li H, Li Y, Liu Y, Zhao Z (2018) Blood mercury, arsenic, cadmium, and lead in children with autism spectrum disorder. Biol Trace Elem Res 181(1):31–37. https://doi.org/10.1007/s12011-017-1002-6

Li M-M, Gao Z-Y, Dong C-Y, Wu M-Q, Yan J, Cao J, Ma W-J, Wang J, Gong Y-L, Xu J, Cai S-Z, Chen J-Y, Xu S-Q, Tong S, Tang D, Zhang J, Yan C-H (2020) Contemporary blood lead levels of children aged 0–84 months in China: a national cross-sectional study. Environ Int 134:105288. https://doi.org/10.1016/j.envint.2019.105288

Lidsky TI, Schneider JS (2003) Lead neurotoxicity in children: basic mechanisms and clinical correlates. Brain 126(1):5–19

Liu J, Han D, Li Y, Zheng L, Gu C, Piao Z, Au WW, Xu X, Huo X (2010) Lead affects apoptosis and related gene XIAP and Smac expression in the hippocampus of developing rats. Neurochem Res 35:473–479

Liu J, Chen L, Cui H, Zhang J, Zhang L, Su C-Y (2014) Applications of metal–organic frameworks in heterogeneous supramolecular catalysis. Chem Soc Rev 43(16):6011–6061. https://doi.org/10.1039/C4CS00094C

Loikkanen JJ, Naarala J, Savolainen KM (1998) Modification of glutamate-induced oxidative stress by lead: the role of extracellular calcium. Free Radical Biol Med 24(2):377–384. https://doi.org/10.1016/S0891-5849(97)00219-0

Lucchini RG, Zoni S, Guazzetti S, Bontempi E, Micheletti S, Broberg K, Parrinello G, Smith DR (2012) Inverse association of intellectual function with very low blood lead but not with manganese exposure in Italian adolescents. Environ Res 118:65–71. https://doi.org/10.1016/j.envres.2012.08.003

Luo Y, Zhu D-M, Ruan D-Y (2011) Galantamine rescues lead-impaired synaptic plasticity in rat dentate gyrus. Toxicology 289(1):45–51

Lyall A, Swanson J, Liu C, Blumenthal TD, Turner CP (2009) Neonatal exposure to MK801 promotes prepulse-induced delay in startle response time in adult rats. Exp Brain Res 197:215–222

Maiti S, Acharyya N, Ghosh TK, Ali SS, Manna E, Nazmeen A, Sinha NK (2017) Green tea (*Camellia sinensis*) protects against arsenic neurotoxicity via antioxidative mechanism and activation of superoxide dismutase activity. Cent Nerv Syst Agents Med Chem (Curr Med Chem Cent Nerv Syst Agents) 17(3):187–195

Mărginean CO, Meliţ LE, Moldovan H, Lupu VV, Mărginean MO (2016) Lead poisoning in a 16-year-old girl: a case report and a review of the literature (CARE compliant). Medicine 95(38)

Markovac J, Goldstein GW (1988) Picomolar concentrations of lead stimulate brain protein kinase C. Nature 334(6177):71–73

Mason LH, Harp JP, Han DY (2014) Pb neurotoxicity: neuropsychological effects of lead toxicity. BioMed Res Int 2014

Masoud AM, Bihaqi SW, Machan JT, Zawia NH, Renehan WE (2016) Early-life exposure to lead (Pb) alters the expression of microRNA that target proteins associated with Alzheimer's disease. J Alzheimer's Dis 51(4):1257–1264

McCue HV, Haynes LP, Burgoyne RD (2010) The diversity of calcium sensor proteins in the regulation of neuronal function. Cold Spring Harb Perspect Biol 2(8):a004085

McMichael AJ, Baghurst PA, Wigg NR, Vimpani GV, Robertson EF, Roberts RJ (1988) Port Pirie Cohort Study: environmental exposure to lead and children's abilities at the age of four years. N Engl J Med 319(8):468–475

Mitra P, Sharma S, Purohit P, Sharma P (2017) Clinical and molecular aspects of lead toxicity: an update. Crit Rev Clin Lab Sci 54(7–8):506–528

Moos T, Skjoerringe T, Gosk S, Morgan EH (2006) Brain capillary endothelial cells mediate iron transport into the brain by segregating iron from transferrin without the involvement of divalent metal transporter 1. J Neurochem 98(6):1946–1958

Murata K, Iwata T, Dakeishi M, Karita K (2009) Lead toxicity: does the critical level of lead resulting in adverse effects differ between adults and children? J Occup Health 51(1):1–12. https://doi.org/10.1539/joh.K8003

Needleman H (2004) Lead poisoning. Annu Rev Med 55:209–222

Nihei MK, Guilarte TR (2001) Molecular changes in glutamatergic synapses induced by Pb^{2+}: association with deficits of LTP and spatial learning. Neurotoxicology 22(5):635–643

Patra RC, Rautray AK, Swarup D (2011) Oxidative stress in lead and cadmium toxicity and its amelioration. Vet Med Int 2011:1–9. https://doi.org/10.4061/2011/457327

Pears C (1995) Structure and function of the protein kinase C gene family. J Biosci 20:311–332

Pocock SJ, Smith M, Baghurst P (1994) Environmental lead and children's intelligence: a systematic review of the epidemiological evidence. BMJ 309(6963):1189–1197

Rocha A, Trujillo KA (2019) Neurotoxicity of low-level lead exposure: history, mechanisms of action, and behavioral effects in humans and preclinical models. Neurotoxicology 73:58–80. https://doi.org/10.1016/j.neuro.2019.02.021

Rogers JT, Venkataramani V, Washburn C, Liu Y, Tummala V, Jiang H, Smith A, Cahill CM (2016) A role for amyloid precursor protein translation to restore iron homeostasis and ameliorate lead (Pb) neurotoxicity. J Neurochem 138(3):479–494. https://doi.org/10.1111/jnc.13671

Sanders T, Liu Y, Buchner V, Tchounwou PB (2009) Neurotoxic effects and biomarkers of lead exposure: a review. Rev Environ Health 24(1):15–46

Sanders AP, Claus Henn B, Wright RO (2015) Perinatal and childhood exposure to cadmium, manganese, and metal mixtures and effects on cognition and behavior: a review of recent literature. Curr Environ Health Rep 2(3):284–294. https://doi.org/10.1007/s40572-015-0058-8

Sandhir R, Gill KD (1994) Calmodulin and cAMP dependent synaptic vesicle protein phosphorylation in rat cortex following lead exposure. Int J Biochem 26(12):1383–1389. https://doi.org/10.1016/0020-711X(94)90181-3

Schneider JS, Kidd SK, Anderson DW (2013) Influence of developmental lead exposure on expression of DNA methyltransferases and methyl cytosine-binding proteins in hippocampus. Toxicol Lett 217(1):75–81

Schwartz BS, Stewart WF, Bolla KI, Simon D, Bandeen-Roche K, Gordon B, Links JM, Todd AC (2000) Past adult lead exposure is associated with longitudinal decline in cognitive function. Neurology 55(8):1144–1150

Schwartz BS, Lee B-K, Bandeen-Roche K, Stewart W, Bolla K, Links J, Weaver V, Todd A (2005) Occupational lead exposure and longitudinal decline in neurobehavioral test scores. Epidemiology 106–113

Shafiq-ur-Rehman (2003) Lead-exposed increase in movement behavior and brain lipid peroxidation in fish. J Environ Sci Health Part A 38(4):631–643

Shirran SL, Barran PE (2009) The use of ESI-MS to probe the binding of divalent cations to calmodulin. J Am Soc Mass Spectrom 20(6):1159–1171. https://doi.org/10.1016/j.jasms.2009.02.008

Singh N, Gupta VK, Kumar A, Sharma B (2017) Synergistic effects of heavy metals and pesticides in living systems. Front Chem 5:70

Stewart WF, Schwartz BS (2007) Effects of lead on the adult brain: a 15-year exploration. Am J Ind Med 50(10):729–739

Stollery BT, Broadbent DE, Banks HA, Lee WR (1991) Short term prospective study of cognitive functioning in lead workers. Occup Environ Med 48(11):739–749

Valko M, Morris H, Cronin MTD (2005) Metals, toxicity and oxidative stress. Curr Med Chem 12(10):1161–1208

Van der Zee EA, Compaan JC, De Boer M, Luiten PG (1992) Changes in PKC gamma immunoreactivity in mouse hippocampus induced by spatial discrimination learning. J Neurosci 12(12):4808–4815

Wang L, Xu T, Lei W, Liu D, Li Y, Xuan R, Ma J (2011) Cadmium-induced oxidative stress and apoptotic changes in the testis of freshwater crab, *Sinopotamon henanense*. PLoS ONE 6(11):e27853

Wei Z, Wei M, Yang X, Xu Y, Gao S, Ren K (2022) Synaptic secretion and beyond: targeting synapse and neurotransmitters to treat neurodegenerative diseases. Oxid Med Cell Longev 2022

Westerink RH, Klompmakers AA, Westenberg HG, Vijverberg HP (2002) Signaling pathways involved in Ca^{2+}- and Pb^{2+}-induced vesicular catecholamine release from rat PC12 cells. Brain Res 957(1):25–36

Winneke G, Lilienthal H, Krämer U (1996) The neurobehavioural toxicology and teratology of lead. In: Toxicology—from cells to man: proceedings of the 1995 EUROTOX congress meeting held in Prague, Czech Republic, 27–30 Aug 1995, pp 57–70

Wu J, Basha MR, Brock B, Cox DP, Cardozo-Pelaez F, McPherson CA, Harry J, Rice DC, Maloney B, Chen D (2008) Alzheimer's disease (AD)-like pathology in aged monkeys after infantile exposure to environmental metal lead (Pb): evidence for a developmental origin and environmental link for AD. J Neurosci 28(1):3–9

Xu C, Liu Q-Y, Alkon DL (2014) PKC activators enhance GABAergic neurotransmission and paired-pulse facilitation in hippocampal CA1 pyramidal neurons. Neuroscience 268:75–86

Yuede CM, Wozniak DF, Creeley CE, Taylor GT, Olney JW, Farber NB (2010) Behavioral consequences of NMDA antagonist-induced neuroapoptosis in the infant mouse brain. PLoS ONE 5(6):e11374. https://doi.org/10.1371/journal.pone.0011374

Zeng X, Xu X, Boezen HM, Huo X (2016) Children with health impairments by heavy metals in an e-waste recycling area. Chemosphere 148:408–415. https://doi.org/10.1016/j.chemosphere.2015.10.078

Zhang J, Cai T, Zhao F, Yao T, Chen Y, Liu X, Luo W, Chen J (2012) The role of α-synuclein and tau hyperphosphorylation-mediated autophagy and apoptosis in lead-induced learning and memory injury. Int J Biol Sci 8(7):935–944. https://doi.org/10.7150/ijbs.4499

Part III
Sustainable Mitigation Strategies of Lead Toxicity

Chapter 7
Phytoremediation of Lead Present in Environment: A Review

Gisela Adelina Rolón-Cárdenas and Alejandro Hernández-Morales

Abstract Lead (Pb) is a heavy metal used in various industrial processes, so its levels are considerably increased in the soil, sediments, surface water, and groundwater. Pb is a non-biodegradable and persistent environmental pollutant that causes toxicity in humans, plants, animals, and microorganisms. Phytoremediation is a clean, eco-friendly, and cost-effective technology to remove Pb from aquatic and terrestrial environments. This technology uses plants to remove, immobilize, and contain Pb through phytoextraction, phytostabilization, and rhizofiltration. This chapter describes the characteristics of plants used in phytoremediation, focusing on the mechanisms employed by the plants to assist in the removal or immobilization of Pb. Moreover, it shows the state of the art on phytoremediation assisted by microorganisms for enhancing phytoremediation of Pb-polluted soils.

Keywords Phytoextraction · Phytostabilization · Rhizofiltration · Microbial-assisted phytoremediation

7.1 Introduction

Lead (Pb) is a soft, malleable, bluish-gray metal located in group IV of the periodic table of elements (Al-Fartusie and Mohssan 2017). Pb occurs naturally in the soil at a concentration from 0.002 to 0.2 mg/kg, while in fresh waters, from 0.001 to 0.010 mg/L, worldwide (Carrillo-Chávez et al. 2006). The presence of Pb in the environment can be from natural or anthropogenic sources (Li et al. 2012). The primary natural sources of Pb are weathering and erosion of lead-rich rocks, forest fires, and

G. A. Rolón-Cárdenas · A. Hernández-Morales (✉)
Facultad de Estudios Profesionales Zona Huasteca, Universidad Autónoma de San Luis Potosí, Ciudad Valles, San Luis Potosí, Mexico
e-mail: alejandro.hernandez@uaslp.mx

A. Hernández-Morales
Facultad de Ciencias Químicas, Centro de Investigación y Estudios de Posgrados, Universidad Autónoma de San Luis Potosí, San Luis Potosí, Mexico

N. Kumar and A. K. Jha (eds.), *Lead Toxicity Mitigation: Sustainable Nexus Approaches*, Environmental Contamination Remediation and Management,
https://doi.org/10.1007/978-3-031-46146-0_7

volcanic eruptions (Zhang et al. 2019). While anthropogenic Pb results from mining, smelting, leaded gasoline combustion, coal burning, and industrial production of pigments, lead-acid batteries, cable sheathing, ammunition, alloys, solder, and pipes (Ballantyne et al. 2018; Eichler et al. 2015). Both natural and anthropogenic Pb sources have caused a significant increase in their levels in the environment (Zhang et al. 2019). For example, high Pb concentration has been detected in arable lands of Ireland (Nag and Cummins 2022), urban agricultural soils of Cameroon (Aboubakar et al. 2021), and in the arable soil in Southwest China (Wu et al. 2018).

Pb is a non-biodegradable and persistent pollutant that exerts toxic effects on plants, animals, and microorganisms at low concentrations (Wong and Li 2004; Abdelkrim et al. 2020; Rahman and Singh 2019). Pb ranks second on the list of hazardous substances according to the U.S. Agency for Toxic Substances and Disease Registry (ATSDR 2022). In humans, Pb causes haematological and cardiovascular disorders, kidney dysfunction, gastrointestinal disease, and central nervous system damage (Rahimpoor et al. 2020; Khanam et al. 2020; Rahman and Singh 2019). Lead also affects brain development in children, causing behavioural changes and lowering IQ score (Heidari et al. 2022).

Different physical and chemical methods have been developed to remove Pb from contaminated sites. These methods include soil washing, landfilling, vitrification, electrokinetic treatments, surface capping, encapsulation, and soil flushing (Liu et al. 2018; Song et al. 2017). Besides, bioremediation has been proposed as an eco-sustainable alternative for removing natural and anthropogenic Pb from soils, sediments, surface water, and groundwater (Bala et al. 2022). In this technology, living organisms such as bacteria, fungi, microalgae, or plants are used to degrade or transform environmental contaminants into non-toxic forms (Vidali 2001). According to the application site, bioremediation techniques have been classified as in situ or ex situ (He et al. 2021). In situ techniques are carried out directly at the contaminated site, while ex situ techniques are applied outside the contaminated site (Boopathy 2000).

7.2 Phytoremediation: Definition and Strategies

Phytoremediation is an in situ bioremediation technology that uses plants to degrade, immobilize, neutralize, and contain environmental contaminants (Wang et al. 2017). Phytoremediation has been successfully applied to clean up heavy metals, radionuclides, petroleum hydrocarbons, explosives, pesticides, pharmaceutical and personal care products (PPCPs) from aquatic and terrestrial environments (Jee 2016; Kurade et al. 2021). Plants perform phytoextraction, phytovolatilization, phytostabilization, rhizofiltration, phytodegradation, and phytostimulation to remove xenobiotics (Alsafran et al. 2022).

Phytoextraction or phytoaccumulation is used to remove heavy metals and other inorganic compounds from soil and sediment to the aerial parts of plants. In this process occurs the absorption of contaminants by the plant roots, translocation through stems, and their accumulation in shoots and leaves (Mahar et al. 2016; Etim 2012).

Phytostabilization or phytoimmobilization is a strategy in which a plant species reduces the mobility of pollutants and decreases its bioavailability for other plants or microorganisms (Alsafran et al. 2022). In this strategy, heavy metals and other inorganic compounds are adsorbed on root cell walls, absorbed within root tissues, or immobilized as non-toxic forms in the rhizosphere through mechanisms including sorption, precipitation, complexation, or metal valence reduction (Rai et al. 2021; Etim 2012; Yan et al. 2020).

Rhizofiltration is a hydroponic-based phytoremediation technique that uses plant roots to eliminate pollutants from the impacted aquatic environments (Srivastava et al. 2021). In this strategy, heavy metals and other inorganic compounds are adsorbed or precipitated on the root surface or absorbed in the roots (Kristanti et al. 2021).

Phytovolatilization is a strategy in which plants uptake environmental contaminants, transform them into volatile forms, and release them into the atmosphere through transpiration (Etim 2012). This process is applied for the treatment of some metals and metalloids such as arsenic (As), selenium (Se), and mercury (Hg) (Rai et al. 2021). This strategy is controversial because the pollutants are not destroyed but transferred to the atmosphere, where they can be redeposited (Mahar et al. 2016).

Phytodegradation is when plants uptake, store, metabolize or mineralize organic contaminants in their tissues (Rai et al. 2021). The phytodegradation process requires degrading enzymes involved in various metabolic processes and enzymes such as nitrilase, nitroreductase, peroxidase, dehalogenase, oxygenase, and laccases (Chatterjee et al. 2013).

Phytostimulation is a strategy in which plants and microorganisms localized in the rhizosphere degrade organic contaminants (Rai et al. 2021). In this strategy, plants secrete root exudates and metabolites that stimulate the growth of degrading microorganisms (Favas et al. 2014).

7.3 Phytoremediation of Lead

Rhizofiltration, phytostabilization, and phytoextraction are the main strategies for Pb removal from polluted environments. Phytostabilization and phytoextraction are applicable for the remediation of Pb in soils and sediments, while rhizofiltration is used for the remediation of surface water, groundwater, and wastewater (Yan et al. 2020; Otte and Jacob 2006).

7.3.1 Phytostabilization of Lead

Phytostabilization reduces the Pb migration from contaminated to non-contaminated soils (Alsafran et al. 2022; Bolan et al. 2011). In this process, the immobilization of Pb can occur by either adsorption on root cell walls or Pb precipitation in the rhizosphere (Ashraf et al. 2015; Arshad et al. 2016).

The bioavailability of Pb in the soil depends on its speciation, which is influenced by various factors such as pH, redox potential, organic matter, sulphur, and carbonate contents (Olaniran et al. 2013; John and Leventhal 1995). Besides, plant roots play an essential role in the metal and nutrient solubility in the soil (Wenzel et al. 1999). The metabolic activity of the plant roots can change the pH, the redox conditions, concentrations of Dissolved Organic Matter (DOM), and microbial activity in the rhizosphere, which enhance the uptake of nutrients such as iron (Fe), phosphorus (P), and zinc (Zn) or immobilize non-essential metals (Li et al. 2021; Seshadri et al. 2015; Rai et al. 2021). For example, *Pelargonium* × *hortorum* L.H. Bailey increased the DOM content and acidified the rhizosphere soil in response to Pb (Manzoor et al. 2020). Root exudation is one of the most critical factors affecting the physicochemical characteristics of the soil (Li et al. 2011b). The root exudates are a mixture of metabolites, including sugars, amino acids, and organic acids, produced by plants and secreted to the soil (Vives-Peris et al. 2020). These compounds can affect the bioavailability of Pb in the rhizosphere (Li et al. 2021; Seshadri et al. 2015). For example, the waterlogged (*Oenanthe javanica* DC.) and yellow melon (*Cucumis melo* L.) roots release citric acid and others organic acids to the rhizosphere which form soluble Pb–organic complexes (Liu and Luo 2019; Irias Zelaya et al. 2020). Similarly, organic acids secreted by pea plants (*Pisum sativumpea* L.) favour the formation of stable metal complexes in the root region (Austruy et al. 2014).

Pb precipitates have also been observed in the root of *Pelargonium* cultivars, Indian mustard (*Brassica juncea*), and two poplar species (Arshad et al. 2016; Yang et al. 2021; Shi et al. 2021). In *Pelargonium* cultivars, Pb precipitates on the root surface in the form of αPbO, $PbOH^+$, carbonates, and ferrite derivatives (Arshad et al. 2016). On the other hand, in the Indian mustard root cells, Pb precipitates as lead phosphate $Pb_3(PO_4)_2$, pyromorphite $Pb_5(PO_4)_3(OH, Cl)$, and other Pb phosphates (Yang et al. 2021). Similarly, *Populus* × *canescens* and *P. nigra* precipitate the Pb as phosphates and oxalates in their roots (Shi et al. 2021).

Pb immobilization in the root can be due to the interaction between the heavy metal ions and the components of the root cell wall (Krzesłowska 2011). The cell wall is the first structure of root cells to come in contact with heavy metals and is involved in ions metal binding (Parrotta et al. 2015). The cell wall's main components are polysaccharide such as cellulose, hemicellulose (HC), and pectin which play an important role in Pb binding in cell walls (Zhang et al. 2021a; Sumranwanich et al. 2018). Pb adsorption by different cell wall components has been reported previously in tea (*Camellia sinensis* L.) and *Athyrium wardii* (Hook.) roots. In the cell wall of tea plant roots, the most significant amount of Pb is adsorbed mainly by cellulose and lignin (68.42%), followed by pectin (20%) and HC2 (5.26%) (Wang et al. 2015). On the other hand, pectin and HC are the primary binding sites for Pb in root cell walls of *A. wardii* (Zhan et al. 2020). The Pb-binding capacity of cell wall polysaccharides is attributed mainly to the presence of carboxyl (–COOH) and hydroxyls (–OH) groups (Sumranwanich et al. 2018; Wang et al. 2015; Zhan et al. 2020).

Previous studies have reported that Pb increases the biosynthesis of polysaccharides in some plant species' root cell walls. For instance, Pb induced pectin and hemicellulose production in root cell walls of tall fescue (*Festuca arundinacea* Schreb) (Zhang et al. 2020), *A. wardii* (Zhan et al. 2020), *Populus* × *canescens*, and *P. nigra* (Shi et al. 2021), which may be a mechanism of tolerance to Pb stress of these plants.

In the phytostabilization process, the plant roots take up Pb ions or Pb-soluble complexes from the rhizosphere and accumulate them internally (Yan et al. 2020). The Pb within root tissue can be associated with the cell wall in apoplastic space or immobilized intracellularly in vacuoles, limiting Pb translocation from roots to shoots (Yan et al. 2020; Rahman et al. 2022; Wierzbicka 1998).

In signal grass (*Brachiaria decumbens*), Indian mustard (*B. juncea*), and *Neyraudia reynaudiana* it have been observed that most Pb precipitates in the cell wall as insoluble deposits inside the roots (Kopittke et al. 2008; Zhou et al. 2016; Yang et al. 2021). Pb mainly exists as lead phosphate precipitates [$Pb_5(PO_4)_3(OH, Cl)$, and $Pb_3(PO_4)_2$] in the Indian mustard roots cells (Yang et al. 2021). Insoluble deposits of chloropyromorphite [$Pb_5(PO_4)_3Cl$] in root cells have also been observed in signal grass (*B. decumbens*) roots (Kopittke et al. 2008).

Vacuolar sequestration of Pb in radicular cells limits its translocation within plants (Sharma and Dubey 2005). Vacuoles are the largest organelle of plant cells and play an essential function in the heavy metal detoxification (Sharma et al. 2016). In this organelle, intracellular Pb is stored by complexation with organic acids and sulfur-rich peptides known as phytochelatins (Zhang et al. 2018; Singh et al. 2017; Zhao et al. 2015). In addition to cell walls, the vacuoles are one of the main storage sites of Pb in *Allium sativum* and *N. reynaudiana* roots (Jiang and Liu 2010; Zhou et al. 2016). Approximately, 31.2–41.3% of total Pb is stored in the vacuoles of roots *A. wardii* (Hook.) (Zhao et al. 2015).

7.3.2 Phytoextraction of Lead

Phytoextraction is a method used to reduce Pb levels in soil and sediments. This method requires Pb uptake by plant roots, root-to-shoot translocation, and intracellular compartmentalization of Pb in aerial tissues (Yan et al. 2020). These processes are dependent on plant species, variety, genotype, environmental conditions, and Pb bioavailability in the soil (Asare et al. 2023).

The first step in Pb accumulation is the Pb uptake by the root (Gong et al. 2022). The Pb ions from the soil are absorbed by the root epidermal cells and can be transported inside the root by apoplastic or symplastic pathways (He et al. 2023; Zhou et al. 2018). In the apoplastic pathway, Pb in the extracellular fluid is transferred from one cell wall to another, whereas in the symplastic pathway, the Pb ions cross the plasma membrane and transfer cell to cell through channels called plasmodesmata (Pasricha et al. 2021). As Pb is not an essential element, plants do not have a specific channel for Pb uptake, so it has been suggested that Pb enter the plant cells via channels or transporters for other essential cations (Peralta-Videa et al. 2009; Gong et al. 2022). Different proteins have been associated with Pb transport across the membrane, such as AtCNGC1 (cyclic nucleotide-gated channel 1), NtCBP4 (Plasma membrane Calmodulin-Binding Protein 4), and OsNRAMP5 (Natural Resistance-Associated Macrophage Proteins 5) (Arazi et al. 1999; Sunkar et al. 2000; Chang et al. 2022). In tobacco (*Nicotiana tabacum*) and *Arabidopsis thaliana*, two cation channels for K^+ and Ca^{2+} called NtCBP4 and AtCNGC1, respectively, have been associated with Pb uptake across the plant plasma membrane and Pb accumulation (Arazi et al. 1999; Sunkar et al. 2000). On the other hand, the OsNRAMP5, a divalent metal transporter, is associated with transporting intracellular Pb in rice (*Oryza sativa*) roots (Chang et al. 2022). Once inside the root cells, Pb is associated with amino acids and organic acids and can be translocated to shoots and leaves by the xylem (Gall and Rajakaruna 2013; Pourrut et al. 2013).

At the shoot level, intracellular Pb is detoxified by metal-binding ligands such as phytochelatins and metallothioneins (Mitra et al. 2014; Pourrut et al. 2011; Eapen and D'Souza 2005). Phytochelatins (PC) are oligopeptides that contain glutamic acid (Glu), cysteine (Cys), and glycine (Gly) $[(\gamma\text{–Glu–Cys})_n\text{–Gly } (n = 2 - 11)]$, whose synthesis is catalyzed by phytochelatin synthase (PCS) from glutathione (Scarano and Morelli 2002; Gupta et al. 2013b). While metallothionein (MT) are low-molecular-weight proteins (7–10 kDa) with 9–16 cysteine residues that are encoded by a family of MT genes (Eapen and D'Souza 2005; Cobbett and Goldsbrough 2002). Previous studies have reported that the plants synthesize PC and MT in response to Pb stress. For example, it has been observed that Pb exposure induces the synthesis of PC in *Salvinia minima* Baker, Dwarf bamboo (*Sasa argenteostriata*), and coontail (*Ceratophyllum demersum* L.) (Jiang et al. 2020; Estrella-Gómez et al. 2009; Mishra et al. 2006), and increases the expression of MT genes in tomato (*Lycopersicon esculentum*), *Bruguiera gymnorrhiza*, and rice (*O. sativa*) plants (Kim and Kang 2018; Kisa et al. 2017; Huang and Wang 2009).

In the cytoplasm of shoot cells, PC and MT binding to intracellular Pb and form stable complexes. The PC–Pb complex is finally transported into the vacuole, where it is stored (Andra et al. 2009; Inouhe et al. 2012).

7.3.3 *Rhizofiltration of Lead*

Rhizofiltration is a technique used to remove Pb from surface water, groundwater, and effluents with low levels of contaminants (Ekta and Modi 2018; Jadia and Fulekar 2009). Similarly, to the phytostabilization process, in the rhizofiltration, the Pb ions can be absorbed within root tissues (Kristanti et al. 2021; Rawat et al. 2012), adsorbed by root cell walls (Ho et al. 2021), or immobilized in the root surface (Delgado-González et al. 2021; Dushenkov et al. 1995).

In aquatic and wetland plants, iron plaque plays an essential role in the sequestration of heavy metals in the roots (Tripathi et al. 2014). The Fe plaques are deposits of different iron oxides and hydroxides on the root surface (Tripathi et al. 2014; Khan et al. 2016). The presence of ferrihydrite $[Fe_{4-5}(OH,O)_{12}]$, lepidocrocite $[\gamma\text{–FeOOH}]$, siderite [$FeCO_3$], and goethite [FeO(OH)] has been observed in the Fe plaques of *Oenanthe javanica*, *Phalaris arundinacea*, and *Vallisneria americana* (Liu and Luo 2019; Hansel et al. 2001; St-Cyr et al. 1993). These Fe (hydr)oxides are result of oxidation of ferrous iron (Fe^{2+}) in the rhizosphere by the oxygen loss from the roots, and the biological activity of microorganisms (Tripathi et al. 2014; Khan et al. 2016). Previous studies have reported that iron plaque can sequester Pb on the root surface (Zandi et al. 2022). In rice (*O. sativa*), the most significant amount of plant taken Pb (> 60%) is stored in the iron plaque of the root (Cheng et al. 2014; Ma et al. 2013). Similarly, most of the total Pb uptake (50–60%) performed by *Phalaris arundinacea* L. and *Carex cinerascens* Kukenth. plants was found in the iron plaque, and only small amounts was found in roots and shoots (Liu et al. 2015, 2016). The Pb–binding capacity of Fe plaques is attributed mainly to the Pb's specific and high affinity for iron (hydr)oxides (Hansel et al. 2001).

7.3.4 Potential Plants for Phytoremediation of Pb

The concentration of heavy metals in plants determines the success of phytoremediation; therefore, selecting suitable plant species is crucial in phytoextraction, phytostabilization, and rhizofiltration efficiency (Yan et al. 2020; Gupta et al. 2013a).

Hyperaccumulator plants can potentially remove Pb from the soil through phytoextraction (Lone et al. 2008). Pb hyperaccumulators are plants able to grow in contaminated soils with heavy metals and accumulate more than 1000 mg/kg of Pb in aerial organs without show phytotoxicity signs (Sytar et al. 2021; Manara et al. 2020). However, Pb hyperaccumulation in plants is uncommon because Pb ions are easily precipitated in the rhizosphere, limiting their uptake by roots and the translocation to shoots (Baker and Brooks 1989). In the Global Hyperaccumulator Database (http://hyperaccumulators.smi.uq.edu.au/collection/), *Alyssum wulfenianum*, *Noccaeae rotundifolium* subsp. *cepaeifolium*, *Polycarpaea synandra*, *Sesbania drummondii*, *Armeria maritima* var. *Halleri*, *Dactyloctenium aegyptium*, *Microstegium ciliatum*, *Polygala umbonata*, and *Spermacoce mauritiana* plants, belonging to seven families, have been identified as Pb hyperaccumulators (Reeves et al. 2018). Although these plants accumulate high concentrations of Pb, hyperaccumulator plants are small and present slow growth, which limits their use in the phytoremediation process (Saifullah et al. 2009; Yan et al. 2020).

Different fast-growing crops with high biomass production, like sorghum (*Sorghum bicolor* L.), sunflower (*Helianthus annuus* L.), and corn (*Zea mays*) have been studied to remove Pb from lead contaminated soil under field conditions (Cheng et al. 2015; Zehra et al. 2020; Yuan et al. 2019). Despite the lower concentrations of Pb in their tissues, the total metal remotion exerted by these plants can be like levels reached by hyperaccumulator plants (Van Slycken et al. 2008).

Native plants grown on heavy metal contaminated sites are another option for Pb remediation. These plant species can survive, grow, and reproduce under metal stress better than plants introduced from other environments (Midhat et al. 2019;

Yoon et al. 2006). Several studies have evaluated the phytoremediation potential of native plants growing in heavy metal-contaminated sites (Table 7.1). For example, Salazar and Pignata (2014) studied the vegetal community growing around a lead smelter plant in Argentina. On the other hand, Mahdavian et al. (2017) and Nouri et al. (2011) investigated plants colonizing a lead–zinc mining area in Iran. In Marocco, Midhat et al. (2019) and Hasnaoui et al. (2020) identified metal-tolerant native plant species from three abandoned mining sites and a contaminated site near a Pb/Zn mining area, respectively. In these sites, some plants belonging to the Asparagaceae, Asteraceae, Brassicaceae, Cucurbitaceae, Cyperaceae, Euphorbiaceae, Fabaceae, Gramineae, Lamiaceae, Liliaceae, Resedaceae, and Tamaricaceae families have been observed (Table 7.1).

The phytoremediation potential of plants can be estimated using Bioconcentration Factors (BCF) and Translocation Factors (TF) (Rolón-Cárdenas et al. 2022). The BCF is the ratio between the heavy metal content in the plant roots and the substrate (Zou et al. 2012; Lorestani et al. 2011). Various native plant species like *Artemisia sieberi* Besser, *Fortuynia bungei* Boiss., *Astragalus durandianus* Aitch. & Baker, *Mentha longifolia* L., and *Allium umbilicatum* Boiss. have showed BCF > 1 for Pb (Table 7.1), indicating the potential of these plants to be used in Pb phytostabilization (Lorestani et al. 2013).

On the other hand, TF is the ratio of heavy metal content in the shoots and the roots (Midhat et al. 2019; Lorestani et al. 2011). Values TF > 1 for Pb has been reported in different native plants such as *Lactuca viminea* (L.) J. Presl & C. Presl, *Scariola orientalis* (Boiss.) Sojak, *Scolymus hispanicus*, *Cyperus iria*, *Juncellus serotinus*, *Euphorbia macroclada* Boiss., *Echinophora platyloba* DC., *Paspalum paspaloides*, *Phragmites australis*, *Reseda alba*, and *Tamarix ramosissima* Ledeb. (Table 7.1). This TF value indicates high efficiency in Pb translocation from the roots to the shoots and, therefore, their potential to be used in phytoextraction (Midhat et al. 2019).

Rhizofiltration uses heavy metal tolerant plants with a fibrous root system and large surface areas to 'filter' Pb ions in solution (Chatterjee et al. 2013; Nedjimi 2021). Different aquatic plants have been studied to remove Pb from water through rhizofiltration (Kafle et al. 2022). For example, *Alternanthera sessilis*, *Enhydra fluctuans*, *Pistia stratiotes*, *Salvinia cucullata*, *Typha latifolia*, and *Vetiveria zizanioides* can remove between 84 and 99% of Pb from solution and accumulate into root and shoot (Das et al. 2021; Parven et al. 2022; Veselý et al. 2011; Alonso-Castro et al. 2009; Singh et al. 2015). Some terrestrial plants such as Indian mustard (*B. juncea*), *Cosmos sulphureus* Cav., sunflower (*H. annuus* L.), *Iris lactea* var. *chinensis*, and *Talinum paniculatum* are also suitable for rhizofiltration due to they remove high amount of Pb from the hydroponic medium and accumulate it in roots and shoots (Liu et al. 2000; Aftab et al. 2021; Seth et al. 2011; Han et al. 2008; dos Reis et al. 2022).

The plants used for rhizofiltration are first cultivated in hydroponic conditions to favour the root system development, later transferred to the contaminated water source, and finally harvested when the root of plants are saturated with contaminants (Mansoor et al. 2022; Yan et al. 2020).

Table 7.1 Native plants growing on heavy metal contaminated sites

Family	Plant	Pb concentration (mg/kg)			Factors		References
		Soil	Root	Shoot	TF	BCF	
Asparagaceae	*Asparagus horridus* L.	10,813.1	45.2	15.2	0.3	0.004	Midhat et al. (2019)
Asteraceae	*Artemisia sieberi* Besser	127.4	345.2	106.2	0.3	2.71	Mahdavian et al. (2017)
Asteraceae	*Bidens pilosa* L.	11,936.0	741.0	59.8	0.1	0.06	Salazar and Pignata (2014)
Asteraceae	*Lactuca viminea* (L.) J. Presl & C. Presl	1077.5	149.2	201.7	1.4	0.14	Midhat et al. (2019)
Asteraceae	*Scariola orientalis* (Boiss.) Sojak	1204.0	9017.0	9140.0	1.0	7.49	Nouri et al. (2011)
Asteraceae	*Scolymus hispanicus*	7792.9	798.7	972.7	1.2	0.10	Hasnaoui et al. (2020)
Asteraceae	*Tagetes minuta* L.	2645.0	30.7	20.0	0.7	0.01	Salazar and Pignata (2014)
Brassicaceae	*Fortuynia bungei* Boiss.	127.0	1720.1	73.0	0.0	13.54	Mahdavian et al. (2017)
Cucurbitaceae	*Citrullus colocynthis* (L.) Schrader	6156.7	94.4	38.4	0.4	0.02	Midhat et al. (2019)
Cyperaceae	*Cyperus iria*	40.8	35.5	54.0	1.5	0.87	Li et al. (2011a)
Cyperaceae	*Juncellus serotinus*	155.0	67.9	91.0	1.3	0.44	Li et al. (2011a)
Euphorbiaceae	*Euphorbia gedrosiaca*	564.9	332.1	52.1	0.2	0.59	Mahdavian et al. (2017)
Euphorbiaceae	*Euphorbia hirta*	283.0	30.6	13.9	0.5	0.11	Li et al. (2011a)
Euphorbiaceae	*Eupatorium inulifolium* Kunth	11,936.0	441.0	52.2	0.1	0.04	Salazar and Pignata (2014)
Euphorbiaceae	*Euphorbia macroclada* Boiss.	9451.0	3809.0	8095.0	2.1	0.40	Nouri et al. (2011)
Fabaceae	*Astragalus durandianus* Aitch. & Baker	241.0	1189.5	137.0	0.1	4.94	Mahdavian et al. (2017)

(continued)

Table 7.1 (continued)

Family	Plant	Pb concentration (mg/kg)			Factors		References
		Soil	Root	Shoot	TF	BCF	
Fabaceae	*Hedysarum spinosissimum*	6445.1	983.6	253.9	0.3	0.15	Hasnaoui et al. (2020)
Fabaceae	*Lotus corniculatus*	12,223.8	1493.0	832.4	0.6	0.12	Hasnaoui et al. (2020)
Gramineae	*Echinophora platyloba* DC.	10,426.0	1421.0	10,121.0	7.1	0.14	Nouri et al. (2011)
Gramineae	*Paspalum paspaloides*	186.0	18.0	50.5	2.8	0.10	Li et al. (2011a)
Gramineae	*Phragmites australis*	174.0	1.3	8.3	6.3	0.01	Li et al. (2011a)
Lamiaceae	*Mentha longifolia* L.	58.0	2168.9	125.0	0.1	37.39	Mahdavian et al. (2017)
Liliaceae	*Allium umbilicatum* Boiss.	25.0	1257.6	80.0	0.1	50.30	Mahdavian et al. (2017)
Resedaceae	*Reseda alba* L.	9535.0	1743.0	703.0	0.4	0.18	Nouri et al. (2011)
Resedaceae	*Reseda alba*	13,487.6	322.7	1607.5	5.0	0.02	Hasnaoui et al. (2020)
Tamaricaceae	*Tamarix ramosissima* Ledeb.	10,401.0	130.0	2010.0	15.5	0.01	Nouri et al. (2011)

7.4 Microbial-Assisted Pb Phytoremediation

Soil microorganisms fulfill essential ecosystem processes since they regulate biogeochemical cycles and decompose organic matter to maintain soil fertility (Basu et al. 2021). Plants establish associations with different types of soil microorganisms like bacteria and fungi which contribute to the host adaptation to environmental conditions (Gan et al. 2017; Narula et al. 2012). The rhizosphere is the zone of the soil around the plants' roots where occurs intense biological activity during the plant-soil-microorganism interactions (More et al. 2019; Pathan et al. 2020).

Rhizobacteria and epiphytic bacteria are a broad group of soil bacteria that colonize the area around the roots, and the root surface, respectively (Taulé et al. 2021). While the endophytic bacteria colonize the internal plant tissues without causing adverse effects on their host plants (Ma et al. 2011).

Plant-associated microorganisms play an essential role in the metal phytoremediation process. These microorganisms can promote plant growth, reduce metal phytotoxicity, modify metal uptake and accumulation in the plant, and increase metal bioavailability in soil or water (Ma et al. 2016; Rajkumar et al. 2012). For example, two rhizospheric bacteria identified as *Bacillus proteolyticus* and *B. licheniformis*, increased the biomass of *Solanum nigrum* plants growing in heavy metal-contaminated soil, and the total Pb content in roots and shoots (He et al. 2020). Under axenic conditions, a rhizospheric arsenic-resistant bacteria also increased Pb concentration in the root of *Pteris vittata* (Manzoor et al. 2019). An endophytic microbial consortium isolated from three native plants increased Pb accumulation in roots and shoots of *B. juncea*, and Pb concentration in sunflower (*H. annuus*) roots (Pietrini et al. 2021).

In recent years, it has been demonstrated the potential of in situ plant-bacteria interaction for promote plant growth under Pb stress and Pb removal from water and contaminated sites. In an agricultural field contaminated with Pb and Cd, a consortium of four heavy metals resistant bacteria (*Rhizobium leguminosarum*, *Bacillus simplex*, *Luteibacter* sp. and, *Variovorax* sp.) increased plant length, dry biomass, nodule number of *Lathyrus sativus* plants, and enhance Pb accumulation in roots in comparison with uninoculated plants (Abdelkrim et al. 2020). Endophyte bacteria *Pseudomonas putida* RE02 reduced the mortality percentage of *Trifolium repens* seedlings under metal stress and improved Pb uptake by *T. repens* plants grown in heavy metal contaminated tailings (Liu et al. 2021).

In a constructed wetland, a consortium of five rhizobacteria (*Bacillus cereus*, *B. pumilus*, *B. subtilis*, *Brevibacillus choshinensis*, and *Rhodococcus rhodochrous*) increased Pb sorption by *Scirpus grossus* plants from contaminated water (Tangahu et al. 2022).

Bacterial communities can improve growth and tolerance to metal stress in host plants by producing phytohormones and enzyme 1–aminocyclopropane–1–carboxylate (ACC) deaminase which reduces ethylene production (Kong and Glick 2017; Sharma 2021). These bacteria also promote plant growth and favour nutritional status by improving the absorption of water and nutritive elements such as nitrogen (N), phosphorus (P), and iron (Fe) through mechanisms like nitrogen fixation, P-solubilization, and siderophores production (Ma et al. 2011; Etesami 2018; Manoj et al. 2020).

Like bacteria, fungi have also been evaluated to increase phytoremediation efficiency. Arbuscular Mycorrhizae (AM) are fungal endophytes that colonize the internal root tissues of higher plants (Deng and Cao 2017; Gaur and Adholeya 2004). AM fungi have also been shown to promote plant growth under Pb stress and increase Pb accumulation in plants. For example, *Funneliformis mosseae*, *Claroideoglomus etunicatum*, and *Rhizophagus intraradices*, promote the growth of the soybean (*Glycine max* L.) exposed to 100 and 300 mg/kg Pb, and increase Pb accumulation in the roots compared to non-inoculated plants (Adeyemi et al. 2021). *F. mosseae* inoculation also increased Pb accumulation in root and dry weights of *Bidens parviflora* under Pb stress (Yang et al. 2022). Similarly, *Rhizophagus irregularis* increases the shoot biomass of *Medicago truncatula* under Pb stress (800 mg/kg), and enhances the Pb concentration and content in its roots (Zhang et al. 2021b).

Arbuscular mycorrhizae inoculation may increase the Pb tolerance of the host plant and accumulation in the roots through immobilization of Pb ions by the root cell wall or by the fungal cells (Zhang et al. 2010, 2021c). *R. irregularis* inoculation induced pectin and hemicellulose production in root cell walls of *M. truncatula*, which increases the Pb immobilization (Zhang et al. 2021c). On the other hand, in maize (*Z. mays*) plants inoculated with AM fungi, the most significant amount of Pb in roots is localized in the hyphal wall and within fungal cells (Zhang et al. 2010).

7.5 Conclusion

Rhizofiltration, phytostabilization, and phytoextraction are the main phytoremediation strategies for Pb removal from polluted environments. In the phytostabilization and rhizofiltration process, the Pb ions can be precipitated in the rhizosphere, immobilized on root cell walls, or sequestered on the root surface. In contrast, in phytoextraction, the Pb is uptake by plant roots, translocated from roots to shoots, and accumulated in aerial tissues. The Pb hyperaccumulator plants, fast-growing crops with high biomass production, and the native plants growing on heavy metal contaminated sites have been used to remove Pb from the soil, while the aquatic and terrestrial plants with fibrous root systems are suitable for Pb removal of surface water, and groundwater. The plant-associated microorganisms like bacteria and fungi could be used as an alternative to improve the Pb phytoextraction efficiency.

Acknowledgements This work was funded by grants from CONAHCYT, Programa Presupuestario F003. Fondo Sectorial de Investigación para la Educación, CB2017-2018 A1-S-40454, to Alejandro Hernández-Morales.

References

Abdelkrim S, Jebara SH, Saadani O et al (2020) *In situ* effects of *Lathyrus sativus*- PGPR to remediate and restore quality and fertility of Pb and Cd polluted soils. Ecotoxicol Environ Saf 192:110260

Aboubakar A, Douaik A, Mewouo YCM et al (2021) Determination of background values and assessment of pollution and ecological risk of heavy metals in urban agricultural soils of Yaoundé, Cameroon. J Soils Sediments 21:1437–1454

Adeyemi NO, Atayese MO, Sakariyawo OS et al (2021) Arbuscular mycorrhizal fungi species differentially regulate plant growth, phosphorus uptake and stress tolerance of soybean in lead contaminated soil. J Plant Nutr 44:1633–1648

Aftab N, Saleem K, Khan AHA et al (2021) *Cosmos sulphureus* Cav. is more tolerant to lead than copper and chromium in hydroponics system. Int J Environ Sci Technol 18:2325–2334

Al-Fartusie FS, Mohssan SN (2017) Essential trace elements and their vital roles in human body. Indian J Adv Chem Sci 5:127–136

Alonso-Castro AJ, Carranza-Álvarez C, Alfaro-De La Torre MC et al (2009) Removal and accumulation of cadmium and lead by *Typha latifolia* exposed to single and mixed metal solutions. Arch Environ Contam Toxicol 57:688–696

Alsafran M, Usman K, Ahmed B et al (2022) Understanding the phytoremediation mechanisms of potentially toxic elements: a proteomic overview of recent advances. Front Plant Sci 13

Andra SS, Datta R, Sarkar D et al (2009) Induction of lead-binding phytochelatins in vetiver grass [*Vetiveria zizanioides* (L.)]. J Environ Qual 38:868–877

Arazi T, Sunkar R, Kaplan B, Fromm H (1999) A tobacco plasma membrane calmodulin-binding transporter confers Ni^{2+} tolerance and Pb^{2+} hypersensitivity in transgenic plants. Plant J 20:171–182

Arshad M, Merlina G, Uzu G et al (2016) Phytoavailability of lead altered by two Pelargonium cultivars grown on contrasting lead-spiked soils. J Soils Sediments 16:581–591

Asare MO, Száková J, Tlustoš P (2023) The fate of secondary metabolites in plants growing on Cd-, As-, and Pb-contaminated soils—a comprehensive review. Environ Sci Pollut Res 30:11378–11398

Ashraf U, Kanu AS, Mo Z et al (2015) Lead toxicity in rice: effects, mechanisms, and mitigation strategies—a mini review. Environ Sci Pollut Res 22:18318–18332

ATSDR (2022) Agency for toxic substances and disease registry substance priority list. [Online]. Available: https://www.atsdr.cdc.gov/spl/index.html#2022spl. Accessed 05 Apr 2023

Austruy A, Shahid M, Xiong T et al (2014) Mechanisms of metal-phosphates formation in the rhizosphere soils of pea and tomato: environmental and sanitary consequences. J Soils Sediments 14:666–678

Baker AJM, Brooks RR (1989) Terrestrial higher plants which hyperaccumulate metallic elements—a review of their distribution, ecology and phytochemistry. Biorecovery 1:81–126

Bala S, Garg D, Thirumalesh BV et al (2022) Recent strategies for bioremediation of emerging pollutants: a review for a green and sustainable environment. Toxics 10

Ballantyne AD, Hallett JP, Riley DJ et al (2018) Lead acid battery recycling for the twenty-first century. R Soc Open Sci 5:171368

Basu S, Kumar G, Chhabra S, Prasad R (2021) Role of soil microbes in biogeochemical cycle for enhancing soil fertility. In: Verma JP, Macdonald CA, Gupta VK et al (eds) New and future developments in microbial biotechnology and bioengineering. Elsevier

Bolan NS, Park JH, Robinson B et al (2011) Phytostabilization: a green approach to contaminant containment. In: Sparks DL (ed) Advances in agronomy. Academic Press

Boopathy R (2000) Factors limiting bioremediation technologies. Bioresour Technol 74:63–67

Carrillo-Chávez A, González-Partida E, Morton-Bermea O et al (2006) Heavy metal distribution in rocks, sediments, mine tailings, leaching experiments, and groundwater from the Mineral de Pozos historical mining site, North-Central Mexico. Int Geol Rev 48:466–478

Chang J-D, Gao W, Wang P, Zhao F-J (2022) OsNRAMP5 is a major transporter for lead uptake in rice. Environ Sci Technol 56:17481–17490

Chatterjee S, Mitra A, Datta S, Veer V (2013) Phytoremediation protocols: an overview. In: Gupta DK (ed) Plant-based remediation processes. Springer Berlin Heidelberg, Berlin, Heidelberg

Cheng H, Wang M, Wong MH, Ye Z (2014) Does radial oxygen loss and iron plaque formation on roots alter Cd and Pb uptake and distribution in rice plant tissues? Plant Soil 375:137–148

Cheng S-F, Huang C-Y, Lin Y-C et al (2015) Phytoremediation of lead using corn in contaminated agricultural land—an in situ study and benefit assessment. Ecotoxicol Environ Saf 111:72–77

Cobbett C, Goldsbrough P (2002) Phytochelatins and metallothioneins: roles in heavy metal detoxification and homeostasis. Annu Rev Plant Biol 53:159–182

Das S, Das A, Mazumder PET et al (2021) Lead phytoremediation potentials of four aquatic macrophytes under hydroponic cultivation. Int J Phytoremediation 23:1279–1288

Delgado-González CR, Madariaga-Navarrete A, Fernández-Cortés JM et al (2021) Advances and applications of water phytoremediation: a potential biotechnological approach for the treatment of heavy metals from contaminated water. Int J Environ Res Public Health [Online] 18

Deng Z, Cao L (2017) Fungal endophytes and their interactions with plants in phytoremediation: a review. Chemosphere 168:1100–1106
dos Reis PE, De Souza KRD, Romão GF et al (2022) Potential of *Talinum paniculatum* cuttings in lead and manganese rhizofiltration. Water Air Soil Pollut 233:243
Dushenkov V, Kumar PBAN, Motto H, Raskin I (1995) Rhizofiltration: the use of plants to remove heavy metals from aqueous streams. Environ Sci Technol 29:1239–1245
Eapen S, D'Souza SF (2005) Prospects of genetic engineering of plants for phytoremediation of toxic metals. Biotechnol Adv 23:97–114
Eichler A, Gramlich G, Kellerhals T et al (2015) Pb pollution from leaded gasoline in South America in the context of a 2000-year metallurgical history. Sci Adv 1:e1400196
Ekta P, Modi NR (2018) A review of phytoremediation. J Pharmacogn Phytochem 7:1485–1489
Estrella-Gómez N, Mendoza-Cózatl D, Moreno-Sánchez R et al (2009) The Pb-hyperaccumulator aquatic fern *Salvinia minima* Baker, responds to Pb^{2+} by increasing phytochelatins via changes in SmPCS expression and in phytochelatin synthase activity. Aquat Toxicol 91:320–328
Etesami H (2018) Bacterial mediated alleviation of heavy metal stress and decreased accumulation of metals in plant tissues: mechanisms and future prospects. Ecotoxicol Environ Saf 147:175–191
Etim E (2012) Phytoremediation and its mechanisms: a review. Int J Environ Bioenergy 2:120–136
Favas P, Pratas J, Varun M et al (2014) Phytoremediation of soils contaminated with metals and metalloids at mining areas: potential of native flora. In: Hernández-Soriano (ed) Environmental risk assessment of soil contamination. InTech
Gall JE, Rajakaruna N (2013) The physiology, functional genomics, and applied ecology of heavy metal-tolerant Brassicaceae. In: Lang M (ed) Brassicaceae: characterization, functional genomics and health benefits. Nova Science Pub Inc.
Gan H, Churchill ACL, Wickings K (2017) Invisible but consequential: root endophytic fungi have variable effects on belowground plant–insect interactions. Ecosphere 8:e01710
Gaur A, Adholeya A (2004) Prospects of arbuscular mycorrhizal fungi in phytoremediation of heavy metal contaminated soils. Curr Sci 86
Gong X, Yang F, Pan X, Shao JF (2022) Accumulation of silicon in shoots is required for reducing lead uptake in rice. Crop J
Gupta DK, Huang HG, Corpas FJ (2013a) Lead tolerance in plants: strategies for phytoremediation. Environ Sci Pollut Res 20:2150–2161
Gupta DK, Vandenhove H, Inouhe M (2013b) Role of phytochelatins in heavy metal stress and detoxification mechanisms in plants. In: Gupta DK, Corpas FJ, Palma JM (eds) Heavy metal stress in plants. Springer Berlin Heidelberg, Berlin, Heidelberg
Han Y-L, Huang S-Z, Gu J-G et al (2008) Tolerance and accumulation of lead by species of *Iris* L. Ecotoxicology 17:853–859
Hansel CM, Fendorf S, Sutton S, Newville M (2001) Characterization of Fe plaque and associated metals on the roots of mine-waste impacted aquatic plants. Environ Sci Technol 35:3863–3868
Hasnaoui SE, Fahr M, Keller C et al (2020) Screening of native plants growing on a Pb/Zn mining area in eastern Morocco: perspectives for phytoremediation. Plants [Online] 9
He X, Xu M, Wei Q et al (2020) Promotion of growth and phytoextraction of cadmium and lead in *Solanum nigrum* L. mediated by plant-growth-promoting rhizobacteria. Ecotoxicol Environ Saf 205:111333
He J, Chen X, Achal V (2021) Bioremediation: concepts and application. In: Sharma N, Sodhi AS, Batra N (eds) Basic concepts in environmental biotechnology. CRC Press, Boca Raton
He W, Wang S, Wang Y et al (2023) Pathways and characteristics of lead uptake and transportation in *Rhus chinensis* Mill. Forests 14:90
Heidari S, Mostafaei S, Razazian N et al (2022) The effect of lead exposure on IQ test scores in children under 12 years: a systematic review and meta-analysis of case-control studies. Syst Rev 11:106
Ho LB, Huynh BV, Bui TC (2021) Accumulation and distribution of lead (Pb) in different tissues of Lucky bamboo plants (*Dracaena sanderiana*). JDAE 20:50–60

Huang G-Y, Wang Y-S (2009) Expression analysis of type 2 metallothionein gene in mangrove species (*Bruguiera gymnorrhiza*) under heavy metal stress. Chemosphere 77:1026–1029

Inouhe M, Huang H, Chaudhary SK, Gupta DK (2012) Heavy metal bindings and their interactions with thiol peptides and other biological ligands in plant cells. In: Gupta DK, Sandalio LM (eds) Metal toxicity in plants: perception, signaling and remediation. Springer Berlin Heidelberg, Berlin, Heidelberg

Irias Zelaya CR, Gadelha JC, Hernandez FFF, Ortiz Escobar ME (2020) Heavy metals behavior in the presence of organic acids and proteins in the rhizosphere and plant tissues of yellow melon var. natal (*Cucumis melo* L.) cultivated in a Quartzarenic neosol. Water Air Soil Pollut 231:572

Jadia CD, Fulekar M (2009) Phytoremediation of heavy metals: recent techniques. Afr J Biotechnol 8

Jee C (2016) Advances in phytoremediation and rhizoremediation. Octa J Environ Res 4:18–32

Jiang W, Liu D (2010) Pb-induced cellular defense system in the root meristematic cells of *Allium sativum* L. BMC Plant Biol 10:40

Jiang M, Cai X, Liao J et al (2020) Different strategies for lead detoxification in dwarf bamboo tissues. Ecotoxicol Environ Saf 193:110329

John DA, Leventhal JS (1995) Bioavailability of metals. In: Bray EAD (ed) Preliminary compilation of descriptive geoenvironmental mineral deposit models. U.S. Geological Survey, Denver, Colorado

Kafle A, Timilsina A, Gautam A et al (2022) Phytoremediation: mechanisms, plant selection and enhancement by natural and synthetic agents. Environ Adv 8:100203

Khan N, Seshadri B, Bolan N et al (2016) Root iron plaque on wetland plants as a dynamic pool of nutrients and contaminants. In: Sparks DL (ed) Advances in agronomy. Academic Press

Khanam R, Kumar A, Nayak AK et al (2020) Metal(loid)s (As, Hg, Se, Pb and Cd) in paddy soil: bioavailability and potential risk to human health. Sci Total Environ 699:134330

Kim Y-O, Kang H (2018) Comparative expression analysis of genes encoding metallothioneins in response to heavy metals and abiotic stresses in rice (*Oryza sativa*) and *Arabidopsis thaliana*. Biosci Biotechnol Biochem 82:1656–1665

Kisa D, Öztürk L, Doker S, Gökçe İ (2017) Expression analysis of metallothioneins and mineral contents in tomato (*Lycopersicon esculentum*) under heavy metal stress. J Sci Food Agric 97:1916–1923

Kong Z, Glick BR (2017) The role of plant growth-promoting bacteria in metal phytoremediation. In: Poole RK (ed) Advances in microbial physiology. Academic Press

Kopittke PM, Asher CJ, Blamey FPC et al (2008) Localization and chemical speciation of Pb in roots of signal grass (*Brachiaria decumbens*) and rhodes grass (*Chloris gayana*). Environ Sci Technol 42:4595–4599

Kristanti RA, Ngu WJ, Yuniarto A, Hadibarata T (2021) Rhizofiltration for removal of inorganic and organic pollutants in groundwater: a review. Biointerface Res Appl Chem 4:12326–12347

Krzesłowska M (2011) The cell wall in plant cell response to trace metals: polysaccharide remodeling and its role in defense strategy. Acta Physiol Plant 33:35–51

Kurade MB, Ha Y-H, Xiong J-Q et al (2021) Phytoremediation as a green biotechnology tool for emerging environmental pollution: a step forward towards sustainable rehabilitation of the environment. Chem Eng J 415:129040

Li G-Y, Hu N, Ding D-X et al (2011a) Screening of plant species for phytoremediation of uranium, thorium, barium, nickel, strontium and lead contaminated soils from a uranium mill tailings repository in South China. Bull Environ Contam Toxicol 86:646–652

Li T, Di Z, Islam E et al (2011b) Rhizosphere characteristics of zinc hyperaccumulator *Sedum alfredii* involved in zinc accumulation. J Hazard Mater 185:818–823

Li F-L, Liu C-Q, Yang Y-G et al (2012) Natural and anthropogenic lead in soils and vegetables around Guiyang city, southwest China: a Pb isotopic approach. Sci Total Environ 431:339–347

Li X, Huang S, Mcbride MB (2021) Rhizosphere effect on Pb solubility and phytoavailability in Pb-contaminated soils. Environ Pollut 268:115840

Liu J, Luo L (2019) Uptake and transport of Pb across the iron plaque of waterlogged dropwort (*Oenanthe javanica* DC.) based on micro-XRF and XANES. Plant Soil 441:191–205

Liu D, Jiang W, Liu C et al (2000) Uptake and accumulation of lead by roots, hypocotyls and shoots of Indian mustard [*Brassica juncea* (L.)]. Bioresour Technol 71:273–277

Liu CY, Gong XF, Tang YP, Chen CL (2015) Lead sequestration in iron plaques developed on *Phalaris arundinacea* Linn. and *Carex cinerascens* Kukenth. from Poyang Lake (China). Aquat Bot 122:54–59

Liu C, Gong X, Chen C et al (2016) The effect of iron plaque on lead translocation in soil-*Carex cinerascens* Kukenth. system. Int J Phytoremediation 18:1–9

Liu L, Li W, Song W, Guo M (2018) Remediation techniques for heavy metal-contaminated soils: principles and applicability. Sci Total Environ 633:206–219

Liu C, Lin H, Li B et al (2021) Endophyte *Pseudomonas putida* enhanced *Trifolium repens* L. growth and heavy metal uptake: a promising in-situ non-soil cover phytoremediation method of nonferrous metallic tailing. Chemosphere 272:129816

Lone MI, He Z-L, Stoffella PJ, Yang X-E (2008) Phytoremediation of heavy metal polluted soils and water: progresses and perspectives. J Zhejiang Univ Sci B 9:210–220

Lorestani B, Cheraghi M, Yousefi N (2011) Phytoremediation potential of native plants growing on a heavy metals contaminated soil of copper mine in Iran. Int J Geol 5:299–304

Lorestani B, Yousefi N, Cheraghi M, Farmany A (2013) Phytoextraction and phytostabilization potential of plants grown in the vicinity of heavy metal-contaminated soils: a case study at an industrial town site. Environ Monit Assess 185:10217–10223

Ma Y, Prasad MNV, Rajkumar M, Freitas H (2011) Plant growth promoting rhizobacteria and endophytes accelerate phytoremediation of metalliferous soils. Biotechnol Adv 29:248–258

Ma X, Liu J, Wang M (2013) Differences between rice cultivars in iron plaque formation on roots and plant lead tolerance. Adv J Food Sci Technol 5:160–163

Ma Y, Rajkumar M, Zhang C, Freitas H (2016) Beneficial role of bacterial endophytes in heavy metal phytoremediation. J Environ Manage 174:14–25

Mahar A, Wang P, Ali A et al (2016) Challenges and opportunities in the phytoremediation of heavy metals contaminated soils: a review. Ecotoxicol Environ Saf 126:111–121

Mahdavian K, Ghaderian SM, Torkzadeh-Mahani M (2017) Accumulation and phytoremediation of Pb, Zn, and Ag by plants growing on Koshk lead–zinc mining area, Iran. J Soils Sediments 17:1310–1320

Manara A, Fasani E, Furini A, Dalcorso G (2020) Evolution of the metal hyperaccumulation and hypertolerance traits. Plant Cell Environ 43:2969–2986

Manoj SR, Karthik C, Kadirvelu K et al (2020) Understanding the molecular mechanisms for the enhanced phytoremediation of heavy metals through plant growth promoting rhizobacteria: a review. J Environ Manage 254:109779

Mansoor S, Khan NF, Farooq I et al (2022) Phytoremediation at molecular level. In: Bhat RA, Tonelli FMP, Dar GH et al (eds) Phytoremediation. Academic Press

Manzoor M, Abid R, Rathinasabapathi B et al (2019) Metal tolerance of arsenic-resistant bacteria and their ability to promote plant growth of *Pteris vittata* in Pb-contaminated soil. Sci Total Environ 660:18–24

Manzoor M, Gul I, Manzoor A et al (2020) Lead availability and phytoextraction in the rhizosphere of Pelargonium species. Environ Sci Pollut Res 27:39753–39762

Midhat L, Ouazzani N, Hejjaj A et al (2019) Accumulation of heavy metals in metallophytes from three mining sites (Southern Centre Morocco) and evaluation of their phytoremediation potential. Ecotoxicol Environ Saf 169:150–160

Mishra S, Srivastava S, Tripathi RD et al (2006) Lead detoxification by coontail (*Ceratophyllum demersum* L.) involves induction of phytochelatins and antioxidant system in response to its accumulation. Chemosphere 65:1027–1039

Mitra A, Chatterjee S, Datta S et al (2014) Mechanism of metal transporters in plants. In: Kumar Gupta D, Chatterjee S (eds) Heavy metal remediation: transport accumulation in plants. Nova Science Publishers, Inc.

More SS, Shinde SE, Kasture MC (2019) Root exudates a key factor for soil and plant: an overview. Pharm Innov J 8:449–459
Nag R, Cummins E (2022) Human health risk assessment of lead (Pb) through the environmental-food pathway. Sci Total Environ 810:151168
Narula N, Kothe E, Behl RK (2012) Role of root exudates in plant-microbe interactions. J Appl Bot Food Qual 82:122–130
Nedjimi B (2021) Phytoremediation: a sustainable environmental technology for heavy metals decontamination. SN Appl Sci 3:286
Nouri J, Lorestani B, Yousefi N et al (2011) Phytoremediation potential of native plants grown in the vicinity of Ahangaran lead–zinc mine (Hamedan, Iran). Environ Earth Sci 62:639–644
Olaniran AO, Balgobind A, Pillay B (2013) Bioavailability of heavy metals in soil: impact on microbial biodegradation of organic compounds and possible improvement strategies. Int J Mol Sci 14:10197–10228
Otte ML, Jacob DL (2006) Constructed wetlands for phytoremediation: rhizofiltration, phytostabilisation and phytoextraction. In: Mackova M, Dowling D, Macek T (eds) Phytoremediation rhizoremediation. Springer
Parrotta L, Guerriero G, Sergeant K et al (2015) Target or barrier? The cell wall of early- and later-diverging plants vs cadmium toxicity: differences in the response mechanisms. Front Plant Sci 6
Parven S, De A, Gupta A (2022) Cu and Pb accumulation and removal from aqueous medium by *Enydra fluctuans* Lour. (Asteraceae)—a medicinal plant with potential for phytoremediation. Environ Sci Pollut Res 29:66902–66912
Pasricha S, Mathur V, Garg A et al (2021) Molecular mechanisms underlying heavy metal uptake, translocation and tolerance in hyperaccumulators-an analysis: heavy metal tolerance in hyperaccumulators. Environ Chall 4:100197
Pathan SI, Ceccherini MT, Sunseri F, Lupini A (2020) Rhizosphere as hotspot for plant-soil-microbe interaction. In: Datta R, Meena RS, Pathan SI et al (eds) Carbon and nitrogen cycling in soil. Springer Singapore, Singapore
Peralta-Videa JR, Lopez ML, Narayan M et al (2009) The biochemistry of environmental heavy metal uptake by plants: implications for the food chain. Int J Biochem Cell Biol 41:1665–1677
Pietrini I, Grifoni M, Franchi E et al (2021) Enhanced lead phytoextraction by endophytes from indigenous plants. Soil Syst 5:55
Pourrut B, Shahid M, Dumat C et al (2011) Lead uptake, toxicity, and detoxification in plants. In: Whitacre DM (ed) Reviews of environmental contamination and toxicology, vol 213. Springer New York, New York, NY
Pourrut B, Shahid M, Douay F et al (2013) Molecular mechanisms involved in lead uptake, toxicity and detoxification in higher plants. In: Gupta DK, Corpas FJ, Palma JM (eds) Heavy metal stress in plants. Springer Berlin Heidelberg, Berlin, Heidelberg
Rahimpoor R, Rostami M, Assari MJ et al (2020) Evaluation of blood lead levels and their effects on hematological parameters and renal function in Iranian lead mine workers. Health Scope 9:e95917
Rahman Z, Singh VP (2019) The relative impact of toxic heavy metals (THMs) (arsenic (As), cadmium (Cd), chromium (Cr)(VI), mercury (Hg), and lead (Pb)) on the total environment: an overview. Environ Monit Assess 191:419
Rahman MM, Swaraz AM, El-Shehawi AM et al (2022) The mechanistic basis of sulfur-mediated alleviation of Pb toxicity in wheat. Gesunde Pflanz 74:571–581
Rai GK, Bhat BA, Mushtaq M et al (2021) Insights into decontamination of soils by phytoremediation: a detailed account on heavy metal toxicity and mitigation strategies. Physiol Plant 173:287–304
Rajkumar M, Sandhya S, Prasad MNV, Freitas H (2012) Perspectives of plant-associated microbes in heavy metal phytoremediation. Biotechnol Adv 30:1562–1574
Rawat K, Fulekar M, Pathak B (2012) Rhizofiltration: a green technology for remediation of heavy metals. Int J Innov Bio-Sci 2:193–199

Reeves RD, Baker AJM, Jaffré T et al (2018) A global database for plants that hyperaccumulate metal and metalloid trace elements. New Phytol 218:407–411

Rolón-Cárdenas GA, Martínez-Martínez JG, Arvizu-Gómez JL et al (2022) Enhanced Cd-accumulation in *Typha latifolia* by interaction with *Pseudomonas rhodesiae* GRC140 under axenic hydroponic conditions. Plants 11:1447

Saifullah, Meers E, Qadir M et al (2009) EDTA-assisted Pb phytoextraction. Chemosphere 74:1279–1291

Salazar MJ, Pignata ML (2014) Lead accumulation in plants grown in polluted soils. Screening of native species for phytoremediation. J Geochem Explor 137:29–36

Scarano G, Morelli E (2002) Characterization of cadmium- and lead-phytochelatin complexes formed in a marine microalga in response to metal exposure. BioMetals 15:145–151

Seshadri B, Bolan NS, Naidu R (2015) Rhizosphere-induced heavy metal(loid) transformation in relation to bioavailability and remediation. J Soil Sci Plant Nutr 15:524–548

Seth CS, Misra V, Singh RR, Zolla L (2011) EDTA-enhanced lead phytoremediation in sunflower (*Helianthus annuus* L.) hydroponic culture. Plant Soil 347:231–242

Sharma P (2021) Efficiency of bacteria and bacterial assisted phytoremediation of heavy metals: an update. Bioresour Technol 328:124835

Sharma P, Dubey R (2005) Lead toxicity in plants. Braz J Plant Physiol 17:35–52

Sharma SS, Dietz K-J, Mimura T (2016) Vacuolar compartmentalization as indispensable component of heavy metal detoxification in plants. Plant Cell Environ 39:1112–1126

Shi W, Zhou J, Li J et al (2021) Lead exposure-induced defense responses result in low lead translocation from the roots to aerial tissues of two contrasting poplar species. Environ Pollut 271:116346

Singh V, Thakur L, Mondal P (2015) Removal of lead and chromium from synthetic wastewater using *Vetiveria zizanioides*. Clean (Weinh) 43:538–543

Singh H, Verma A, Kumar M et al (2017) Phytoremediation: a green technology to clean up the sites with low and moderate level of heavy metals. Austin Biochem 2:1012

Song B, Zeng G, Gong J et al (2017) Evaluation methods for assessing effectiveness of in situ remediation of soil and sediment contaminated with organic pollutants and heavy metals. Environ Int 105:43–55

Srivastava S, Anand V, Singh P et al (2021) Microbial systems as a source of novel genes for enhanced phytoremediation of contaminated soils. In: Kumar A, Singh VK, Singh P et al (eds) Microbe mediated remediation of environmental contaminants. Woodhead Publishing

St-Cyr L, Fortin D, Campbell PGC (1993) Microscopic observations of the iron plaque of a submerged aquatic plant (*Vallisneria americana* Michx). Aquat Bot 46:155–167

Sumranwanich T, Boonthaworn K, Singh A (2018) The roles of plant cell wall as the first-line protection against lead (Pb) toxicity. KMUTNB IJAST 11:239–245

Sunkar R, Kaplan B, Bouché N et al (2000) Expression of a truncated tobacco NtCBP4 channel in transgenic plants and disruption of the homologous Arabidopsis CNGC1 gene confer Pb^{2+} tolerance. Plant J 24:533–542

Sytar O, Ghosh S, Malinska H et al (2021) Physiological and molecular mechanisms of metal accumulation in hyperaccumulator plants. Physiol Plant 173:148–166

Tangahu BV, Sheikh Abdullah SR, Basri H et al (2022) Lead (Pb) removal from contaminated water using constructed wetland planted with *Scirpus grossus*: optimization using response surface methodology (RSM) and assessment of rhizobacterial addition. Chemosphere 291:132952

Taulé C, Vaz-Jauri P, Battistoni F (2021) Insights into the early stages of plant–endophytic bacteria interaction. World J Microbiol Biotechnol 37:13

Tripathi RD, Tripathi P, Dwivedi S et al (2014) Roles for root iron plaque in sequestration and uptake of heavy metals and metalloids in aquatic and wetland plants. Metallomics 6:1789–1800

Van Slycken S, Meers E, Meiresonne L et al (2008) The use of bio-energy crops for phytoremediation of metal enriched soils in the Campine region. Commun Agric Appl Biol Sci 73:19–22

Veselý T, Tlustoš P, Száková J (2011) The use of water lettuce (*Pistia Stratiotes* L.) for rhizofiltration of a highly polluted solution by cadmium and lead. Int J Phytoremediation 13:859–872

Vidali M (2001) Bioremediation. An overview. Pure Appl Chem 73:1163–1172
Vives-Peris V, De Ollas C, Gómez-Cadenas A, Pérez-Clemente RM (2020) Root exudates: from plant to rhizosphere and beyond. Plant Cell Rep 39:3–17
Wang M, Duan D, Xu C et al (2015) Adsorption ability of cell wall (CW) components in roots of tea plant (*Camellia sinensis* L.) to Pb and FTIR spectra of their functional groups. Acta Ecol Sin 35:1743–1751
Wang L, Ji B, Hu Y et al (2017) A review on in situ phytoremediation of mine tailings. Chemosphere 184:594–600
Wenzel WW, Lombi E, Adriano DC (1999) Biogeochemical processes in the rhizosphere: role in phytoremediation of metal-polluted soils. In: Prasad MNV, Hagemeyer J (eds) Heavy metal stress in plants: from molecules to ecosystems. Springer Berlin Heidelberg, Berlin, Heidelberg
Wierzbicka M (1998) Lead in the apoplast of *Allium cepa* L. root tips—ultrastructural studies. Plant Sci 133:105–119
Wong CSC, Li XD (2004) Pb contamination and isotopic composition of urban soils in Hong Kong. Sci Total Environ 319:185–195
Wu J, Long J, Liu L et al (2018) Risk assessment and source identification of toxic metals in the agricultural soil around a Pb/Zn mining and smelting area in southwest China. Int J Environ Res Public Health 15:1838
Yan A, Wang Y, Tan SN et al (2020) Phytoremediation: a promising approach for revegetation of heavy metal-polluted land. Front Plant Sci 11
Yang W, Luo L, Bostick BC et al (2021) Effect of combined arsenic and lead exposure on their uptake and translocation in Indian mustard. Environ Pollut 274:116549
Yang Y, Huang B, Xu J et al (2022) Heavy metal domestication enhances beneficial effects of arbuscular mycorrhizal fungi on lead (Pb) phytoremediation efficiency of *Bidens parviflora* through improving plant growth and root Pb accumulation. Environ Sci Pollut Res 29:32988–33001
Yoon J, Cao X, Zhou Q, Ma LQ (2006) Accumulation of Pb, Cu, and Zn in native plants growing on a contaminated Florida site. Sci Total Environ 368:456–464
Yuan X, Xiong T, Yao S et al (2019) A real filed phytoremediation of multi-metals contaminated soils by selected hybrid sweet sorghum with high biomass and high accumulation ability. Chemosphere 237:124536
Zandi P, Yang J, Darma A et al (2022) Iron plaque formation, characteristics, and its role as a barrier and/or facilitator to heavy metal uptake in hydrophyte rice (*Oryza sativa* L.). Environ Geochem Health
Zehra A, Sahito ZA, Tong W et al (2020) Assessment of sunflower germplasm for phytoremediation of lead-polluted soil and production of seed oil and seed meal for human and animal consumption. J Environ Sci 87:24–38
Zhan J, Huang H, Yu H et al (2020) The combined effects of Cd and Pb enhanced metal binding by root cell walls of the phytostabilizer *Athyrium wardii* (Hook.). Environ Pollut 258:113663
Zhang H-H, Tang M, Chen H et al (2010) Effect of inoculation with AM fungi on lead uptake, translocation and stress alleviation of *Zea mays* L. seedlings planting in soil with increasing lead concentrations. Eur J Soil Biol 46:306–311
Zhang J, Martinoia E, Lee Y (2018) Vacuolar transporters for cadmium and arsenic in plants and their applications in phytoremediation and crop development. Plant Cell Physiol 59:1317–1325
Zhang Y, Hou D, O'Connor D et al (2019) Lead contamination in Chinese surface soils: source identification, spatial-temporal distribution and associated health risks. Crit Rev Environ Sci Technol 49:1386–1423
Zhang J, Qian Y, Chen Z et al (2020) Lead-induced oxidative stress triggers root cell wall remodeling and increases lead absorption through esterification of cell wall polysaccharide. J Hazard Mater 385:121524
Zhang B, Gao Y, Zhang L, Zhou Y (2021a) The plant cell wall: biosynthesis, construction, and functions. J Integr Plant Biol 63:251–272

Zhang H, Ren W, Zheng Y et al (2021b) Arbuscular mycorrhizal fungi increase Pb uptake of colonized and non-colonized *Medicago truncatula* root and deliver extra Pb to colonized root segment. Microorganisms 9:1203

Zhang X, Hu W, Xie X et al (2021c) Arbuscular mycorrhizal fungi promote lead immobilization by increasing the polysaccharide content within pectin and inducing cell wall peroxidase activity. Chemosphere 267:128924

Zhao L, Li T, Yu H et al (2015) Changes in chemical forms, subcellular distribution, and thiol compounds involved in Pb accumulation and detoxification in *Athyrium wardii* (Hook.). Environ Sci Pollut Res 22:12676–12688

Zhou C, Huang M, Li Y et al (2016) Changes in subcellular distribution and antioxidant compounds involved in Pb accumulation and detoxification in *Neyraudia reynaudiana*. Environ Sci Pollut Res 23:21794–21804

Zhou J, Zhang Z, Zhang Y et al (2018) Effects of lead stress on the growth, physiology, and cellular structure of privet seedlings. PLoS ONE 13:e0191139

Zou T, Li T, Zhang X et al (2012) Lead accumulation and phytostabilization potential of dominant plant species growing in a lead–zinc mine tailing. Environ Earth Sci 65:621–630

Chapter 8
Application of Nanoadsorbents for Lead Decontamination in Water

Nitish Dhingra

Abstract Clean water availability is a primary requisite of all living organisms. The rapid rise in population has led to significant development and industrial growth all over the globe to fulfil the increasing demands. This has resulted in the contamination of water bodies due to the release of heavy metals and metalloids caused by sudden mine tailings, gasoline, leaded paints, usage of fertilizers inland, animal manures, pesticides, sewage sludge, wastewater irrigation, coal, etc. The contamination of water bodies has caused severe environmental concerns. As a limited natural resource, the water preservation and its quality maintenance are of fundamental importance to ensure its availability for future generations. Therefore, eliminating heavy metals and other pollutants from contaminated streams is a primary concern due to their ability to cause toxic chaos that can affect the metabolism of flora and fauna. However, the existing decontamination techniques, such as ion exchange and reverse osmosis, suffer many disadvantages; hence, the focus has been shifted to developing novel, efficient techniques to remove heavy metals such as lead from the water. Out of these, the adsorption based on nanoadsorbents has gained popularity due to its ease of operation and cost-effectiveness. This chapter highlights the recent advances in water decontamination methods using nanoadsorbents involving carbon nanotubes, graphene, polymer-based, metal oxide nanoparticles, zeolites, and nano-clays.

Keywords Decontamination · Heavy metal · Lead · Nanoadsorbent · Water

8.1 Introduction

There are a few prime necessities for the survival of living beings. Among these, access to pure and safe water is one of the most challenging needs for the multifarious growth of a nation and a flourishing economy. Earth is popularly termed a blue

N. Dhingra (✉)
Electron Microscopy and Nanoscience Laboratory, Department of Soil Science, Punjab Agricultural University, Ludhiana, Punjab 141004, India
e-mail: nitishdhingra@pau.edu

N. Kumar and A. K. Jha (eds.), *Lead Toxicity Mitigation: Sustainable Nexus Approaches*, Environmental Contamination Remediation and Management,
https://doi.org/10.1007/978-3-031-46146-0_8

planet because the water spans about one-third of the entire earth's surface, of which about 97.5% is saline water, while the remaining 2.5% is freshwater. Out of this freshwater fraction, about 68.9% of water is available in the form of ice, permanent snow, and glaciers. Groundwater accounts for 30.8% of freshwater and only 0.3% is easily accessible (UNESCO 2023). The rapid rise in the population level, growing industrialization, urbanization, and extensive agriculture practices are contributing to the persistent deterioration of the water resources quality, hence becoming a matter of grave concern (Kamali et al. 2019; Olvera et al. 2017). Globally, around 1.2 billion population lacks access to pure drinking water, around 2.6 billion people struggle to meet primary sanitation, and millions of children have died due to communicable diseases spread through contaminated water (Kumar et al. 2014a, b; Yamamura et al. 2011).

The quantum of usable water is reducing every day due to rapid industrialization, leading to severe environmental pollution and contamination of drinking and groundwater due to diverse organic, inorganic, and biological pollutants (Ali 2014). Figure 8.1 gives an overview of all possible categories of water contamination. Among these pollutants, the pollution caused by heavy metal ions is critical due to their non-biodegradability and accumulation inside living organism's bodies. Lead (Pb) is one of the highly toxic metals (Shanmugalingam and Murugesan 2018; Engida and Chandravanshi 2017; Benabdallah et al. 2017). Lead is a naturally occurring metal that often forms compounds in combination with two or more elements. It reacts with air and water to produce lead carbonates, lead oxide, or lead sulfate. Such compounds can prevent corrosion by acting as a protective barrier. Although the occurrence of lead is suggested in nature, anthropogenic activities are the primary source of rising lead content in the environment (Shahid et al. 2015). Inadvertent exposure to lead can occur in many ways, such as the corrosion of pipes or faucets, household plumbing systems containing lead, printing, old paints, storage/automobile battery manufacturing, and other industrial waste. These potential sources result in the introduction of lead at elevated concentrations into the water cycle (Salam 2013; Mubarak et al. 2014; Ozlem Kocabas-Ataklı and Yurum 2013; Manzoor et al. 2013). Table 8.1 summarizes the standard limits of selected heavy metals in drinking water.

Lead released into the air from industrial activities is removed by the rain and shifted to soil or surface water. Besides, lead is utilized as a pesticide for vegetable and fruit cultivation (Gall et al. 2015). Consequently, humans can be exposed to lead through ingestion or skin breaks, resulting in lead accumulation through absorption in the blood, bones, and soft tissues. Lead accumulation severely influences the central nervous system and can cause short-term memory loss and other neurological, renal, gastrointestinal, and cardiovascular disorders. Hence, there is a dire necessity to invent decontamination techniques to eradicate heavy metals such as lead from the water.

Conventionally, several physical and chemical techniques are used for water remediation. These include reverse osmosis, chemical precipitation, coagulation or flocculation, ion exchange, electrolysis, ultrafiltration, and adsorption. The applicability of these techniques depends on various factors. Table 8.2 summarizes various water

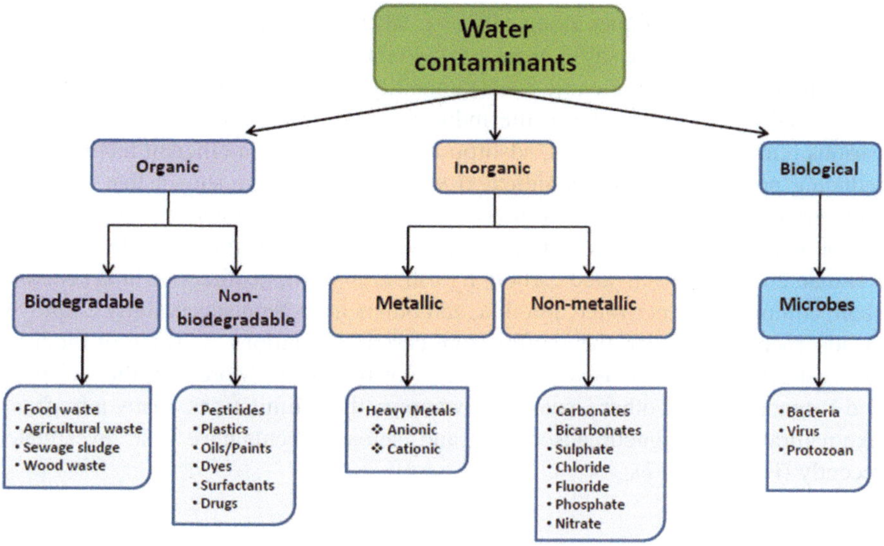

Fig. 8.1 Pictorial representation of the possible categories of water contaminants

Table 8.1 Reference limit of heavy metal contamination in drinking water (WHO 2017)

Contaminant	Environmental protection agency (EPA) maximum permissible limit (mg/L)	Value as per WHO guidelines	Ill-effects on human health
Arsenic	0.01	0.01	Vascular disease, skin manifestations, and visceral cancers
Cadmium	0.005	0.003	Kidney damage, renal disorder, human and carcinogen
Chromium	0.1	0.05	Headache, diarrhea, nausea, vomiting, and carcinogenic
Copper	1.3	2	Liver damage, Wilson disease, and insomnia
Lead	0.015	0.01	Damage the fetal brain, diseases of the kidneys, circulatory system, and nervous system
Mercury	0.002	0.006	Rheumatoid arthritis, and diseases of the kidneys, circulatory system, and nervous system
Nickel	–	0.07	Dermatitis, nausea, chronic asthma, coughing, and human carcinogen
Zinc	5	3	Depression, lethargy, neurological signs, and increased thirst

decontamination techniques along with their advantages and limitations. These techniques are effective but suffer from unavoidable issues such as energy intensiveness, high cost, less efficiency, unsustainability, and tediousness, thereby making them challenging to execute at the industrial scale (Tahoon et al. 2020). Out of various remediation techniques, adsorption is considered an efficient technique for reducing pollutants from contaminated water. Studies have shown that industrial and agricultural wastes can be utilized to remove various kinds of pollutants from synthetic aqueous solutions and industrial wastewater (Nazar et al. 2018; Fazal-ur-Rehman 2018). The activated carbon, fly ash, crab-shell, zeolite, rice husk, coconut shell, commercially activated alumina, and others have been successfully employed as adsorbents. However, they suffer from certain limitations, as some of them are not cost-effective, whereas some require prior treatment. Moreover, their disposal and regeneration are other issues. To overcome these limitations, many adsorbents like natural clay, magnetic adsorbents, and nanoadsorbents have been investigated recently (Pandey 2017).

8.2 Nanotechnology for Water Remediation

In the contemporary era, nanotechnology has unfolded as a promising solution in diverse sectors involving research and development. Some of these are the agriculture sector (Acharya and Pal 2020), bioanalytical sciences (Keçili et al. 2019), the food sector (Palit 2020), and water purification (Puri et al. 2021). With the help of technological advances, it is feasible to employ nanoscale materials (<100 nm) in addressing water remediation problems owing to their remarkable properties such as higher surface-to-volume ratio, effortless functionalization ability for enhanced selectivity and affinity, and high sorbent capabilities (Zhang et al. 2019).

Adsorption is a surface phenomenon and refers to the ability of a solid substance to attach to its surface the molecules of gases or solutions with which it is in close contact. There may be variations from the micron to nanometer range in the size of adsorbent molecules participating in such interactions. The nanomaterials can offer a relatively larger surface area for adsorption, yielding higher decontamination efficiency than their bulk counterparts. This process of separating the contaminants with nanoscale adsorbents is termed nanoadsorption. Nanoparticles are favored over other adsorbents due to their unique features, such as numerous sorption sites, porosity, increased specific surface area, surface functionalities, low-temperature modification, little intraparticle diffusion distance, and enhanced capabilities for ion binding (Singh et al. 2018). In addition, other physicochemical properties like dimensions, shape, chemical constitution, physicochemical stability, crystal structure and surface traits like roughness, energy, and area also affect the efficiency and properties of nanosorbent materials. The reactiveness of nanomaterials could be enhanced by reducing the size further, thereby improving the surface area-to-volume ratio. The nanoparticle's toxicity is strongly influenced by the surface charge because it controls various characteristics of nanomaterials, including its colloidal behavior, selective

Table 8.2 Summary of various water decontamination techniques

Technique	Description	Advantages	Limitations
Adsorption	Formation of a molecular or atomic film by accumulation of a gas or liquid solute on the adsorbent surface	• Removes most pollutants • High efficiency, cost-effective • Easy operation	• Requires regeneration • Removal of adsorbents from the treated water is cumbersome
Chemical precipitation	Precipitation of metal ions by the addition of coagulants (organic polymers) like lime, alum, etc	• Cost-effective • Can remove most metals • Easy to operate	• Sludge formation • Additional cost for sludge disposal
Coagulation/ flocculation	Addition of a coagulant to the water results in the formation of small aggregates called "flocs"	• High efficiency • Require limited investment	• High operational cost • Formation of sludge
Electrodialysis	An electrical potential between two electrodes causes the separation of cations and anions, leading to the formation of cells of concentrated and dilute salts	• High recovery rate • Limited pretreatment is required	• Membrane fouling occurs • High operational cost • Energy consumption
Ion exchange	The ions held by electrostatic forces get exchanged with the metal ions	• Faster removal • Highly effective • Materials are regenerated	• Too much expensive • Complete removal of ions is not possible
Reverse osmosis	The separation of metal ions caused by the dissolved solids when external pressure becomes greater than the internal osmotic pressure	• Environment friendly • Good quality water	• Removal of minerals • Time-consuming • High cost
Ultrafiltration	Passage of fluid through a semipermeable membrane, whereas the suspended solids retain on the other side of the membrane	• Possible to remove a variety of pollutants • Possible to regenerate • Produces the highest quality water	• Ineffective against inorganic pollutants

adsorption, integrity of blood–brain barrier, binding of plasma protein, and trans-membrane permeability. Moreover, the crystalline structure, composition, surface coating, and surface roughness also play a critical role in determining the toxicity of nanoparticles.

Based on above-said physicochemical characteristics, a wide variety of nanomaterials have been synthesized recently, such as carbon nanotubes, metal oxide nanoparticles, polymeric nanoparticles, and nanowires. The physicochemical properties of nanomaterials may also be influenced by intrinsic compositions, inherent surface properties, external functional groups attached to nanosorbents, and sizes. The reason for such extraordinary properties and behavior of nanosorbents was explained in terms of the nature of active sites and their arrangement on the surface of these materials (Neyaz et al., 2014). Nanoparticles alone are readily oxidized by atmospheric oxygen leading to the formation of aggregates in aqueous systems. Therefore, it is essential to do surface modification of these nanoparticles to stabilize them and subsequently employ them as nanoadsorbents. To eradicate heavy metal pollutants, nanomaterials were recently surface-modified to enhance their properties, like efficiency, stability, and adsorption capacity. The performance of nanoadsorbents was successfully improved by modifying their surface using various methods.

8.3 Nanoadsorbents Used for Wastewater Remediation

With the advent of new technological advancements, a wide range of nanomaterials have been synthesized that find applications in various fields and treatments. Some of these include nanobots, nanoelectronics, nanofertilizers, nanotubes, nanoparticles, nanowires, quantum dots, etc. Nanomaterials are effectively employed in water/wastewater remediation to eliminate heavy metals and other toxic pollutants because of their extraordinary properties, the foremost of which is high absorption capacity. The subsequent subsections summarize several general and newly found nanomaterials in wastewater remediation, such as carbon nanotubes (CNTs) or carbon-based nanoadsorbents, metal-based nanoadsorbents, polymer-based nanoadsorbants, and zeolites.

8.3.1 Carbon Nanotubes

Carbon nanotubes (CNTs) are macromolecules with cylindrical geometry containing carbon atoms organized as a hexagonal lattice in the separations of the tubes with ends capped via the support of a semi-fullerene-like structure (Iijima 1991). Their classification is primarily based on the carbon atom's hybridization in the layers of CNT. As a result, CNTs can exist as single-walled carbon nanotubes (SWCNTs) and multi-walled carbon nanotubes (MWCNTs). The diameter of SWCNT is typically close to 1 nm, whereas its length can be several thousand times the diameter. When rolled into a seamless cylinder, a single layer of graphite (graphene layer) may form SWCNTs. The MWCNTs can be viewed as a queue of compressed SWCNTs with variable diameters. The SWCNTs possess diameters in the 0.3–3 nm range, whereas the diameters for MWCNTs can reach up to 100 nm.

Various methods, like arc discharge, chemical vapor deposition (CVD), laser ablation, and pyrolysis, can be used to synthesize CNTs. The carbon nanotubes have several superior characteristics, such as high adsorption capability, large surface area, and tunable surface chemistry, that make them effective in treating heavy metals in wastewater. CNTs are unusal because they have multiple distinctive features, such as high adsorption capacity, morphology, high permeability, and physicochemical properties. The electron donor–acceptor (EDA) interactions are considered as the primary adsorption mechanism in CNTs. It is possible to modulate the surface affinity of CNTs to a broader spectrum of contaminants in the remediation of wastewater. The adsorption capability of CNTs toward heavy metals can be significantly improved by introducing functional groups, such as –COOH, $-NH_2$, –OH, onto the surface of CNTs through oxidation of CNTs using acids such as HCl, H_2O_2, HNO_3, H_2SO_4, $KMnO_4$, and NaOCl (Kumar et al. 2014a, b).

CNTs have also been reported to display antimicrobial properties and induce oxidative stress in bacteria, causing the removal of cell wall (Das et al. 2018). Carbon nanotubes functionalized with silver nanoparticles showed a remarkable ability to inactivate microorganisms. Despite the multiple advantages of CNTs, there are some areas for improvement in using CNTs. Firstly, the cost associated with CNTs is relatively high and hinders their usage at the commercial scale. Still, much work is needed to make them cost-effective. Moreover, the isolation of CNTs from the water after adsorption is a challenging task. This further adds to the cost associated with this remediation technique in addition to causing secondary pollution. Consequently, the toxicological investigation of CNTs is also in high demand.

8.3.2 Graphene Nanomaterials

Graphene is yet another noteworthy nanomaterial involving carbon. It is the first two-dimensional atomic crystal used for decontaminating heavy metals from water. Due to its remarkable properties like elasticity, electrical conductivity, mechanical strength, stiffness, and thermal conductivity, it finds extensive applications in many fields. Two kinds of graphene-based nanomaterials, graphene oxide (GO) and reduced graphene oxide (RGO) are frequently employed for the remediation of wastewater contaminated with heavy metals. Oxidation of graphene yields GO. It possesses various functional groups (hydroxyl, carboxyl, epoxide, and carbonyl) containing oxygen, driving it a fair candidate for removing heavy metals (Gao et al. 2011). RGO, on the other hand, results from the reduction of GO and generally contains higher defects compared to pristine graphene. It is possible to readily modify RGO with functional groups, such as –OH and –COOH.

The logic behind graphene-based nanomaterials for heavy metal decontamination lies in their ability to offer extensive specific surface areas and other outstanding characteristics, such as ample functional groups, highly hydrophilic attributes, and high negative charge density. Additionally, several studies have been carried out on graphene-based nanocomposites for eliminating heavy metals from water. However, separating GO from the aqueous solution is problematic as GO is well dispersed in water. A study by Arshad et al. (2018) revealed the synthesis of a novel graphene-modified absorbent that can tackle this problem. Moreover, studies performed on the reusability of the adsorbent showed that Pb(II) removal efficiency remained at

75–80% even after five operation cycles. However, much of the research on graphene-based nanomaterials is still immature to be implemented practically for wastewater remediation multiple pollutants.

8.3.3 Polymer-Based Nanoadsorbents

Traditional adsorbents suffer some limitations, such as deficiency in specificity, low recyclability, and less adsorption capacity (Siddiqui and Chaudhry 2017; Burakov et al. 2018). Accordingly, multiple organic–inorganic hybrid polymers with more robust adsorption capability, better thermal resilience, and higher recyclability have been synthesized to resolve the problems of traditional adsorbents (Lofrano et al. 2016). Polymer-based nanoadsorbent offers a large specific surface area and permeable structure. Furthermore, functional groups attached to the surface lead to an enhanced binding capability toward heavy metal ions such as arsenic, cadmium, lead, and zinc from wastewater and organic dyes (Lofrano et al. 2016; Baruah et al. 2019).

Various kinds of polymer-based nanoadsorbents have been synthesized based on the material used. Dendrimers are another category of organic polymer-based nanoadsorbents containing highly branched and star-shaped macromolecules with nanoscale dimensions.

8.3.4 Hematite (Fe_2O_3) Nanoparticles

Studies showed that Pb^{2+} has the highest affinity toward hematite than other heavy metal ions such as Zn^{2+}, Cd^{2+}, and Cu^{2+} (Shipley et al. 2013). For an initial Pb contamination level of 500 $\mu g\ L^{-1}$, 100% decontamination efficiency was achieved using hematite nanoparticles with the particle size of 37 nm with different adsorbent concentrations of 0.05, 0.1, to 0.5 $g\ L^{-1}$ at pH 8.0 and 120 min of contact time. The higher adsorbent concentration increased the adsorption rate due to increased metal adsorbing sites.

8.3.5 ZnO Nanoparticles

Hua et al. (2012) reported that zinc oxide (ZnO) nanoparticles possess high adsorption capability for copper metal ions but are equally effective in removing lead ions. Some other studies suggest that ZnO nanoparticles result in better remediation of wastewater from heavy metal ions than titanium (Mahdavi et al. 2012). The study reported the lead decontamination efficiency of rod-like ZnO nanoparticles having particle size of 25 nm with homogeneous morphology. As per the reported study, optimum operating conditions are pH 6 with 2–3 h of contact time. The sorbent's (ZnO nanoparticles) basic nature (pH 9.2) readily influenced the lead ions possessing positive charge. The optimum concentration of adsorbent was observed to be 2 gL^{-1}. The study demonstrated that electrostatic attraction and chemisorption were the primary mechanisms responsible for the adsorption. An efficiency of 80% was reported, which was promising compared to iron oxide and copper oxide nanoparticles.

8.3.6 Copper Oxide Nanoparticles

Mahmoud et al. (2021) synthesized copper oxide nanoparticles following a green approach and studied their efficiency in eliminating lead, nickel, and cadmium from contaminated water. The study demonstrated that decontamination efficiency depends on the concentration of nanosorbents and the efficiency enhanced with increased doses of nanosorbents. Such a trend is due to the increased number of available binding sites on the surface of nanosorbents. The decontamination efficiency was observed to be highest for Pb^{2+}, followed by Ni^{2+} and the least for Cd^{2+}. The optimum removal efficiency of Pb^{2+}, Ni^{2+}, and Cd^{2+} was observed to be 84, 52.5, and 18%, respectively, obtained at a pH of 6 for wastewater remediation under standard environmental conditions.

8.3.7 Zeolites

Zeolites are aluminosilicate minerals composed of silicon, aluminum, and oxygen, with a framework structure containing pores, also called molecular sieves, where water, alkali, and alkali earth cations may reside. The typical structure of zeolites involves a tetrahedral linkage between the silica and aluminum atoms via shared oxygen atoms. Zeolites may be both naturally occurring and laboratory-synthesized materials. Naturally occurring zeolites are low-cost materials that occur in abundance with remarkable ion exchange and sorption properties. Therefore, they are preferred over other common cation exchange materials, like organic resins, for removing metal cations from wastewater. Zeolites are compatible with current water decontamination methods because it is possible to employ them as fixed adsorbents in the form of pellets and beads.

Since the past decades, zeolites have been widely applied in separation and purification techniques as adsorbents. Multiple analyses have demonstrated high-silica zeolites' efficacy for removing organic micro-pollutants (OMPs) from wastewater/water (Wasielewski et al. 2018). Nano-zeolites having dimensions varying from 10 to 500 nm have exhibited exceptional performance in wastewater treatment because of their high surface area, water stability, low-cost production and, most notably, their affinity with the natural environment (Shepard et al. 2020). Zeolites possess electrostatic pores that can trap nanoparticles such as silver ions and exchange them with other cations. Various materials having nano-silver, including zeolites, were also observed to display antimicrobial behavior to inhibit microbe's growth due to the presence of silver.

8.3.8 Nano-Clay Adsorbents

Nano-clay is another excellent candidate for heavy metal elimination from water due to its non-toxic nature, low cost, high ion-exchange potential, and abundance in nature. Nano-clay finds its origin in the naturally existing clays, primarily composed of fine mineral particles. The primary segment of clay minerals is layered silicates, made up of silicon and oxygen bonds along with some additional elements. They can be seen as 2D single, double, or multilayers placed in a stacked manner, assembled by corner-linked silicate (SiO_4) tetrahedrons (Kennedy 1990). The lack of chemical bonding between different clay geometry layers improves the stacked layers' adsorption capacity, resulting in better adsorption. Nano-clay minerals possess extremely small molecular size, high porosity, and a large surface area, resulting in better reactivity for physical and chemical interactions on the surface of clay minerals.

Recent studies have demonstrated that nano-clay follows a surface sorption mechanism which enables the adsorption of various contaminants from water. Nano-clay offers increased specific surface area and higher sorption capacity that furnishes chemical and structural stability for the adsorption of organic and inorganic pollutants. The adsorption of pollutants occurs due to availability of charge on the surface of nano-clay. A pictorial representation of pollutants adsorption at different locations of the nano-clay is given in Fig. 8.2. The adsorption of pollutant can proceed as surface attachment, edge attachment, inter-laminar spaces, and inter-particle spaces.

8.4 Limitations of Nanoadsorbents

Although nanoadsorbents display remarkable performance for the remediation of water and wastewater owing to their extraordinary physicochemical characteristics, they do have certain limitations. One is their ecotoxicity in the treated water, which can ruin aquatic ecosystems. Another limitation is the huge price and upkeep associated with nanoadsorbents; e.g., CNTs are admiringly efficient as nanoadsorbents though they are pretty costly. Therefore, making cost-effective nanomaterials is a vital aspect to consider when choosing nanoadsorbents. For instance, photocatalysis, another technique used for water decontamination, can maintain its activity by restoring nanoadsorbent materials (Zhu et al. 2019). Some investigations have also conveyed the unwanted impacts of nanomaterials caused by adding substances in water for its purification; e.g., chlorine mixed with water to eliminate pathogens, resulted in the production of cancer-causing by-products (Srivastav et al. 2020). Moreover, the small size of nanoparticles enables them to quickly penetrate the

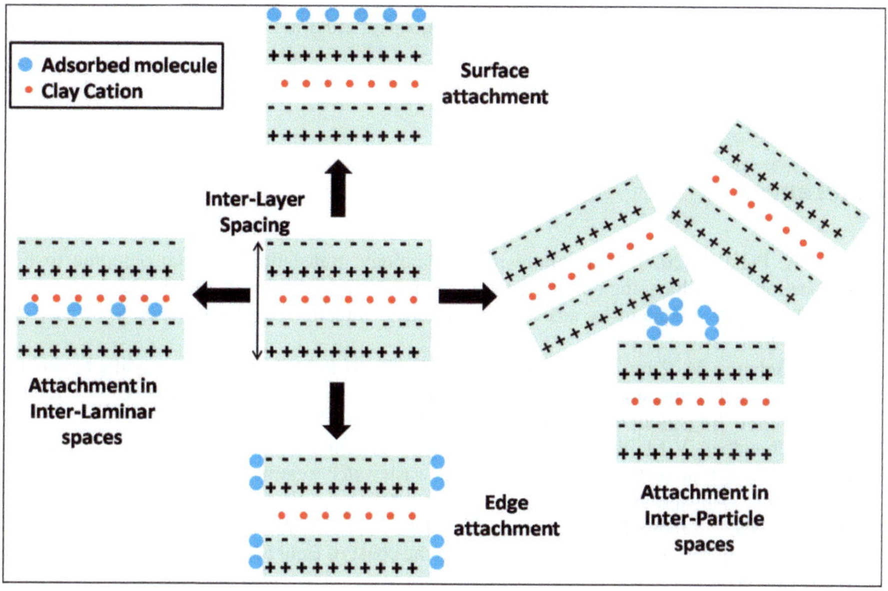

Fig. 8.2 Pollutants adsorption at different locations of the nano-clay

lymph and blood through epithelial and endothelial barriers. From there, they can proceed further into the brain, heart, liver, and other organs and tissues. Addressing such issues will enable nanotechnology to furnish the expected results, such as highly pure water and low-cost remedies.

8.5 Conclusions

To summarize, numerous water decontamination techniques have evolved; however, the adsorption-based remediation technique has emerged as the most powerful and popular technique. It can be applied to efficiently reduce various kinds of inorganic and organic pollutants without significant side effects. On account of their exceptional properties, the nanomaterials are broadly utilized to eradicate heavy metals in water/wastewater. In this context, nanoadsorbent materials are gaining wider recognition in water remediation due to their extraordinary adsorption potential compared to traditional bulk-scale adsorbents. Therefore, nanoadsorbents can be named as next-generation adsorbents beneficial for the sanctification of the environment and controlling water/wastewater pollution. The chapter presents an overview of using nanomaterials as adsorbents either independently or with altered surfaces giving supplementary functional groups for more promising sequestration of lead ions in wastewater.

Acknowledgements The author is thankful to Punjab Agricultural University, Ludhiana, India, for the support during the work.

References

Acharya A, Pal PK (2020) Agriculture nanotechnology: translating research outcome to field applications by influencing environmental sustainability. NanoImpact 19:100232

Ali I (2014) Water treatment by adsorption columns: evaluation at ground level. Sep Purif Rev 43:175–205

Arshad F, Selvaraj M, Zain J, Banat F, Haija MA (2018) Polyethylenimine modified graphene oxide hydrogel composite as an efficient adsorbent for heavy metal ions. Separat Purif Technol 209:870–880. https://doi.org/10.1016/j.seppur.2018.06.035

Baruah A, Chaudhary V, Malik R, Tomer VK (2019) Nanotechnology based solutions for wastewater treatment. In: Nanotechnology in water and wastewater treatment: theory and applications. Elsevier Inc., Amsterdam

Benabdallah NK, Harrache D, Mir A, Guardia MDL, Benhachem FZ (2017) Bioaccumulation of trace metals by red alga *Corallina elongata* in the coast of Beni Saf, west coast, Algeria. Chem Int 3:320–331

Burakov AE, Galunin EV, Burakova IV, Kucherova AE, Agarwal S, Tkachev AG, Gupta VK (2018) Adsorption of heavy metals on conventional and nanostructured materials for wastewater treatment purposes: a review. Ecotoxicol Environ Saf 148:702–712

Das R, Leo BF, Murphy F (2018) The toxic truth about carbon nanotubes in water purification: a perspective view. Nanoscale Res Lett 13:183. https://doi.org/10.1186/s11671-018-2589-z

Engida AM, Chandravanshi BS (2017) Assessment of heavy metals in tobacco of cigarettes commonly sold in Ethiopia. Chem Int 3:212–218

Fazal-ur-Rehman M (2018) Methodological trends in preparation of activated carbon from local sources and their impacts on production: a review. Chem Int 4(2):109–119

Gall JE, Boyd RS, Rajakaruna N (2015) Transfer of heavy metals through terrestrial food webs: a review. Environ Monit Assess 187:201. https://doi.org/10.1007/s10661-015-4436-3

Gao W, Majumder M, Alemany LB, Narayanan TN, Ibarra MA, Pradhan BK, Ajayan PM (2011) Engineered graphite oxide materials for application in water purification. ACS Appl Mater Interfaces 3:1821–1826. https://doi.org/10.1021/am200300u

Hua M, Zhang S, Pan B, Zhang W, Lv L, Zhang Q (2012) Heavy metal removal from water/wastewater by nanosized metal oxides: a review. J Hazard Mater 211:317–331

Iijima S (1991) Helical microtubules of graphitic carbon. Nature 354:56–58

Kamali M, Persson KM, Costa ME, Capela I (2019) Sustainability criteria for assessing nanotechnology applicability in industrial wastewater treatment: current status and future outlook. Environ Int 125:261–276

Keçili R, Büyüktiryaki S, Mustansar Hussain C (2019) Advancement in bioanalytical science through nanotechnology: past, present and future. TrAC Trends Anal Chem 110:259–276

Kennedy BA (1990) Surface mining. In: Society for mining, metallurgy, and exploration, 2nd edn. Port City Press, Pikesville

Kumar R, Khan MA, Haq N (2014a) Application of carbon nanotubes in heavy metals remediation. Crit Rev Environ Sci Technol 44:1000–1035

Kumar S, Ahlawat W, Bhanjana G, Heydarifard S, Nazhad MM, Dilbaghi N (2014b) Nanotechnology-based water treatment strategies. J Nanosci Nanotechnol 14:1838–1858

Lofrano G, Carotenuto M, Libralato G, Domingos RF, Markus A, Dini L, Gautam RK, Baldantoni D, Rossi M, Sharma SK, Chattopadhyaya MC, Giugni M, Meric S (2016) Polymer functionalized nanocomposites for metals removal from water and wastewater: an overview. Water Res 92:22–37

Mahdavi S, Jalali M, Afkhami A (2012) Removal of heavy metals from aqueous solutions using Fe_3O_4, ZnO, and CuO nanoparticles. J Nanopart Res 14:1–18

Mahmoud AED, Al-Qahtani KM, Alflaij SO, Al-Qahtani SF, Alsamhan FA (2021) Green copper oxide nanoparticles for lead, nickel, and cadmium removal from contaminated water. Sci Rep 11(1):12547. https://doi.org/10.1038/s41598-021-91093-7

Manzoor Q, Nadeem R, Iqbal M, Saeed R, Ansari TM (2013) Organic acids pretreatment effect on *Rosa bourbonia* phyto-biomass for removal of Pb(II) and Cu(II) from aqueous media. Bioresour Technol 132:446–452

Mubarak N, Sahu J, Abdullah E, Jayakumar N (2014) Removal of heavy metals from wastewater using carbon nanotubes. Sep Purif Rev 43:311–338

Nazar N, Bibi I, Kamal S, Iqbal M, Nouren S, Jilani K, Umair M, Ata S (2018) Cu nanoparticles synthesis using biological molecule of *P. granatum* seeds extract as reducing and capping agent: growth mechanism and photo-catalytic activity. Int J Biol Macromol 106:1203–1210. https://doi.org/10.1016/j.ijbiomac.2017.08.126

Neyaz N, Siddiqui W, Nair KK (2014) Application of surface functionalized iron oxide nanomaterials as a nanosorbents in extraction of toxic heavy metals from ground water: a review. Int J Environ Sci 4:472–483

Olvera RC, Silva SL, Robles-Belmont E, Lau EZ (2017) Review of nanotechnology value chain for water treatment applications in Mexico. Resour Technol 3:1–11

Ozlem Kocabas-Ataklı Z, Yurum Y (2013) Synthesis and characterization of anatase nanoadsorbent and application in removal of lead, copper and arsenic from water. Chem Eng J 225:625–635

Palit S (2020) Recent advances in the application of nanotechnology in food industry and the vast vision for the future. In: Grumezescu AM, Holban AM (eds) Nanoengineering in the beverage industry. Academic Press, New York, pp 1–34

Pandey S (2017) A comprehensive review on recent developments in bentonite-based materials used as adsorbents for wastewater treatment. J Mol Liq 241:1091–1113. https://doi.org/10.1016/j.molliq.2017.06.115

Puri N, Gupta A, Mishra A (2021) Recent advances on nano-adsorbents and nanomembranes for the remediation of water. J Clean Prod 322:129051

Salam MA (2013) Coating carbon nanotubes with crystalline manganese dioxide nanoparticles and their application for lead ions removal from model and real water. Colloids Surf A 419:69–79

Shahid M, Khalid M, Abbas G, Shahid N, Nadeem M, Aslam M (2015) Heavy metal stress and crop productivity. In: Hakeem K (ed) Crop production and global environmental issues. Springer, Cham

Shanmugalingam A, Murugesan A (2018) Removal of hexavalent chromium by adsorption on microwave assisted activated carbon prepared from stems of *Leucas Aspera*. Z Phys Chem 232:489–506. https://doi.org/10.1515/zpch-2017-0998

Shepard ZJ, Lux EM, Oyanedel-Craver VA (2020) Performance of silver nanoparticle-impregnated ovoid ceramic water filters. Environ Sci Nano 7:1772–1780. https://doi.org/10.1039/D0EN00115E

Shipley HJ, Engates KE, Grover VA (2013) Removal of Pb(II), Cd(II), Cu(II), and Zn(II) by hematite nanoparticles: effect of sorbent concentration, pH, temperature, and exhaustion. Environ Sci Pollut Res 20:1727–1736

Siddiqui SI, Chaudhry SA (2017) Arsenic: toxic effects and remediation. Advanced Materials for Wastewater Treatment 1–28, Scrivener Publishing LLC

Singh NB, Nagpal G, Agrawal S, Rachna S (2018) Water purification by using adsorbents: a review. Environ Technol Innov 11:187–240. https://doi.org/10.1016/j.eti.2018.05.006

Srivastav AL, Patel N, Chaudhary VK (2020) Disinfection by-products in drinking water: occurrence, toxicity and abatement. Environ Pollut 267:115474. https://doi.org/10.1016/j.envpol.2020.115474

Tahoon MA, Siddeeg SM, Salem Alsaiari N, Mnif W, Ben Rebah F (2020) Effective heavy metals removal from water using nanomaterials: a review. Processes 8:645

UNESCO (2023) Definition of freshwater resources. https://web.archive.org/web/20160411064155/, http://webworld.unesco.org/water/ihp/publications/waterway/webpc/definition.html. Accessed 22 Feb 2023

Wasielewski S, Rott E, Minke R, Steinmetz H (2018) Evaluation of different Clinoptilolite Zeolites as adsorbent for Ammonium removal from highly concentrated synthetic wastewater. Water 10:584. https://doi.org/10.3390/w10050584

WHO (2017) Drinking water quality guidelines. World Health Organization, Geneva

Yamamura H, Hashino M, Kubota N (2011) Membrane filtration: principle and applications in water treatment. SEN-I GAKKAISHI 67:3

Zhang W, Xu W, Lan Y, He X, Liu K, Liang Y (2019) Antitumor effect of hyaluronic-acid-modified chitosan nanoparticles loaded with siRNA for targeted therapy for non-small cell lung cancer. Int J Nanomed 14:5287–5301

Zhu Y, Liu X, Hu Y, Wang R, Chen M, Wu J, Wang Y, Kang S, Sun Y, Zhu M (2019) Behavior, remediation effect and toxicity of nanomaterials in water environments. Environ Res 174:54–60. https://doi.org/10.1016/j.envres.2019.04.014

Chapter 9
Microbial Tolerance Strategies Against Lead Toxicity

Saurabh Gupta, Manjot Kaur, Amrit Kaur, Amanpreet Kaur, Ravindra Kumar, Vijay Singh, and Bhairav Prasad

Abstract Lead (Pb) is a toxic element that is not required for any known function in living organisms; hence, it is not considered essential. It does not break down and will persist in the environment for an undetermined amount of time, including in the soil, water, and air. Because PB is the cause of serious health and environmental problems, we need to develop remediation solutions that are both effective and efficient. Microorganisms that are native to Pb-affected places have developed processes that are unique to them to live and even thrive in an environment that is contaminated with lead. Bacteria protect themselves from lead in many different ways, including biosorption, efflux, and the synthesis of metal chelators like siderophores and metallothioneins, the production of exopolysaccharides, extracellular sequestration, and intracellular bioaccumulation. Bacteria also produce metal chelators like siderophores and metallothioneins. One interesting potential alternative for lead contamination removal is the use of microbes. The employment of transgenic bacterial strains that possess metal-binding properties, metal chelating proteins, or higher metal adsorption ability, along with the utilisation of bacterial activity, such as the integration of plant growth-promoting rhizobacteria to enhance Pb resistance, exopolysaccharide and siderophores, and metallothionein-mediated immobilisation, has the potential to effectively accomplish bioremediation and phytoremediation objectives.

S. Gupta (✉) · M. Kaur · A. Kaur · A. Kaur
Department of Microbiology, Mata Gujri College, Fatehgarh Sahib, Punjab 140406, India
e-mail: sau27282@gmail.com

R. Kumar
Department of Agriculture, Mata Gujri College, Fatehgarh Sahib, Punjab 140406, India

V. Singh
Department of Botany, Mata Gujri College, Fatehgarh Sahib, Punjab 140406, India

B. Prasad
Department of Biotechnology, Chandigarh Group of Colleges, Landran, Mohali, Punjab 140307, India

N. Kumar and A. K. Jha (eds.), *Lead Toxicity Mitigation: Sustainable Nexus Approaches*, Environmental Contamination Remediation and Management,
https://doi.org/10.1007/978-3-031-46146-0_9

Keywords Biosorption · Exopolysaccharides · Metallothionein · Phytoremediation

9.1 Introduction

Heavy metals above 4.5 g/cm^3 inhibit biota growth and metabolism (Etesami 2018; Li et al. 2019). Due to its tremendous toxicity, Group 14 element lead (Pb) is the most heinous of these elements (Sorrentino et al. 2018). Mining, smelting, refining, acid industry, battery production, industrial effluent, pesticides, petrol additives, and other 900 + industries use a lead and its compounds (Mohammadi et al. 2008; Karrariet al. 2012; Hu et al. 2019). Pb is non-biodegradable, widespread, and persistent in soil, air, and water, causing serious environmental issues (Carocci et al. 2016). Agriculture using lead-contaminated soil and water raises the danger of lead entering the food chain and harming human health and longevity (Ma et al. 2017; Nonget al. 2019; Gupta 2020). According to Mielke and Reagan (1998), lead-based paint and soil from both anthropogenic and natural sources can cause higher blood lead levels or severe lead poisoning in newborns and adults. However, acute toxicity only occurs after extensive exposure. Chronic Lead (Pb) poisoning begins around 40–60 g/−1 blood levels (Flora et al. 2012). Lead is a mutagenic and teratogenic metal that causes hypertension, baldness, anaemia (it inhibits porphobilinogen synthase and ferrochelatase enzyme activities), dementia, cognitive problems, and death (it is an analogue to calcium, and at low concentrations, Pb selectively blocks voltage-dependent calcium channels; see also Wani et al. 2015; Chen et al. 2015).

Lead (Pb) damages DNA, protein, and lipids and replaces essential metal ions like Zn, Ca, and Fe from enzymes important in microbial development, according to Asmusset al. (2000), Hartwig et al. (2002), and Roanne (1999). Because of this, the US Environmental Protection Agency (EPA) has classified lead as a contaminant of concern (Vilchez et al. 2011), and the WHO advises lead levels in drinking water be fewer than 10 g/L (Watt et al. 2000). Thus, Pb(II) clean-up is essential. Pb(II) remediation methods include activated carbon adsorption, ion exchange, chemical precipitation, reverse osmosis, foam flotation, and others. The low Pb(II) concentration makes these physicochemical methods expensive and inefficient (Gavrilescu 2004; Srivastava and Majumder 2008). Bioremediation of Pb(II)-contaminated wastes is promising due to its high removal capacity, eco-friendliness, and cost-effectiveness. Microbes tolerate and immobilise Pb(II) well (Akar and Tunali 2006; Cho and Kim 2003; Sari and Tuzen 2009; Tuzunet al. 2005). These bacteria reduce Pb(II)'s mobility and biological toxicity through adsorption/chelating, active transport, extracellular precipitation, intracellular absorption, and biomineralisation (Naik and Dubey, 2013). Higher Pb concentrations negatively affect photosynthesis, respiration, food uptake, enzyme activity, water balance, and plasma membrane integrity (Huang et al. 2008; Gupta et al. 2009, 2010; Shahid et al. 2014; Mitra et al. 2020). Most of these negative consequences are caused by Pb-induced oxidative stress, which accelerates oxidation and alters DNA, RNA, protein, and lipid molecules (Chanu and Gupta

2016; Mittler, 2017; Khan et al. 2018). Redox-driven protein function changes affect transcription, phosphorylation, and other cellular signalling processes (Gupta et al., 2013; Schieber and Chandel, 2014; Reczek and Chandel, 2015). Rhizosphere soil is rich in carbon and nitrogen, and plants emit a variety of chemicals that support a diverse microbiome (Compant et al. 2010). To survive, microbes must advance and root adsorb, link to the root surface via cell surface polymers, propagate into the root, infiltrate plant tissues (endophytes), etc. (Wang et al. 2016). Plant-allied microbes influence both symbiotic partners' ability to adapt and thrive in biotic and abiotic environments (Gopalakrishnan et al. 2015). Rhizobacteria, endobacteria, and arbuscular mycorrhizal fungus (PGPR, EBMF) assist plants to adapt to metal stress in metal-loaded soil. The PGPR secretes enzymes, phytohormones such as IAA, CK, and GA, and heavy metal chelating agents like metallophores, organic acids, siderophores, and biosurfactants. These metabolites directly regulate the availability of minerals like nitrogen, phosphorus, potassium, calcium, iron, manganese, copper, and zinc, which affect plants' response to HM stress (Glick 2010; Gupta et al. 2015; Ma 2016). Pb exposure also endangers microorganisms. Microorganisms have evolved ways to grow and operate normally in Pb-contaminated environments. Pb-resistant microorganisms are many, such as *Acidithiobacillus ferrooxidans*, *Citrobacter* spp., *Bacillus pumilus*, and *Pseudomonas aeruginosa*. Thus, Lead (Pb) contamination is a significant environmental issue, and bioremediation using Pb-resistant bacterial strains can be a promising approach to mitigate its impact.

9.2 Sources of Lead

Lead (Pb) is a prevalent naturally occurring constituent of rocks, soils, and particulate matter, typically exhibiting concentrations that span from 2 to 200 parts per million (ppm). According to Gupta et al. (2020), the release of this substance from rocks occurs through natural weathering processes. The Earth's crust typically harbours Pb at an average concentration of approximately 15 mg kg^{-1}. However, it is noteworthy that certain rock categories, such as metamorphic and igneous rocks, exhibit elevated levels of Pb, whereas sedimentary rocks demonstrate relatively lower levels of Pb and related minerals. For instance, potassium and silicate minerals are frequently encountered in sedimentary rocks (Lovering 1969). Granites that contain minerals with radionuclides, such as allanite, uraninite, thorite, monazite, zircon, xenotime, and titanite, have been found to exhibit higher concentrations of Pb (Kushwaha et al. 2018). Further, Sammut et al. (2010) have reported that lead (Pb) may exist in the soil in various forms, including as a free metal, an inorganic complex (with SO_4^{2-}, Cl^-, CO_3^{2-}, or HCO^{3-}), or an organic acid (with humic, amino, or fulvic acids). The dispersion of Pb in soils is governed by various processes such as adsorption, desorption, ion exchange, precipitation, dissolution, aqueous complexion, biological immobilisation, and mobilisation. The processes in question are influenced by various regional factors, such as climate, topography, organic matter, soil pH, soil type, and biota (Wuana and Okieimen 2011; Amundson et al. 2015; Kushwaha et al.

2018). The significant rise in environmental Pb concentrations, exceeding 1000 times the previous levels, can be attributed to human activities and the excessive utilisation of materials containing Pb, as stated by Naik and Dubey (2013). Lead-based paints, smelting processes, glazed ceramics, and solders are extensively utilised across diverse applications. The presence of tetraethyl Pb in gasoline has been found to have a notable impact on the release of Pb into the environment. Additionally, the use of Pb-containing pesticides, such as lead arsenate, in agricultural and horticultural practices has been identified as a source of contamination for both soil and natural water sources. Lead, being a non-degradable contaminant, can endure in the environment and accumulate through processes such as deposition, leaching, and erosion in various ecological systems such as soils, aquatic systems, and sediments (Brännvall et al. 1999). According to Schock (1990), the primary cause of lead contamination in drinking water may be the process of lead leaching from household plumbing systems that incorporate lead-containing PVC in the form of pipes, solder, fittings, or service connections. Lead may also be released through the detachment of lead carbonate deposits from lead pipes and the accumulation of iron silt, which has a propensity to accumulate lead, in aged galvanised plumbing systems (Schock 1990). According to Cosgrove et al. (1989), it has been demonstrated that soldered connections on copper pipework can release significant amounts of lead (210–390 g/l), which can result in intoxication among children. Magill et al. (2012) have reported that 210Pb, a beta emitter with a half-life of 22 years, is a naturally occurring radioactive isotope of lead that is derived from 238U and is ubiquitously present in nature, particularly in regions with high concentrations of uranium in the soil and rocks. The concentration of 210 Pb is influenced by both natural and anthropogenic sources, as reported by Cook et al. (2018). Nevertheless, there exists a notable regional escalation in 210Pb concentration owing to the actions of naturally occurring radioactive materials (NORMs). Some of the activities that have been identified as potential sources of environmental contamination include oil and gas extraction, the generation of waste from phosphoric acid production, the utilisation of geothermal energy and fracking practices, as well as uranium mining and milling operations. The stable isotopes of lead exhibit a significant association with natural radioactivity. It is noteworthy that 204Pb is the only primordial isotope, while 206Pb, 207Pb, and 208Pb represent the terminal points of the three natural decay series, namely 238U, 235U, and 232U, respectively (Magill et al. 2012). Consequently, regions that contain minerals with high concentrations of U or Th exhibit modifications in both lead concentrations and isotopic compositions. Understanding the distribution of metals necessitates a comprehension of the distribution of mineral rocks and processing materials (Cook et al. 2018).

9.3 Microbes Play an Important Role in the Migration and Transformation of Pb (II)

The significance of microbes in the cycling of metals and formation of minerals in surface environments of the Earth has been widely acknowledged by scholars (Lowenstam 1981; Reith et al. 2009). In recent years, there has been a growing body of research on the effects of microbes on the transportation and alteration of Pb. The research conducted by Smeaton et al. (2009) has demonstrated that *Shewanella putrefaciens* CN32 has the ability to produce intracellular crystalline precipitates that exhibit an improved Pb and P mineral phase. Furthermore, this microorganism can facilitate the dissolution of Pb-jarosite through dissimilatory Fe reduction, particularly under anaerobic circumneutral conditions. The study conducted by Park and Bolan (2013) has shown that *Enterobacter cloacae* can enhance the solubilisation of P and facilitate the immobilisation of Pb, leading to decreased mobility, bioavailability, and toxicity of Pb. This effect is achieved through the release of P from insoluble P additions. According to Govarthanan et al. (2013), Bacillus sp. KK1 has the ability to transform Pb $(NO_3)_2$ into PbS and $PbSiO_3$. Additionally, the presence of KK1 bacteria in soil leads to an augmentation of the carbonate fraction, as the bacteria are capable of inducing calcite precipitation of Pb. The bacterium Bacillus sp. MN3-4, which resides within plant tissues, has been found to possess a remarkable ability to resist lead toxicity. This is attributed to its capacity to sequester lead extracellularly and accumulate it intracellularly via the binding of Pb ions. Additionally, the bacterium has been observed to promote the growth of Alnus firma plants by producing plant hormones such as indoleacetic acid and siderophores. These findings suggest that bacteria may have a potential role in mitigating phytotoxicity and enhancing Pb accumulation in A. firma plants, as reported by Shin et al. (2012). In general, microorganisms have a significant impact on the migration and transformation of Pb(II) in natural systems, and studying the mechanism of Pb bioremediation is of great practical importance.

9.4 Lead-Resistant Microbes as a Means to Improve Phytoextraction

Bacteria that are associated with plants have a significant impact on the adaptation of the host to a dynamic environment. The production of plant growth-promoting factors, including siderophores, IAA, and 1-aminocyclopropane-1-carboxylate (ACC) deaminase, by rhizospheric microbial communities has been found to enhance plant tolerance against heavy metal toxicity (Miransari 2011; Lugtenberg and Kamilova 2009). Current studies indicate that plant endophytes could play a significant role in the process of heavy metal phytoremediation. This is supported by various research works, such as those conducted by Waranusantigul et al. (2011), Deng et al. (2011), He et al. (2009), Jiang et al. (2008), Newman and

Reynolds (2005), Sheng et al. (2008), and Weyens et al. (2009). According to Jiang et al. (2008), the *Burkholderia* sp. strain of bacteria, which is resistant to heavy metals, has been found to notably enhance the biomass of maize and tomato plants. Additionally, the strain has been observed to increase the levels of Pb and Cd in plant tissue by a range of 38–192%. In a separate instance, the use of multiple diazotrophs (nitrogen-fixing bacteria) for co-inoculation with maize has been observed to yield superior results in achieving Pb phytoextraction and promoting plant growth and biomass as compared to the use of single culture inoculations. According to Arias et al. (2010), the utilisation of arbuscular mycorrhiza (AM) fungi has been found to enhance the tolerance and accumulation of Pb in mesquite plants. The research conducted by Zaier et al. (2010) has demonstrated that *Sesuvium portulacastrum*, a halophytic plant species, exhibits a greater potential for phytoextraction of Pb compared to *Brassica juncea*, which is commonly employed for Pb phytoextraction. Additionally, Gisbert et al. (2003) have observed that the seedlings of genetically modified plants grown in mining soils with elevated levels of Pb (1572 ppm) accumulate twice the concentration of this heavy metal as compared to the non-modified plants. The present study reports that the application of 400 and 800 mg Pb kg1 in soil resulted in an increase in Pb uptake from 18 to 46% in the cotton (*Gossypium hirsutum*) rhizosphere. This was achieved through the isolation of *Bacillus edaphicus* (strain NBT) from the rhizosphere of cotton, which was cultured with *B. juncea* L. (Indian mustard). According to Sheng et al. (2008a), strain NBT can synthesise IAA, siderophores, and 1-aminocyclopropane-1-carboxylate deaminase. The present study reports a significant increase in $CaCl_2$-extractable Pb in soil contaminated with heavy metals, after the inoculation of two Cd-resistant bacteria, namely Pseudomonas sp. RJ10 and Bacillus sp. RJ16. The observed increase ranged between 69 and 93%. According to He et al. (2009), there was a range of 73–79% increase in Pb concentrations in the above-ground tissues of inoculated plants when compared to the control group that was not inoculated. *Pseudomonas fluorescens* G10 and *Microbacterium* sp. G16 was identified as facultative endophytes that were extracted from rape plants cultivated in areas with high levels of heavy metal contamination. These endophytes exhibited the ability to establish themselves within the rhizosphere soils and internal plant tissues. According to Sheng et al. (2008b), the presence of endophytic bacteria in the rhizosphere soils and plant tissues of rape may facilitate the growth of the plant and the uptake of Pb through various mechanisms such as the production of IAA, siderophore, or ACC deaminase or by solubilising Pb in soils. Two endophytic bacteria (Q2BJ2 and Q2BG1) that are resistant to Pb and exhibit increased ACC deaminase activity were extracted from *Commelina communis* plants that were cultivated on mine tailings containing lead and zinc. The study conducted by Zhang et al. (2011) involved rape plants that were grown in quartz sand with a Pb concentration of 100 mg kg1. The results showed that inoculation with the isolates led to a significant increase in above-ground Pb contents, ranging from 58 to 62% in the inoculated rape plants as compared to the uninoculated control. The fungal endophytes, specifically Mucor sp. CBRF59, were extracted from the roots of rape plants that were cultivated in soil that was heavily contaminated with metals. According to Deng et al. (2011), the presence of active mycelia from CBRF59 has been observed to enhance the availability of soil Pb by 77% in Pb-contaminated soils. The study reports that a strain of Bacillus bacteria obtained from the cadmium hyperaccumulator Solanum nigrum exhibited a high degree of selectivity in absorbing 80% of Pb^{2+} ions when exposed to an initial concentration of 10 mgL1. The observed hormetic response appears to be a consequence of an atypically heightened ATPase activity, which is intended to furnish energy to facilitate EBL14 and mitigate the deleterious effects of heavy metal toxicity by exporting cations, as noted by Guo et al. (2010). According to Guo et al. (2010), subcellular fractionation analyses indicate that the majority of Pb uptake by cells, approximately 77%, is localised in membrane fractions. Conversely, the cytoplasm and cell wall contain only 7.4% and 16.1%, respectively, of the total Pb uptake. The colonisation of Vetiver grass with arbuscular mycorrhizal (AM) fungi, specifically Glomus mosseae, has been observed to result in an increase in chlorophyll content and a decrease in GSH levels. This suggests that the plant is better equipped to withstand stress induced by metal exposure. According to Punamiya et al. (2010), the treatment with a concentration of 1200 mg kg1 resulted in the highest observed level of Pb in shoot tissue, which was measured at 2179 mg kg1 when adjusted for dry weight. Additionally, the inoculation of plants with G. mosseae was found to increase the translocation of Pb to the shoot. The findings of hydroponic research indicate that the introduction of *Ochrobactrum intermedium* BN-3 has a notable impact on both the biomass and Pb accumulation of *Eucalyptus camaldulensis*, surpassing the results of the control group, which was not inoculated. The findings indicate that natural rhizospheric bacteria present on the root surfaces of E. camaldulensis play a significant role in enhancing both plant growth and Pb accumulation, as per the study conducted by Waranusantigul et al. in 2011. The successful remediation of Pb through phytoremediation may be achieved by utilising bacteria with the ability to dissolve Pb and enhance the growth of plants in soils that have been contaminated.

9.5 Mechanisms of Pb (II) Resistance and Bioremediation by Microbes

Numerous microbial species exhibit the ability to mitigate the toxicity of lead. The microbes utilise a variety of mechanisms to achieve this, such as extracellular binding, intracellular sequestration, active transport, exclusion by forming a permeable barrier (such as exopolysaccharide), precipitation, and biomineralisation, as described in previous studies (Naik and Dubey 2013; Park et al. 2011b). Figure 9.1, depicting the various techniques for immobilising microorganisms, is available for reference at the provided location.

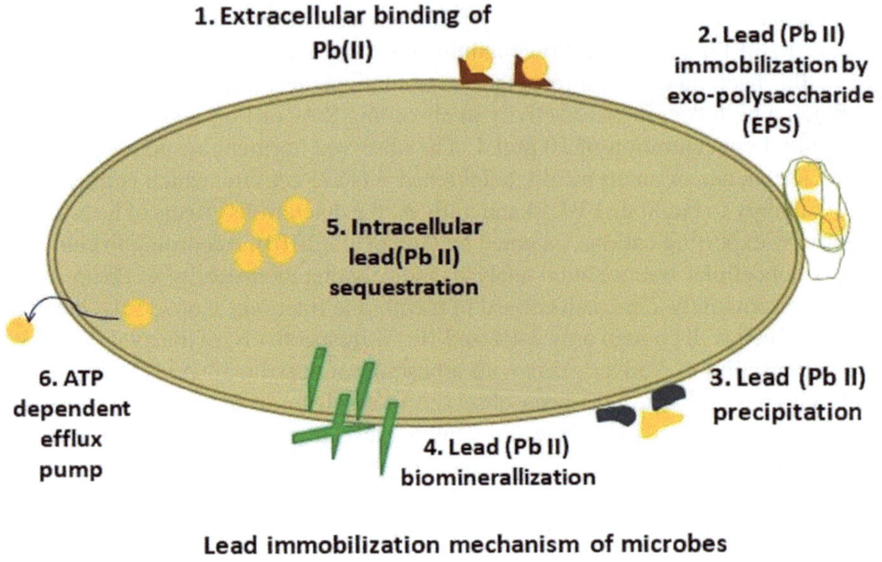

Fig. 9.1 Lead immobilisation mechanism of microbes [1] Extracellular binding of Pb (II) [2] Pb (II) immobilisation by exopolysaccharides [3] Pb (II) precipitation [4] Pb (II) biomineralization [5] Intracellular Pb (II) sequestration [6] ATP—dependent efflux pump

9.5.1 Pb (II) Resistance Mechanisms in Microbes

Microbes, including bacteria and fungi, have developed various mechanisms to resist the toxic effects of Pb(II) ions. It's worth noting that different microbial species have evolved distinct ways of resisting Pb(II) toxicity, and that some may use more than one strategy at once. Hynninen et al. (2009) found that efflux systems help reduce cation buildup in both bacterial and eukaryotic organisms. Bacteria generally have a wide variety of resistance mechanisms to heavy metal toxicity. Metal-binding proteins, or metallothioneins (MTs) (Naik and Dubey 2013; Naik et al. 2012a, b; Roane 1999; Sharma et al. 2006), are synthesised by microorganisms to improve the immobilisation of heavy metals and reduce their toxicity. MTs have been found in a variety of bacterial species, including *Pseudomonas aeruginosa* (BmtA), *P. putida* (BmtA), and *Synechococcus* PCC 7942 (SmtA) (Blindauer et al. 2002; Turner et al. 1996). In addition, Roane (1999) reported the detection of a protein in *Bacillus megaterium* that shares properties with an MT protein and displays binding affinity towards Pb(II). It has been discovered that certain lead-resistant bacterial strains carry the genes for P-type ATPase and phosphatase. To carry ions and tiny organic molecules across the cytoplasmic membrane, cells use transmembrane transporters called P-type ATPases, which use ATP as an energy source (Apell 2004; Jaroslawiecka and Piotrowska-Seget 2014; Naik and Dubey 2013). Potentially useful lead-resistance mechanisms have been proposed for the bacterium *Cupriavidus metallidurans* DN440. Hynninen et al. (2009) reported that PbrB generates inorganic phosphate, which is then used to store

Pb(II) as a phosphate salt, while PbrA is in charge of transporting Pb(II) out of the cytoplasm. In addition, the exclusion of lead via the development of a permeable barrier or the active transfer of Pb^{2+} may potentially account for microbial resistance to lead.

9.5.2 Pb(II) Bioremediation Mechanisms of Microbes

The process of microbial biosorption or complexation of Pb(II) is a widely recognised mechanism for bioremediation, as reported by Naik and Dubey (2013) and Templeton et al. (2003). The metal ions can be immobilised by the biomass through various mechanisms such as surface functional groups, ion exchange, or electrostatic adsorption. This leads to a decrease in the toxicity of the metal ions. According to Bueno et al. (2008), the process of Pb(II) uptake by *Rhodococcus opacus* involves the participation of COOH, OH, and NH groups, as indicated by the results of FT-IR and SEM–EDS analyses. The mechanism responsible for this uptake may be attributed to either complexation or electrostatic adsorption. The present study reports the isolation of two distinct bacterial strains, namely *Pseudomonas marginalis* and *Bacillus megaterium*, from soils contaminated with metals. The study further demonstrates that these bacterial strains exhibit resistance to lead and are capable of immobilising and sequestering lead through extracellular exclusion and intracellular accumulation mechanisms, respectively. This finding is consistent with the previous study conducted by Roane in 1999. According to Edris et al. (2014), the pseudo-first-order kinetic model applies to the biosorption of Pb(II) onto dead cells of Chlorella vulgaris biomass. The maximum adsorption capacity for lead has been determined to be 178.5 mg/g. The utilisation of fungi, specifically *Saccharomyces cerevisiae* and *Rhizopus arrhizus*, is a common practice in the process of Pb(II) biosorption. The efficacy of their adsorption capacity is subject to significant influence from factors such as pH, initial metal concentration, and temperature, as evidenced by studies conducted by Ozer and Ozer (2003) and Sag et al. (1995). The Freundlich isotherm model is also applicable to the biosorption process, wherein a metal-tolerant actinomycete (strain 723) has demonstrated a remarkable capacity to eliminate lead ions (116 mg/g) from wastewater.

9.5.3 Bioprecipitation and Biomineralisation

Numerous studies have indicated that bacteria can facilitate the production of insoluble Pb compounds with sulphide, hydroxide, and carbonate as a means of mitigating the toxicity of Pb (II) (Park et al. 2011b). Furthermore, certain microorganisms can facilitate or accelerate the transformation of Pb (II) from its mobile state into highly stable and insoluble minerals. Thus, the microorganisms exhibit potential for on-site remediation of areas contaminated with Pb (II). It is imperative to take into account the pathogenic properties and sanitary significance of microorganisms in practical scenarios. *Bacillus thuringiensis* 016 has been identified as a viable and secure biosorbent for the conversion of Pb(II) into lead-containing minerals on a large scale. This is due to its non-pathogenic properties towards humans, plants, and animals, as reported by Chen et al. (2015). According to Govarthanan et al. (2013), the *Bacillus* sp. KK1 isolate has demonstrated a high level of efficacy in the process of mineralising active Pb ions into inactive Pb. The resulting precipitation has been identified as PbS and $PbSiO_3$. The microorganisms *Klebsiella aerogenes*, *Bacillus iodinium* GP13, and *Bacillus pumilus* S3 have been observed to cause precipitation of Pb(II) into PbS, as reported by Aiking et al. (1985) and De et al. (2008). Additionally, Bacillus cereus 12-2 has been found to facilitate the transformation of Pb(II) into $Ca_{2.5}Pb_{7.5}(OH)_2(PO_4)_6$ through an enzyme-mediated process, as noted by Chen et al. (2016). In addition, studies by Mire et al. (2004) and Naik et al. (2013) have shown that *Vibrio harveyi* and *Providenciaalcalifaciens* strain 2EA can facilitate the transfer of Pb(II) into $Pb_9(PO_4)_6$. Similarly, Enterobacter sp. and *Staphylococcus aureus* have been found to possess the capacity to precipitate Pb(II) into $Pb_3(PO_4)_2$, as reported by Levinson et al. (1996) and Park et al. (2011a). Pyromorphite, which is a Pb mineral found in the Earth's crust, has a high level of stability. Its chemical formula is $Pb_5(PO_4)_3X$, where X can be F, Cl, Br, or OH. According to various studies (Debela et al. 2010; Miretzky and Cirelli 2008; Rhee et al. 2014), pyromorphite has a low solubility, with a Ksp value ranging from 1071.6 to 1084.4. According to previous research conducted by Henry et al. (2015), Scheckel et al. (2013), and Scheckel and Ryan (2002), the synthesis of pyromorphite has the potential to reduce the mobility and bioavailability of soil Pb. This is because the stability of pyromorphite is positively correlated with ageing time, resulting in long-term protection of treated soils. Rhee et al. (2014) have provided evidence that various fungi, such as *Botryotinia fuckeliana*, *Penicillium* sp., *Aureobasidum pullulans*, and *Phomamacrostoma*, are capable of facilitating the conversion of Pb(II) to $Pb_5(PO_4)_3Cl$.

9.5.4 Detoxification of Lead by Microbes

Biosorption, precipitation, efflux, leaky chelating compounds, extracellular sequestration, intracellular bioaccumulation, compartmentalisation, and extracellular sequestration are some of the defensive strategies that bacteria have taken to minimise the harmful effects of lead (Fig. 9.2). Extracellular sequestration is another strategy that bacteria have developed to lessen the harmful effects of lead.

9.5.5 Biosorption

The cell wall of microorganisms serves as the primary defence mechanism, as various macromolecules within the cell wall participate in metal binding, as noted by Fomina

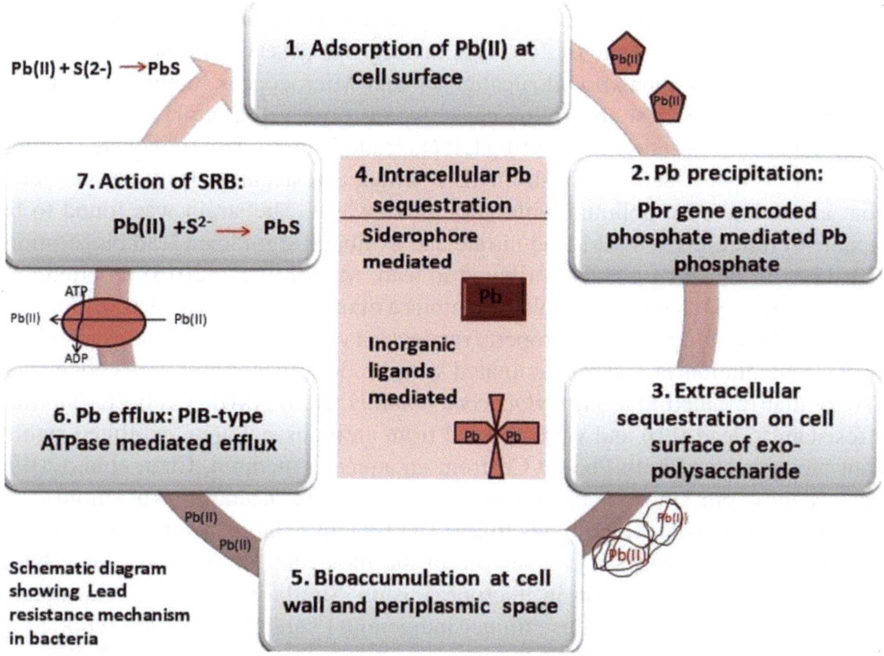

Fig. 9.2 Schematic diagram showing lead-resistance mechanism in bacteria

and Gadd (2014). The cellular walls of gramme-positive and gramme-negative bacteria contain negative ions that can bind with metal cations. This binding process, known as biosorption, can impede the entry of heavy metals into the cells by immobilising the metal on the outer surface. This information is supported by Kushwaha et al. (2018) and Tiquia-Arashiro (2018). The biosorption process involves the participation of various negatively charged groups present on the microbial surface, such as carboxyl, hydroxyl, sulfonate, amine, sulfhydryl, and phosphonate. The absorption of metals by microorganisms is facilitated by diverse mechanisms such as ion exchange, chelation, adsorption, and diffusion through cell membranes and walls, as reported by Chang et al. (1997). Chang et al. (1997) reported the biosorption of lead, cadmium, and copper in an aqueous solution by Pseudomonas aeruginosa in a prior investigation. Gabr et al. (2008) conducted research that revealed that the immobilisation of lead (Pb_{2+}) by a strain of Pseudomonas aeruginosa, specifically ASU6a, is facilitated by the presence of carbonyl, phosphate, hydroxyl, and amino groups located on the cell surface.

Cabuk et al. (2006) reported that the cell surface of "*Bacillus* sp. ATS-2" contains carboxyl and hydroxyl groups, as well as amide and sulphonamide groups, which can bind with Pb_{2+}. Cabuk et al. (2007) reported a comparable outcome in the immobilisation of Pb_{2+} through the utilisation of Saccharomyces cerevisiae. The adsorption capacity of Pb_{2+} on the cell surface is strongly regulated by the primary concentration of lead and the pH. The research conducted by Leung et al. (2000) indicates that the process of metal biosorption is positively influenced by an increase in pH levels, specifically within the range of 2–6. This finding was observed through experimentation with Pseudomonas pseudoalcaligenes and Micrococcus luteus. The authors reported that the highest absorption capacity was observed at a pH of 5, when the initial metal concentration was 100 mg L1. Naik and Dubey (2011) conducted a study that revealed that P. aeruginosa strain 4EA, which is resistant to lead, was obtained from areas that were polluted with automotive waste. The strain was found to be capable of tolerating 0.8 mM lead nitrate and exhibited significant lead biosorption, with 11% of the cell's weight attributed to lead. Munoz et al. (2015) reported that Klebsiella sp. 3S1, a bacterial isolate, exhibits a maximum capacity of 140.19 mg g1 dry cell at 25 °C and pH 5. This property renders it a viable and economical biosorbent for the remediation of soil contaminated with Pb. Rahman et al. (2019) conducted a study that revealed that *Staphylococcus hominis* strain AMB-2, a bacterium that is resistant to lead (Pb) and was obtained from an industrial area, exhibited significant biosorption of both Pb and Cd from an aqueous medium. Chen et al. (2019) reported the expression of a flagellin protein with a high concentration of amino acids containing carbonyls, which binds to Pb. The protein exhibits a substantial affinity for Pb-binding and plays a crucial role in the molecular mechanisms underlying Pb tolerance and biosorption in microorganisms. Similarly, selenium (Se) uptake and immobilisation of Bacillus strains (indigenous) studies have been reported in seleniferous soil of Punjab, India (Gupta et al. 2022). The aforementioned attributes exhibited by bacterial strains that are tolerant to lead indicate their potential suitability as efficacious agents for the process of bioremediation, specifically in the context of soil or water that has been contaminated with lead.

9.5.6 Extracellular Sequestration

Metal-tolerant bacteria can biosorb using exopolysaccharide (EPS)-mediated immobilisation on the outer surface (Bramhacharie et al. 2007; Arashiro 2018). External plastid envelope (EPS) complexes located outside the plastid envelope are composed of polyanionic polymers with high molecular weight, including proteins, polysaccharides, humic acids, lipids, uronic acid, glycoproteins, and nucleic acids. These polymers have the ability to bind with cationic metals with different levels of specificity and affinity, ultimately resulting in the formation of a slime layer surrounding the cell (Sheng et al. 2010). Because of its carboxyl, phosphate, amide, and H-bonding groups, EPS may be able to sequester heavy metals (Vimalnath and Subramanian 2018). Roanne (1999), Salehizadeh (2003), Raungsomboonet (2007), Amoozegaret (2012), Shamim (2013), Kalita and Joshi (2017), and others have discovered lead-resistant bacteria binding Pb_{2+} to EPS. EPS from Azotobacter chroococcum XU1 adsorbs 22.38 mg/g of Pbat at neutral pH (Rasulovet al., 2013). Acinetobacter junii L. EPS adsorbs Pb (Kushwaha et al. 2017). Klebsiellamichiganensis R19, Providenciarettgeri L2, Raoultellaplanticola R3, and Serratia sp. L30 can absorb lead from mono- and mixed-metal solutions (Bowman et al. 2018). EPS may help bioremediate lead in biofilms, activated sludge, and granules by fixing cations (Gupta and Diwan, 2017). Metals interact with biofilm EPS depending on pH, metal concentration, organic matter, biomass and protein ratio, metal ion radius, electronegativity, cell surface sheath shape, temperature, and ph (Nong et al. 2019). Gabr et al. (2008) found that Pb_{2+} biosorption was better than Ni_{2+} biosorption because the Pb ion has a greater radius (1.20 vs. 0.9). Metal electronegativity affects EPS biosorption (Ping et al., 2007). High-electronegativity metals prefer negative organic groups. Bacterial capsules may reduce biosorption. Capsulated K. pneumoniae adsorbs heavy metals faster than decapsulated C. freundii, regardless of pH and metal concentration (Al-Garni 2005). Pb biosorption is optimal at 4–7 (Pardo et al. 2003; Al-Garni 2005). When developing biosorbents, this is significant because EPSs gradually become negatively charged when pH rises, allowing metal cations to be absorbed via carboxyl, phosphate, and amino groups (Pardo et al. 2003). A recent study found that EPSs, citrate, and oxalate can dramatically alter the surface properties of an acidic ultisol (reddish-yellow acid soil) to efficiently mobilise Pb (Nkohetal 2019). Thus, EPSs are recommended for remediating Pb-contaminated environmental samples.

9.5.7 Siderophores as Chelators

In reaction to the heavy metal stress, microorganisms and monocotyledonous plants both release fungi, which are members of the principal class of chelators known as siderophores. Siderophores are low-molecular-weight chelating agents (0.2–2.0 kDa), and they can mediate iron transport into the cell. In addition, they are very effective at binding heavy metals that are found outside the cell. According to Rajkumar et al. (2010), the iron receptor protein that is located on the cell's outer membrane is responsible for recognising Fesiderophore complexes (three siderophores linked with one Fe^{3+}) and directing their translocation into the cell. The presence of lone pair electrons of oxygen and nitrogen atoms on a functional group (such as the hydroxamate group) readily improves the functional group's metal chelation capabilities. This makes it possible for the functional group to form a stable polycyclic bond with the metals, which is what gives siderophores their chelation power. According to Braud et al. (2010), the chelating capacities of the siderophores found in *P. aeruginosa* allow the bacteria to be resistant to a wide variety of toxic metals. These metals include Co, Cu, Ni, Pb, and Zn. Additional evidence for the increased Pb tolerance via siderophore secretion in Pseudomonas aeruginosa 4EA (Naik and Dubey 2011) and P. putida KNP9 (Tripathi et al., 2005) has been obtained. According to Burde et al. (2000), an increase in the production of a mutant strain of the plant-growth-promoting bacteria Kluyveraascorbata SUD165 was connected to lower toxicity of Ni, Pb, and Zn following siderophore treatment.

9.5.8 Efflux of Metals

Numerous bacterial species employ a proficient mechanism for eliminating metals from their cellular structures. Various microorganisms, such as *E. coli* (Fan and Rosen 2002), *Enterococcus hirae* (Bissig et al. 2001), *Candida albicans* (Riggle and Kumamoto 2000), and *Pseudomonas putida* (Adaikkalam and Swarup 2002), have been documented to utilise this mechanism as a defence against the detrimental effects of metals on their cellular processes. The P1B-type ATPase plays a crucial role in the efflux mechanism, facilitating the transmembrane transportation of ions and small organic molecules. Metals can undergo transportation via either active or passive mechanisms. ABC transporters in the active state and P-type ATPases employ the process of ATP hydrolysis to facilitate the transfer of metal ions. The passive pathway, as opposed to the active pathway, involves the exclusion of metals across the proton gradient through the utilisation of resistance-nodulation cell division (RND) proteins, as stated by Ruggerone et al. (2013). The RND proteins have been identified as the primary metal extruders, responsible for exporting superfluous cations, while the CDF (cation diffusion facilitators) have been classified as the secondary metal extruders. The protein classification known as P-type ATPases (Nies 2003) is composed of two subcategories: the "Cu^{1+}-translocating ATPases," which discharge Cu^{1+} and Ag^{1+}, and the "Zn^{2+}-translocating ATPases," which discharge Zn^{2+}, Cd^{2+}, and Pb^{2+}, respectively (Hou et al. 2001; Rensing et al. 1999). PbtA is a type of ATPase that belongs to the family of ATPases. It is responsible for the efflux of Pb and is found in Achromobacter xylosoxidans A8. Furthermore, Kushwaha et al. (2018) reported that the NreB and CnrT proteins are transporter systems that belong to the CHR protein family. Remediation strategies aimed at reducing metal concentrations in the environment are not conducive to the efflux mechanism of microbial cells due to their ineffectiveness. Transporter and regulatory proteins that possess a significant affinity for Pb have the potential to serve as intermediaries in the advancement of precise and sensitive Pb sensors, as suggested by Nong et al. (2019). The pbrR gene has been established as a crucial genetic element in the development of plasmid-based, whole-cell bacterial biosensors for the rapid detection of lead-contaminated water, as demonstrated in this particular context (Bereza-Malcolm et al. 2016). Lead sensors possess the capability to contribute to lead surveillance and identification. Furthermore, due to their biological compatibility, they can be conveniently incorporated into biological entities.

9.5.9 Intracellular Immobilisation of Lead

Various bacterial species frequently utilise a mechanism that involves the proficient expulsion of metallic substances from their cellular structures. Various microorganisms such as *Candida albicans* (Riggle and Kumamoto 2000), *E. coli* (Fan and Rosen 2002), *Enterococcus hirae* (Bissig et al. 2001), and *Pseudomonas putida*

(Adaikkalam and Swarup 2002) have been reported to utilise this mechanism as a defence mechanism against the detrimental effects of metal accumulation within their cellular structures. The P1B-type ATPase is a pivotal component of the efflux mechanism, enabling the transportation of ions and small organic molecules across the cellular membrane. Metals possess the capacity to undergo displacement through either active or passive mechanisms. The transfer of metal ions is facilitated by the utilisation of ATP hydrolysis in ABC transporters in the active state and P-type ATPases. According to Ruggerone et al. (2013), the passive pathway, in contrast to the active pathway, employs resistance-nodulation cell division (RND) proteins to transport metals against the proton gradient. The RND proteins are recognised as the principal metal efflux pumps owing to their capacity to transport excess cations, whereas the CDF (cation diffusion facilitators) function as the secondary metal efflux pumps. The P-type ATPases, as classified by Nies (2003), can be divided into two distinct subgroups. The first subgroup is the "Cu^{1+}-translocating ATPases," which are responsible for releasing Cu^{1+} and Ag^{1+}. The second subgroup is the "Zn^{2+}-translocating ATPases," which are responsible for releasing Zn^{2+}, Cd^{2+}, and Pb^{2+}. These subgroups have been extensively studied by Hou et al. (2001) and Rensing et al. (1999). PbtA belongs to the ATPase protein family and is accountable for the removal of Pb from cells. The aforementioned statement elucidates that the origin of the subject in question can be traced back to Achromobacter xylosoxidans A8, and its primary function is to act as a Zn^{2+}-translocating ATPase. Kushwaha and colleagues (2018) have reported that the transporter systems NreB and CnrT are members of the CHR protein family. The effectiveness of remediation strategies in reducing metal concentrations in the environment is hindered by their incompatibility with the efflux mechanism of microbial cells. Nong et al. (2019) have suggested that transporter and regulatory proteins, possessing a significant affinity for Pb, could potentially serve as intermediaries in the advancement of accurate and sensitive Pb sensors. Bereza-Malcolm et al. (2016) conducted a study that demonstrated the crucial role of the pbrR gene as a genetic component in the formation of bacterial biosensors that employ plasmids to detect lead concentrations in water. This methodology facilitates the prompt identification of lead pollution. Lead sensors have the potential to aid in the surveillance and identification of lead. Moreover, owing to their biological compatibility, they can be conveniently integrated into living organisms.

9.5.9.1 Role of Multimetal Tolerant Bacteria in Regulating Pb Uptake by Plants

Extracellular enzymes and organic acids secreted by bacteria improve Pb availability in soil by dissolving "inaccessible" forms of heavy metal-bearing minerals (Lors et al. 2004; Drewniake et al. 2017). Bacteria isolated from the naturally contaminated heavy metal zone frequently exhibit multimetal tolerance (Pal et al. 2005). Microorganisms resistant to heavy metals play a major role in regulating plant uptake of these contaminants. Examples of plants that metal-resistant bacteria boost absorption include the tomato plant (*Solanum lycopersicum*) and the rapeseed (*Brassica napus*) (Sarathambal et al. 2017; Sheng et al. 2008). Canola plants treated with PGPRPaenibacillus jamilae HTb8 and Pseudomonas sp. GTa5 showed significantly increased above-ground biomass and Pb absorption compared to uninoculated canola plants, according to research by Zhang et al. (2012). Bacterial isolates from the rhizosphere soil of the Mn hyperaccumulator Polygonum pubescense, named Enterobacter sp. JYX7 and Klebsiella sp. JYX10 by Jing et al. (2014), was shown to increase Cd, Pb, and Zn availability in culture solutions and metal-amended soils and to exhibit high levels of multimetal tolerance. Results from a pot experiment showed that PGPR, JYX7, and JYX10 significantly increased the growth and Cd, Pb, and Zn uptake in Brassica napus by either increasing the bioavailability of those metals in soils or by promoting the synthesis of IAA, siderophore, and ACC deaminase. Under Pb stress, Pteris vittata's root and shoot biomass grew by a factor of 2–3, while multimetal (particularly As, Cd, and Pb)-resistant Pseudomonas sp. strain PG-12 dramatically decreased Pb concentrations in the root and shoot. Hynninen et al. (2009) and Jarosawiecka and Piotrowska-Seget (2014) both suggest that the bacterial efflux transporters PbrA and PbrB are to blame for phosphate-based Pb immobilisation in bacterial cells, rendering it unavailable for plant uptake. Inoculating Sedum plumbizincicolum with the RC6B strain of the multimetal-resistant bacteria Phyllobacteriummyrsinacearum decreased Pb uptake and, hence, Pb buildup in root and shoot, as was also shown by Ma et al. (2013).

9.5.9.2 Multi-mechanisms

The bioremediation of Pb (II) contamination is indeed governed by several mechanisms. These mechanisms work together to facilitate the removal, transformation, or immobilisation of Pb (II) ions from the environment. Prior studies have investigated the mechanism of Pb (II) biosorption by Saccharomyces cerevisiae cells, revealing that the elimination of Pb is influenced by both the complexation of Pb (II) with surface functional groups and ion exchange (Cabuk et al. 2007). Urrutia and Beveridge (1993) have noted that mineral production comprises multiple processes. The interaction of metal ions with bacterial surfaces has been observed to exhibit a distinct progression from sorption to nucleation and precipitation processes, as demonstrated by Warren and Ferris in 1998. The biosorption and complexation of metal ions by microbes are believed to be attributed to the surface functional groups of these ions, such as carboxyl, phosphoryl, and amino groups. The adsorbed metal ions are considered to serve as sites for the nucleation and precipitation of minerals. This notion is supported by previous studies conducted by Schultze-Lam et al. (1996), Templeton et al. (2003), and Warren and Ferris (1998). Simultaneous biosorption and biomineralisation of Pb (II) take place in Burkholderiacepacia, leading to an increased accumulation of Pb through the generation of pyromorphite [$Pb_5(PO_4)_3(OH)$] nanocrystals. However, the precise mechanism underlying the formation of lead phosphate by B. cepacia remains unclear, as noted by Templeton et al. (2003).

9.5.9.3 Integrated Approach for the Bioremediation of Lead

The bacteria's Pb resistance, such as precipitation, extracellular or intracellular sequestration, or modification of Pb's oxidation state (Wang and Chen 2009; Kang et al. 2016), may help bioremediate Pb-contaminated soil or water. Plant growth-promoting rhizobacteria (PGPRs) are being utilised in bioremediation to help plants cope with heavy metals and other abiotic stresses (Lucy et al. 2004; Zhuang et al. 2007) (Fig. 9.3). The indirect pathway stimulates the production of enzymes and metabolites like siderophores and ACC deaminase that help plants grow in the presence of metals, unlike the direct pathway, which involves metal immobilisation and biotransformation (Zaidi et al. 2006). Multiple studies have indicated that PGPR applied to seeds, soil, or leaves boosts biomass output and plant tolerance to Pb-contaminated soil, improving phytoremediation (Zulfiqar et al. 2019). Figure 9.2 shows the integrated Pb bioremediation technique. According to Ma et al. (2013), soil nutrients, pH, plant species, and their microbial flora, all have an impact on the plant–microbe connection, which affects heavy metal uptake by metalliferous soil plants. However, metal-tolerant bacteria associated with hyper accumulators can mobilise or immobilise heavy metals by excreting a range of metabolites, which is essential to phytoremediation. Bacterial siderophores boost plant iron content, metal mobility,

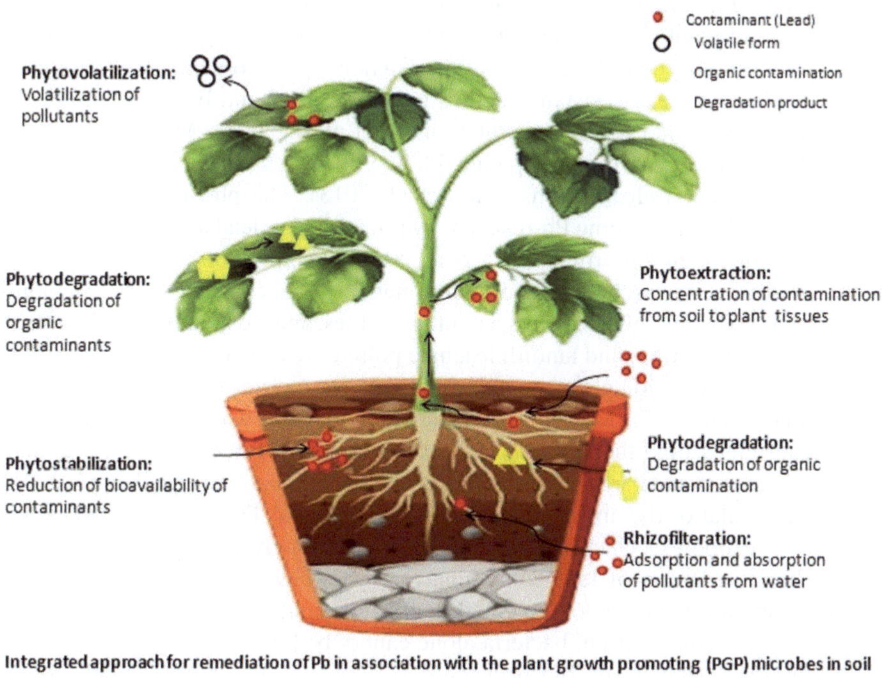

Fig. 9.3 Integrated approach for remediation of Pb in association with the plant-growth-promoting (PGP) microbes in soil

and bioavailability for metal hyper accumulators (Dimkpa et al. 2008; Braud et al. 2009). Pseudomonas aeruginosa-inoculated maize under Cr and Pb stress increased shoot biomass, metal concentration, and metal transfer factor from root to shoot (Braud et al. 2009). Maize's phytoextraction was enhanced by siderophore-mediated metal bioavailability (Braud et al. 2009). Inoculated with a Pb- and Cd-tolerant siderophore-producing strain, Phaseolus vulgaris lowered metal uptake (Tripathi et al. 2005). Siderophore-producing PGPR has potential for Pb bioremediation due to rhizosphere microorganisms' metal mobilisation, plant species, Pb content, soil chemistry, and specific siderophores. To improve phytoremediation with this method, one must understand siderophore-producing bacteria's reactivity to Pbinmultimetal polluted areas. Bacteria that solubilise phosphorus to promote plant growth (Azotobacter, Bacillus, Beijerinckia, Burkholderia, Enterobacter, Erwinia, Flavobacterium, Microbacterium, Pseudomonas, Rhizobium, and Serratia sp.) are promising agents for phytoremediation of heavy metals because they make metals more bioavailable in the soil. Pb^{2+}, the more reactive form of soil lead, forms less soluble complexes with cations including Cl-, $CO_3{}^{2-}$, $SO_4{}^{2-}$, and $PO_{4-}{}^{3}$ and organic ligands such as humic and fulvic acids, EDTA, and amino acids (Ahmad 2019). Phosphate-stabilising bacteria help plants absorb and remove insoluble Pb compounds. Organic acids and H^+ acidify soil, releasing phosphorus from insoluble mineral complexes (Kong and Glick 2017). The plant-growth-promoting endophytic bacterium Rahnella sp. JN6, which solubilises phosphate, increased Cd, Pb, and Zn absorption in aerial and root tissue and plant development in canola plants (He 2013). Precipitating heavy metals into insoluble complexes reduces their toxicity. Microorganisms precipitating lead are unknown. Lead-resistant Bacillus iodinium GP13 and Bacillus pumilus S3 detected lead sulphide (PbS) precipitation (Deet et al. 2008). Mire et al. (2004) reported that Vibrio harveyi precipitates Pb as $Pb_9\ (PO_4)_6$. Same effect with lead-resistant Providentia alcalifaciens (Naik et al. 2013). The phosphate solubilising bacterium *E. cloacae* become Pb-resistant by immobilising lead as the insoluble lead phosphate mineral pyromorphite (Park et al. 2011). Precipitation removes lead from polluted soil effectively, cheaply, and environmentally (Naik and Dubey 2013). Industrial discharge (battery plates, paints, ceramics, cables, and weapons), recycling facilities, car manufacturing, and landfill leachate pollute wastewater with lead. Biofilm-mediated bioremediation cleans metals and treats wastewater safely and effectively (Kalita and Joshi 2017). Microbial EPS is widely used in wastewater bioremediation because they bind metallicions (Pal and Paul 2008). Bacillus sp. ATS-2 binds Pb^{2+} via hydroxyl and carboxyl groups, amide, and sulfonamide (Abuk et al. 2006), and EPS flocculation dynamics affect heavy metal removal. Pseudomonas isolate W6 produces an exopolysaccharide that binds lead more strongly, making it a good option for the bioremediation of lead-contaminated wastewater (Kalita and Joshi 2017). Due to temperature, pH, redox potential, nutritional condition, moisture, and heavy metal chemical composition, bacteria alone cannot be bioremediated (Shukla et al. 2013). Inorganic fertilisers, biosurfactants, bulking agents, compost, and biochar can solve the strain's low competitiveness and high metal content (Wiszniewska et al. 2016). The most promising Pb bioremediation method may be PGPR-mediated siderophores. Siderophores increase phytoextraction, which reduces heavy metals in the food chain (Gaonkar and Borkar 2017). Siderophores show how bacteria react to metals. In *Pseudomonas aeruginosa* strain 4EA, Pb promotes siderophore synthesis, while Cd and Co limit it (Naik and Dubey 2011). Siderophore production increases Pb mobility and plant absorption. Siderophores can bind and solubilise Cu, Ni, Zn, Pb, Cd, and actinides in multimetal-contaminated environments (Schalk et al. 2011). Besides stability, ligand efficiency limits siderophore binding to a metal (Hernlem et al. 1999).

9.6 The Strategy for Effective Pb(II) Immobilisation

The enhancement of Pb(II)-binding proteins, specifically MTs, is a crucial requirement for effective immobilisation of Pb(II) and can be achieved through the genetic modification of microbes. The utilisation of genetically engineered microorganisms is a viable approach for addressing lead contamination in sites that exhibit substantial levels of this toxic metal. This is because molecular techniques have been shown to enhance the precipitation of Pb (III), as demonstrated in studies conducted by Naik and Dubey (2013) and Park et al. (2011b). Furthermore, it has been demonstrated by Chen et al. (2003) that genetic engineering can enhance the production of metal-binding proteins in bacteria, leading to a 33% increase in Pb(II) sorption. The augmented ability of bacteria to biosorb heavy metals is believed to be due to the presence of cysteine-containing transport proteins that are associated with the cell membrane, as suggested by Chang and Hong (1994). Phosphate solubilizing bacteria (PSB) are frequently utilised to facilitate the dissolution of phosphorus, which has prompted numerous research endeavours focused on the use of PSB to assist in the immobilisation of Pb. The study conducted by Park et al. (2010) demonstrated the efficacy of two types of phosphosulfate-reducing bacteria (PSB), namely Pantoea sp. and Enterobacter sp., in immobilising Pb (II) present in soil contaminated with lead from shooting ranges. The PSB was able to increase the solubility of phosphorus from sources that were previously insoluble by producing organic acids. The removal of Pb (II) has been investigated using biomass immobilised on host matrices such as sodium alginate, silica, gelatin, polyacrylamide, polysulfone, and polyvinyl alcohol. These matrices are effective hosts for immobilising or encapsulating cells. Manasi et al. (2014) reported that immobilisation of the Halomonas BVR 1 strain in sodium alginate results in enhanced efficiency of lead removal. This phenomenon is attributed to a physicochemical interaction between the Pb^{2+} and the sodium alginate beads that have been immobilised with the microbe. Furthermore, the immobilised beads can be regenerated using a 0.1 M HCl solution. Likewise, the elimination of Pb(II) (CI) is facilitated through the immobilisation of fungal biomass (*Aspergillus niger*) within a polysulfone matrix and Bacillus sp. within a silica matrix. Genetic engineering can be employed to augment microbes to synthesise specialised Pb(II)-binding proteins, including MTs, that are crucial for the effective immobilisation of Pb(II). The utilisation of genetically engineered microorganisms is a viable approach for addressing lead contamination in sites that exhibit substantial levels of this toxic metal. This is because molecular techniques have been shown to enhance the precipitation of Pb (III), as demonstrated by previous studies conducted by Naik and Dubey (2013) and Park et al. (2011b). Furthermore, it has been demonstrated by Chen et al. (2003) that genetic engineering can enhance the production of metal-binding proteins in bacteria, leading to a 33% increase in Pb(II) sorption. The augmented capacity of bacteria to biosorb heavy metals is believed to be due to the presence of cysteine-containing transport proteins that are associated with the cell membrane, as suggested by Chang and Hong (1994). Phosphate solubilizing bacteria (PSB) are frequently utilised to facilitate the dissolution of phosphorus, resulting in numerous research endeavours focused on PSB-mediated lead immobilisation. The study conducted by Park et al. (2010) demonstrated the efficacy of two types of phosphosulfate-reducing bacteria (PSB), namely Pantoea sp. and Enterobacter sp., in the immobilisation of Pb(II) in soil contaminated with Pb. This was achieved through the enhancement of phosphorus solubility, which was made possible by the production of organic acids. The removal of Pb(II) has been investigated using biomass immobilised on host matrices such as sodium alginate, silica, gelatin, polyacrylamide, polysulfone, and polyvinyl alcohol. These matrices are effective hosts for immobilising or encapsulating cells. Manasi et al. (2014) reported that immobilising the Halomonas BVR 1 strain in sodium alginate results in a higher efficiency of lead removal. This is attributed to a physicochemical interaction between the Pb^{2+} and the sodium alginate beads that have been immobilised with the microbe. Furthermore, the microbe-immobilised sodium alginate beads can be regenerated using a 0.1 M HCl solution. The immobilisation of Bacillus sp. in a silica matrix (Cabuk et al. 2006) and Aspergillus niger biomass in a polysulfone matrix (Cabuk et al. 2006) has been found to facilitate the removal of Pb(II) (Abuk et al. 2006; Kapoor and Viraraghavan 1998).

9.7 Transgenic Approach for Lead Bioremediation

The use of genetic engineering has made it feasible to generate microorganisms with the necessary features, such as tolerance to metal stress, overexpression of metal-chelating proteins and peptides, and the ability to accumulate metals to boost bioremediation (Tiquia-Arashiro 2018). This has made it possible to develop microorganisms with the qualities necessary to increase bioremediation. For example, Wei et al. (2014) built *E. coli* with a lead-specific binding protein (PBrR) and the promoter region for PbrR from Cupriavidus metallidurans CH34 inserted into it. They also included the red fluorescence protein (RFP) in their construct. Because of this, the transgenic bacteria were endowed with the ability to selectively adsorb and immobilise lead in a solution that contained a variety of different heavy metal ions. This method, as stated by Kuroda and Ueda (2010), lessens the load of toxic metals that are carried inside the cells, which results in a greater adsorption capacity. Almaguer-Cant et al. (2011) found that the efficacy of metal biosorption in transgenic E. coli that expressed the mouse metallothionein gene (pMt-Thio) was greatly improved. This enhancement was observed for the ions with the charge of Pb^{2+} as well as

Cd^{2+}. In a similar vein, it was discovered that genetically engineered E. coli BL21 (DE3) containing the metallothionein gene CgMT from Corynebacterium glutamicum has a much increased capacity for absorbing Pb^{2+} and Zn^{+2} when compared to non-engineered *E. coli* BL21 (DE3) (Jafarian and Ghaffari 2017). According to Das et al. (2016), a new bacterial strain can be engineered to have specific metal-binding properties (either by synthesising metal chelating proteins or by having higher metal adsorption ability) by introducing a single gene or operon and/or modifying the gene sequences of already-existing genes using molecular techniques. This can be done either by modifying the gene sequences of already-existing genes or by introducing a new gene. Developing a novel bacterial strain to have certain metal-binding capabilities is one way to achieve this goal. Although there has been a lot of interest in the use of transgenic bacteria for the removal of lead due to their effectiveness, low cost, and environmentally friendly nature, this technology has not yet been deployed outside of the laboratory setting. Instead, you will most frequently come across its application within the confines of a managed setting.

9.8 Future Perspectives

Both bioremediation and microbiological tactics have shown some promise in the past few years as potential methods for cleaning polluted ecosystems of heavy metals. The removal of heavy metals from damaged environmental areas could be accomplished through the use of bioremediation as a feasible strategy. For instance, Alishewanella sp. WH16-1 can decrease the bioavailability of Pb and significantly enhance rice biomass in pot experiments of Pb (II)-contaminated paddy soil, since the bacteria can effectively remove Pb(II) by forming Fe–Mn oxide-bound, organic matter-bound, and PbS precipitate, implying its suitable application of Pb bioremediation in Pb-contaminated soil (Zhou et al. 2016). Together with many phosphate additions (three apatites), the bacteria *Alcaligenes piechaudii* and *Pseudomonas putida* have the potential to both decrease the bioavailability of lead and increase the amount of lead that is immobilised in lead-contaminated soil. The increased level of lead immobilisation may be generated by the kinetics of apatite dissolution. In addition, the incorporation of biological apatite into the soil leads to a significant increase in the average rate of oxygen consumption, as well as an increase in the density of microorganisms; both of these results point to the significance of microbial activity in the metabolic process of the soil (Wilson et al. 2006). Biological apatite is a type of calcium phosphate that is naturally occurring in some rocks and minerals. On the other hand, biological applications of large-scale Pb(II) removal are still uncommon. This is because there are various limitations in the employment of microorganisms as a tool for lead bioremediation (Park et al. 2011). Although microorganisms are capable of immobilising lead (II) via a variety of distinct mechanisms, the use of microbial remediation in real-world settings presents several problems that need to be overcome. The expense of microbial treatment, the efficient recycling of lead (II) from microorganisms, and the differing biosorption behaviours between simulated and real polluted sites are only a few of the problems that need to be overcome. Therefore, further studies need to explore the immobilisation and bioremediation behaviour in lead-contaminated environments. In addition, having an in-depth understanding of the process by which microorganisms remove Pb(II) from the environment might provide crucial foundational data that can be used in the development of Pb(II) bioremediation approaches that are both practical and effective. Further, uses of phosphate solubilizing bacteria (Rhizobacteria) may promote growth of many agricultural crops under lead toxicity by reducing Pb bioavailability (Hareem et al. 2023), and can also help in bioremediation of these toxic elements (Riseh et al. 2023).

References

Adaikkalam V, Swarup S (2002) Molecular characterization of an operon, cueAR, encoding a putative P1-type ATPase and a MerR-type regulatory protein involved in copper homeostasis in *Pseudomonas putida.* MicrobiolSgm 148:2857–2867

Ahemad M (2019) Remediation of metalliferous soils through the heavy metal resistant plant growth promoting bacteria: paradigms and prospects

Aiking H, Govers H, van't Riet J (1985) Detoxification of mercury, cadmium and lead in *Klebsiella aerogenes* NCTC 418 growing in continuous culture. Appl Environ Microbiol 50:1262–1267

Akar T, Tunali S (2006) Biosorption characteristics of *Aspergillus flavus* biomass for removal of Pb(II) and Cu(II) ions from an aqueous solution. Bioresour Technol 97:1780–1787

Al-Garni SM (2005) Biosorption of lead by Gram-ve capsulated and noncapsulated bacteria. Water SA 31:345–349

Almaguer-Cantú V, Morales-Ramos LH, Balderas-Rentería I (2011) Biosorption of lead (II) and cadmium (II) using *Escherichia coli* genetically engineered with mice metallothionein I. Water Sci Technol 63:1607–1613

Amoozegar MA, Ghazanfari N, Didari M (2012) Lead and cadmium bioremoval by *Halomonassp.*an exopolysaccharide producing halophilic bacterium. Prog Biol Sci 2:1–11

Amundson R, Berhe AA, Hopmans JW, Olson C, Sztein AE, Sparks DL (2015) Soil and human security in the 21st century. Science 348:1261071–1261076

Apell HJ (2004) How do P-type ATPases transport ions? Bioelectrochemistry 63:149–156

Asmuss M, Mullenders LH, Hartwig A (2000) Interference by toxic metal compounds with isolated zinc finger DNA repair proteins. Toxicol Lett 112–113:227–231

Bereza-Malcolm L, Aracic S, Franks AE (2016) Development and application of a synthetically-derived lead biosensor construct for use in Gram-negative bacteria. Sensors 16:13

Bissig KD, Voegelin TC, Solioz M (2001) Tetrathiomolybdate inhibition of the *Enterococcus hirae*CopB copper ATPase. FEBS Lett 507:367–370

Blindauer CA, Harrison MD, Robinson AK, Parkinson JA, Bowness PW, Sadler PJ, Robinson NJ (2002) Multiple bacteria encode metallothione in sand SmtA-like fingers. Mol Microbiol 45:1421–1432

Bowman N, Patel D, Sanchez A, Xu W, Alsaffar A, Tiquia-Arashiro SM (2018) Lead-resistant bacteria from Saint Clair River sediments and Pb removal in aqueous solutions. Appl Microbiol Biotechnol 102:2391–2398

Bramhachari PV, Kavi Kishor PB, Ramadevi R, Kumar R, Rao BR, Dubey SK (2007) Isolation and characterization of mucous exopolysaccharide produced by *Vibrio furnissii* VB0S3. J Microbiol Biotechnol 17:44–51

Brännvall ML, Bindler R, Renberg I, Emteryd O, Bartnicki J, Billström K (1999) The nedieval metal industry was the cradle of modern largescale atmospheric lead pollution in northern Europe. Environ Sci Technol 33:4391–4395

Braud A, Geoffroy V, Hoegy F, Mislin GLA, Schalk IJ (2010) Presence of the siderophores pyoverdine and pyochelin in the extracellular medium reduces toxic metal accumulation in *Pseudomonas aeruginosa* and increases bacterial metal tolerance. Environ Microbiol Rep 2:419–425

Braud A, Jezequel K, Bazot S, Lebeau T (2009) Enhanced phytoextraction of an agricultural Cr and Pb-contaminated soil by bioaugmentation with siderophore producing bacteria. Chemosphere 74:280–286

Bueno BYM, Torem ML, Molina F, de Mesquita LMS (2008) Biosorption of lead(II), chromium(III) and copper(II) by *R. opacus*: Equilibrium and kinetic studies. Miner Eng 21:65–75

Çabuk A, Akar T, Tunali S, Gedikli S (2007) Biosorption of Pb (II) by industrial strain of *Saccharomyces cerevisiae* immobilized on the biomatrix of cone biomass of Pinus nigra: equilibrium and mechanism analysis. Chem Eng J 131:293–300

Çabuk A, Akar T, Tunali S, Tabak Ö (2006) Biosorption characteristics of *Bacillus sp.* ATS-2 immobilized in silica gel for removal of Pb (II). J Hazard Mater 136:317–323

Carocci A, Catalano A, Lauria G, Sinicropi MS, Genchi G (2016) Lead toxicity, antioxidant defense and environment. Rev Environ Contam Toxicol 238:45–67

Chang JS, Law R, Chang CC (1997) Biosorption of lead, copper and cadmium by biomass of *Pseudomonas aeruginosa* PU21. Water Res 31:1651–1658

Chen B, Fang L, Yan X, Zhang A, Chen P, Luan T, Hu L, Jiang GA (2019) A unique Pb-binding flagellin as an effective remediation tool for Pb contamination in aquatic environment. J Hazard Mater 363:34–40

Chen M, Ma LQ, Singh SP, Cao RX, Melamed R (2003) Field demonstration of in situ immobilization of soil Pb using P amendments. Adv Environ Res 8:93–102

Chen Z, Pan XH, Chen H, Guan X, Lin Z (2016) Biomineralization of Pb(II) into Pb-hydroxyapatite induced by Bacillus cereus 12–2 isolated from Lead-Zinc mine tailings. J Hazard Mater 301:531–537

Chen Z, Pan X, Chen H, Lin Z, Guan X (2015) Investigation of lead(II) uptake by *Bacillusthuringiensis* 016. World J Microbiol Biotechnol 31:1729–1736

Cho DH, Kim EY (2003) Characterization of Pb^{2+} biosorption from aqueous solution by *Rhodotorula glutinis*. Bioprocess Biosyst Eng 25:271–277

Compant S, Clément C, Sessitsch A (2010) Plant growth-promoting bacteria in the rhizo- and endosphere of plants: their role, colonization, mechanisms involved and prospects for utilization. Soil Biol Biochem 42:669–678

Cook NJ, Ehrig KJ, Rollog M, Ciobanu CL, Lane DJ, Schmandt DS, Owen ND, Hamilton T, Grano SR (2018) ^{210}Pb and ^{210}Po in geological and related anthropogenic materials: implications for their mineralogical distribution in base metal ores. Minerals 8:211

Cosgrove E, McNulty P, McNulty F, Brown MJ, Okonski L, Olkowski L, Madigan P, Schmidt J (1989) Childhood lead poisoning: case study traces source to drinking water. J Environ Health 1:346–349

Das S, Dash HR, Chakraborty J (2016) Genetic basis and importance of metal resistant genes in bacteria for bioremediation of contaminated environments with toxic metal pollutants. Appl Microbiol Biotechnol 100:2967–2984

De J, Ramaiah N, Vardanyan L (2008) Detoxification of toxic heavy metals by marine bacteria highly resistant to mercury. Mar Biotechnol 10:471–477

Debela F, Arocena JM, Thring RW, Whitcombe T (2010) Organic acid-induced release of lead from pyromorphite and its relevance to reclamation of Pb-contaminated soils. Chemosphere 80:450–456

Dimkpa CO, Svatoš A, Dabrowska P, Schmidt A, Boland W, Kothe E (2008) Involvement of siderophores in the reduction of metalinduced inhibition of auxin synthesis in Streptomyces *spp.* Chemosphere 74:19–25

Drewniak Ł, Skłodowska A, Manecki M, Bajda T (2017) Solubilization of Pb-bearing apatite Pb5 (PO4)3Cl by bacteria isolated from polluted environment. Chemosphere 171:302–307

Edris G, Alhamed Y, Alzahrani A (2014) Biosorption of cadmium and lead from aqueous solutions by *Chlorella vulgaris* Biomass: equilibrium and kinetic study. Arab J Sci Eng 39:87–93

Etesami H (2018) Bacterial mediated alleviation of heavy metal stress and decreased accumulation of metals in plant tissues: mechanisms and future prospects. Ecotoxicol Environ Saf 147:175–191

Fan B, Rosen BP (2002) Biochemical characterization of CopA, the *Escherichia coli* Cu(I)-translocating P-type ATPase. J Biol Chem 277:46987–46992

Flora G, Gupta D, Tiwari A (2012) Toxicity of lead: a review with recent updates. Interdiscip Toxicol 5:47–58

Fomina M, Gadd GM (2014) Biosorption: current perspectives on concept, definition and application. Bioresour Technol 160:3–14

Gabr RM, Hassan SHA, Shoreit AAM (2008) Biosorption of lead and nickel by living and non-living cells of *Pseudomonas aeruginosa* ASU 6a. Int Biodeterior Biodegrad 62:195–203

Gaonkar T, Borkar S (2017) Applications of siderophore producing marine bacteria in bioremediation of metals and organic compounds. Marine pollution and microbial remediation. Springer, Singapore, pp 177–187

Gavrilescu M (2004) Removal of heavy metals from the environment by biosorption. Eng Life Sci 4:219–232

Glick BR (2010) Using soil bacteria to facilitate phytoremediation. Biotechnol Adv 28:367–374

Gopalakrishnan S, Sathya A, Vijayabharathi R, Varshney RK, Gowda CL, Krishnamurthy L (2015) Plant growth promoting rhizobia: challenges and opportunities. 3 Biotech 5:355–377

Govarthanan M, Lee KJ, Cho M, Kim JS, Kamala-Kannan S, Oh BT (2013) Significance of autochthonous *Bacillus sp.* KK1 on biomineralization of lead in mine tailings. Chemosphere 90:2267–2272

Gupta DK, Chatterjee S, Walther C (2020) Lead in plants and the environment. Springer, U.S.A. Hamer D H 1986 Metallothioneins. Annu Rev Biochem 55:913–951

Gupta DK, Huang HG, Corpas FJ (2013) Lead tolerance in plants: strategies for phytoremediation. Environ Sci Pollut Res 20:2150–2161

Gupta DK, Nicoloso F, Schetinger MRC, Rossato LV, Pereira LB, Castro GY, Srivastava S, Tripathi RD (2009) Antioxidant defence mechanism in hydroponically grown *Zea mays* seedlings under moderate lead stress. J Hazard Mater 172:479–484

Gupta P, Diwan B (2017) Bacterial exopolysaccharide mediated heavy metal removal: a review on biosynthesis, mechanism and remediation strategies. Biotechnol Rep 13:58–71

Gupta S, Bandana B, Kumar A, Singh V (2022) Selenium uptake and immobilization using indigenous Bacillus strain isolated from seleniferous soils of Punjab. Bioremediat J. https://doi.org/10.1080/10889868.2022.2040414

Gupta VK, Nayak A, Agarwal S (2015) Bioadsorbents for remediation of heavy metals: current status and their future prospects. Environ Engg Res 20:1–18

Hareem M, Zakriya M, Waheed A, Meryem SS, Hirad AH, Ansari SA, Ali MAE, Egeh MA (2023) Mitigating the negative effects of lead toxicity on Vigna mungo: the promising role of rhizobacteria. J King Saud Univ-Sci 35(7):102765

Hartwig A, Asmuss M, Ehleben I, Herzer U, Kostelac D, Pelzer A, Schwerdtle T, Burkle A (2002) Interference by toxic metal ions with DNA repair processes and cell cycle control: molecular mechanisms. Environ Health Perspect 110(Suppl 5):797–799

He H, Ye Z, Yang D, Yan J, Xiao L, Zhong T, Yuan M, Cai X, Fang Z, Jing Y (2013) Characterization of endophytic *Rahnella sp.* JN6 from Polygonum pubescens and its potential in promoting growth and Cd, Pb Zn Uptake by Brassica Napus. Chemosphere 90:1960–1965

Henry H, Naujokas MF, Attanayake C, Basta NT, Cheng Z, Hettiarachchi GM, Maddaloni M, Schadt C, Scheckel KG (2015) Bioavailability-based in situ remediation to meet future lead (Pb) standards in urban soils and gardens. Environ Sci Technol 49:8948–8958

Hernlem BJ, Vane LM, Sayles GD (1999) The application of siderophores for metal recovery and waste remediation: examination of correlations for prediction of metal affinities. Water Res 33:951–960

Hou ZJ, Narindrasorasak S, Bhushan B, Sarkar B, Mitra B (2001) Functional analysis of chimeric proteins of the Wilson Cu (I)-ATPase (ATP7B) and ZntA, a Pb(II)/Zn(II)/Cd(II)-ATPase from *Escherichia coli.* J Biol Chem 276:40858–40863
Hu YM, Zhou J, Du BY, Liu HL, Zhang WT, Liang JN (2019) Health risks to local residents from the exposure of heavy metals around the largest copper smelter in China. Ecotoxicol Environ Saf 171:329–336
Huang H, Li TX, Tian S, Gupta DK, Zhang X, Yang XE (2008) Role of EDTA in alleviating lead toxicity in accumulator species of *Sedum alfredii* H. Bioresour Technol 99:6088–6096
Hynninen A, Touzé T, Pitkänen L, Mengin-Lecreulx D, Virta M (2009) An efflux transporter PbrA and a phosphatase PbrB cooperate in a lead-resistance mechanism in bacteria. Mol Microbiol 74:384–394
Jafarian V, Ghaffari F (2017) A unique metallothionein-engineered in *Escherichia coli* for biosorption of lead, zinc, and cadmium; absorption or adsorption? Microbiology 86:73–81
Jarosławiecka A, Piotrowska-Seget Z (2014) Lead resistance in microorganisms. Microbiology 160:12–25
Jing YX, Jun LY, Huai DH, Dan JY, Li X, Ting Z, Ming Y, Xin DC, Shu BL (2014) Characterization of bacteria in the Rhizosphere soils of *Polygonum pubescen*s and their potential in promoting growth and Cd, Pb, Zn uptake by *Brassica napus.* Int J Phytorem 16:321–333
Kalita D, Joshi SR (2017) Study on bioremediation of lead by exopolysaccharide producing metallophilic bacterium isolated from extreme habitat. Biotechnol Rep 16:48–57
Kang CH, Kwon YJ, So JS (2016) Bioremediation of heavy metals by using bacterial mixtures. Ecol Eng 89:64–69
Kapoor A, Viraraghavan T (1998) Removal of heavy metals from aqueous solutions using immobilized fungal biomass in continuous mode. Water Res 32:1968–1977
Karrari P, Mehrpour O, Abdollahi M (2012) A systematic review on status of lead pollution and toxicity in Iran; guidance for preventive measures. DARU J Pharm Sci 20:2
Khan MM, Islam E, Irem S, Akhtar K, Ashraf MY, Iqbal J, Liu D (2018) Pb induced phytotoxicity in para grass (*Brachiariamutica*) and Castorbean (*Ricinus communis* L.): antioxidant and ultrastructural studies. Chemosphere 200:257–265
Kong Z, Glick BR (2017) The role of plant growth-promoting bacteria in metal phytoremediation. Adv Microbial Physiol 71:97–132
Kuroda K, Ueda M (2010) Engineering of microorganisms towards recovery of rare metal ions. Appl Microbiol Biotechnol 87:53–60
Kushwaha A, Hans N, Kumar S, Rani R (2018) A critical review on speciation, mobilization and toxicity of lead in soil microbe plant system and bioremediation strategies. Ecotoxicol Environ Saf 147:1035–1045
Kushwaha A, Rani R, Kumar S, Thomas T, David AA, Ahmed M (2017) A new insight to adsorption and accumulation of high lead concentration by exopolymer and whole cells of lead-resistant bacterium *Acinetobacter junii* L. Pb1 isolated from coal mine dump. Environ Sci Pollut Res 24:10652–10661
Leung WC, Wong MF, Chua H, Lo W, Yu PHF, Leung CK (2000) Removal and recovery of heavy metals by bacteria isolated from activated sludge treating industrial effluents and municipal wastewater. Water Sci Technol 41:233–240
Levinson HS, Mahler I, Blackwelder P, Hood T (1996) Lead resistance and sensitivity in *Staphylococcus aureus.* FEMS Microbiol Lett 145:421–425
Li YP, Wang SL, Nan ZR, Zang F, Sun HL, Zhang Q (2019) Accumulation, fractionation and health risk assessment of fluoride and heavy metals in soil-crop systems in Northwest China. Sci Total Environ 663:307–314
Lors C, Tiffreau C, Laboudigue A (2004) Effects of bacterial activities on the release of heavy metals from contaminated dredged sediments. Chemosphere 56:619–630
Lovering TG (1969) Distribution of minor elements in samples of biotite from igneous rocks. In: Geological survey research
Lowenstam HA (1981) Minerals formed by organisms. Science 211:1126–1131

Lucy M, Reed E, Glick BR (2004) Applications of free living plant growth-promoting rhizobacteria. Antonie Leeuwenhoek 86:1–25

Ma X, Zheng C, Li W, Ai S, Zhang Z, Zhou X, Pang C, Chen H, Zhou K, Tang M, Li L, Wang Y, Li Y, Guo L, Dong H, Yang D (2017) Potential use of cotton for remediating heavy metal-polluted soils in southern China. J Soils Sediments 17:2866–2872

Ma Y, Oliveira RS, Freitas H, Zhang C (2016) Biochemical and molecular mechanisms of plant-microbe-metal interactions: relevance for phytoremediation. Front Plant Sci 7:918

Ma Y, Rajkumar M, Luo Y, Freitas H (2013) Phytoextraction of heavy metal polluted soils using Sedum plumbizincicola inoculated with metal mobilizing *Phyllobacterium myrsinacearum* RC6b. Chemosphere 93:1386–1392

Magill J, Dreher R, Soti Z, Lasche GP (2012) Nucleonica: web based software tools for simulation and analysis. In: 21st international conference nuclear energy for New Europe, Ljubljana, 5–7 Sept

Manasi RV, Rajesh N (2014) Adsorption isotherms, kinetics and thermodynamic studies towards understanding the interaction between a microbe immobilized polysaccharide matrix and lead. Chem Eng J 248:342–351

Mielke HW, Reagan PL (1998) Soil is an important pathway of human lead exposure. Environ Health Perspect 106:217–229

Mire CE, Tourjee JA, O'Brien WF, Ramanujachary KV, Hecht GB (2004) Lead precipitation by *Vibrio harveyi*: evidence for novel quorum-sensing interactions. Appl Environ Microbiol 70:855–864

Mitra A, Chatterjee S, Voronina AV, Walther C, Gupta DK (2020) Lead toxicity in plants: a review. In: Gupta DK, Chatterjee S, Walther C (eds) Lead in plants and the environment. Springer, Cham, pp 99–116

Mittler R (2017) ROS are good. Trends Plant Sci 22:11–19. Mohseni M, Khosravi F, Mohajerani M, Chaichi MJ (2014) Bioremediation activity of Pb (II) resistance *Citrobacter sp* MKH2 isolated from heavy metal contaminated sites in Iran. J Sci IR Iran 25:105–110

Mohammadi S, Mehrparvar A, Aghilinejad M (2008) Appendectomy due to lead poisoning: a case-report. J Occup Med Toxicol 3:23

Muñoz AJ, Espínola F, Moya M, Ruiz E (2015) Biosorption of Pb (II) ions by *Klebsiella* sp. 3S1 isolated from a wastewater treatment plant: kinetics and mechanisms studies. Bio Med Res Inter 2015:719060

Naik MM, Dubey SK (2011) Lead-enhanced siderophore production and alteration in cell morphology in a Pb-resistant *Pseudomonas aeruginosa* strain 4EA. Curr Microbiol 62:409–414

Naik MM, Dubey SK (2013) Lead resistant bacteria: lead resistance mechanisms, their applications in lead bioremediation and biomonitoring. Ecotoxicol Environ Saf 98:1–7

Naik MM, Pandey A, Dubey SK (2012a) *Pseudomonas aeruginosa* strain WI-1 from Mandovi estuary possesses metallothionein to alleviate lead toxicity and promotes plant growth. Ecotoxicol Environ Saf 79:129–133

Naik MM, Shamim K, Dubey SK (2012b) Biological characterization of lead resistant bacteria to explore role of bacterial metallothionein in lead resistance. Curr Sci 103:1–3

Naik MM, Khanolkar D, Dubey SK (2013) Lead-resistant *Providencia alcalifaciens*strain 2 EA bioprecipitates Pb+2 as lead phosphate. Lett Appl Microbiol 56:99–104

Nies DH (2003) Efflux-mediated heavy metal resistance in prokaryotes. FEMS Microbiol Rev 27:313–339

Nkoh JN, Lu HL, Pan XY, Dong G, Kamran MA, Xu RK (2019) Effects of extracellular polymeric substances of *Pseudomonas fluorescens*, citrate, and oxalate on Pb sorption by an acidic Ultisol. Ecotoxicol Environ Saf 171:790–797

Nong Q, Yuan K, Li Z, Chen P, Huang Y, Hu L, Jiang J, Luan T, Chen B (2019) Bacterial resistance to lead: chemical basis and environmental relevance. J Environ Sci 31:46–55

Pal A, Dutta S, Mukherjee PK, Paul AK (2005) Occurrence of heavy metal-resistance in microflora from serpentine soil of Andaman. J Basic Microbiol 45:207–218

Pal A, Paul AK (2008) Microbial extracellular polymeric substances: central elements in heavy metal bioremediation. Indian J Microbiol 48:49–64

Pardo R, Herguedas M, Barrado E, Vega M (2003) Biosorption of cadmium, copper, lead and zinc by inactive biomass of *Pseudomonas putida.* Anal Bioanal Chem 376:26–32

Park J H, Bolan N, Megharaj M, Naidu R (2011) Concomitant rock phosphate dissolution and lead immobilization by phosphate solubilizing bacteria (*Enterobacter sp.*) J Environ Manag 92:1115–1120

Ping XS, Yenpeng TA, Chen JP (2007) Biosorption of heavy metal ions (Pb, Cu, and Cd) from aqueous solutions by the marine alga *Sargassum sp.* in single- and multiple-metal systems. Ind Eng Chem Res 46:2438–2444

Rahman Z, Thomas L, Singh VP (2019) Biosorption of heavy metals by a lead (Pb) resistant bacterium, *Staphylococcus hominis* strain AMB2. J Basic Microbiol 59:477–486

Rajkumar M, Ae N, Prasad MN, Freitas H (2010) Potential of siderophoreproducing bacteria for improving heavy metal phytoextraction. Trends Biotechnol 28:142–149

Rasulov BA, Yili A, Aisa HA (2013) Biosorption of metal ions by exopolysaccharide produced by *Azotobacter chroococcum* XU1. J Environ Prot 4:989–993

Raungsomboon S, Chidthaisong A, Bunnag B, Inthorn D, Harvey NW (2007) Lead (Pb2+) adsorption characteristics and sugar composition of capsular polysaccharides of cyanobacterium *Calothrix marchica.* Songklanakarin J Sci Technol 29:529–541

Reczek CR, Chandel NS (2015) ROS-dependent signal transduction. Curr Opin Cell Biol 33:8–13

Rensing C, Ghosh M, Rosen BP (1999) Families of soft-metal-iontransporting ATPases. J Bacteriol 181:5891–5897

Riggle PJ, Kumamoto CA (2000) Role of a *Candida albicans* P1-type ATPase in resistance to copper and silver ion toxicity. J Bacteriol 182:4899–4905

Riseh RS, Vazvani MG, Hajabdollahi N, Thakur VK (2023) Bioremediation of heavy metals by Rhizobacteria. Appl Biochem Biotechnol 195(8):4689–4711. https://doi.org/10.1007/s12010-022-04177-z

Roanne TM (1999) Lead resistance in two bacterial isolates from heavy metal-contaminated soils. Microb Ecol 37:218–224

Ruggerone P, Murakami S, Pos KM, Vargiu AV (2013) RND efflux pumps: structural information translated into function and inhibition mechanisms. Curr Top Med Chem 13:3079–3310

Salehizadeh H, Shojaosadati SA (2003) Removal of metal ions from aqueous solution by polysaccharide produced from *Bacillus firmus.* Water Res 37:4231–4235

Sammut M, Noack Y, Rose J, Hazemann J, Proux O, DepouxZiebel M, Fiani E (2010) Speciation of Cd and Pb in dust emitted from sinter plant. Chemosphere 78:445–450

Sarathambal C, Khankhane PJ, Gharde Y, Kumar B, Varun M, Arun S (2017) The effect of plant growth-promoting rhizobacteria on the growth, physiology, and Cd uptake of *Arundo donax* L. Int J Phytorem 19:360–370

Schalk IJ, Hannauer M, Braud A (2011) New roles for bacterial siderophores in metal transport and tolerance. Environ Microbiol 13:2844–2854

Schieber M, Chandel NS (2014) ROS function in redox signaling and oxidative stress. Curr Biol 24:R453–R462

Schock MR (1990) Causes of temporal variability of lead in domestic plumbing systems. Environ Monit Assess 15:59–82

Shahid M, Pourrut B, Dumat C, Nadeem M, Aslam M, Pinelli E (2014) Heavy metal-induced reactive oxygen species: phytotoxicity and physicochemical changes in plants. Rev Environ Contam Toxicol 232:1–44

Shamim K, Naik MM, Pandey A, Dubey SK (2013) Isolation and identification of *Aeromonas caviae* strain KS-1 as TBTC- and leadresistant estuarine bacteria. Environ Monit Assess 185:5243–5249

Sheng GP, Yu HQ, Li XY (2010) Extracellular polymeric substances (EPS) of microbial aggregates in biological wastewater treatment systems: a review. Biotechnol Adv 28:882–894

Sheng XF, Xia JJ, Jiang CY, He LY, Qian M (2008) Characterization of heavy metal-resistant endophytic bacteria from rape (*Brassica napus*) roots and their potential in promoting the growth and lead accumulation of rape. Environ Pollut 156:1164–1170

Shukla KP, Sharma S, Singh NK, Singh V, Bisht S, Kumar V (2013) Rhizoremediation: a promising rhizosphere technology. Appl Bioremed: Active Passive Approaches 2:331

Sorrentino MC, Capozzi F, Amitrano C, Giordano S, Arena C, Spagnuolo V (2018) Performance of three cardoon cultivars in an industrial heavy metal-contaminated soil: effects on morphology, cytology and photosynthesis. J Hazard Mater 351:131–213

Tiquia-Arashiro SM (2018) Lead absorption mechanisms in bacteria as strategies for lead bioremediation. Appl Microbiol Biotechnol 102:5437–5444

Tripathi M, Munot HP, Shouche Y, Meyer JM, Goel R (2005) Isolation and functional characterization of siderophore-producing lead- and cadmiumresistant *Pseudomonas putida* KNP9. Curr Microbio l50:233–237

Turner JS, Glands PD, Samson ACR, Robinson NJ (1996) Zn2+-sensing by the cyanobacterial metallothione in repressor SmtB: different motifs mediate metal-induced protein-DNA dissociation. Nucleic Acids Res 24:3714–3721

Vimalnath S, Subramanian S (2018) Studies on the biosorption of Pb(II) ions from aqueous solution using extracellular polymeric substances (EPS) of *Pseudomonas aeruginosa.* Colloid Surf B-Biointerfac 172:60–67

Wang J, Chen C (2009) Biosorbents for heavy metals removal and their future. Biotechnol Adv 27:195–226

Wang W, Zhai Y, Cao L, Tan H, Zhang R (2016) Endophytic bacterial and fungal microbiota in sprouts, roots and stems of rice (*Oryza sativa* L.). Microbiol Res 188:1–8

Wani AL, Ara A, Usman JA (2015) Lead toxicity: a review. Interdiscip Toxicol 8:55–64

Wei W, Liu X, Sun P, Wang X, Zhu H, Hong M, Mao ZW, Zhao J (2014) Simple whole-cell biodetection and bioremediation of heavy metals based on an engineered lead-specific operon. Environ Sci Technol 48:3363–3371

Wiszniewska A, Hanus-Fajerska E, Muszyńska E, Ciarkowska K (2016) Natural organic amendments for improved phytoremediation of polluted soils: a review of recent progress. Pedosphere 26:1–2

Wuana RA, Okieimen FE (2011) Heavy metals in contaminated soils: a review of sources, chemistry, risks and best available strategies for remediation. ISRN *E.coli* 2011:109557

Zaidi S, Usmani S, Singh BR, Musarrat J (2006) Significance of *Bacillus subtilis* strain SJ-101 as a bioinoculant for concurrent plant growth promotion and nickel accumulation in Brassica juncea. Chemosphere 64:991–997

Zhang WH, Huang Z, He LY, Sheng XF (2012) Assessment of bacterial communities and characterization of lead-resistant bacteria in the rhizosphere soils of metal-tolerant Chenopodium ambrosioides grown on lead-zinc mine tailings. Chemosphere 87:1171–1178

Zhang Y, He L, Chen Z, Zhang W, Wang Q, Qian M, Sheng X (2011) Characterization of lead-resistant and ACC deaminase-producing endophytic bacteria and their potential in promoting lead accumulation of rape. J Hazard Mater 186:1720–1725

Zhou GT, Xia X, Wang H, Li L Q, Wang GJ, Zheng SX, Liao SJ (2016) Immobilization of lead by *Alishewanella sp.* WH16-1 in pot experiments of Pb-contaminated paddy soil.Water Air Soil Pollution

Zhuang X, Chen J, Shim H, Bai Z (2007) New advances in plant growthpromoting rhizobacteria for bioremediation. Environ Int 33:406–413

Zulfiqar U, Farooq M, Hussain S, Maqsood M, Hussain M, Ishfaq M, Ahmad M, Anjum MZ (2019) Lead toxicity in plants: impacts and remediation. J Environ Manag 250:10955

Chapter 10
Effect and Responses of Lead Toxicity in Plants

Mamta Rani, Vikas, Rohtas Kumar, Mamta Lathwal, and Ankush Kamboj

Abstract Heavy metal toxicity is becoming a great concern to living organism due to exponential population growth and futile race of development. As one of the latent contaminants, lead (Pb) contributes about 10% of the trace metal(oids) pollution. It can easily absorb and accumulate in different plants parts but have not any beneficial role in cell metabolism. Increase in Pb content in environment is caused by natural as well as human activities such as industrial waste, mining, and irrigation by sewage water, sewage sludge application, chemical fertilizer, and pesticides. Plants such as crops, vegetables, and fruits grown on highly Pb contaminated soil show some toxic symptoms that may retard their growth (vegetative and reproductive), reduction in photosynthetic rate, seed germination rate blackening of roots, decline in quality, and yield of the plants. Lead also affects the activity of various enzymes that play a significant role in metabolic pathways, i.e., catalase, peroxidase, superoxide dismutase, ATP synthase, RuBP carboxylase/oxygenase, APX, AsA, GPX, and ABA. When the contaminated produce consumed by living being, it causes many life-threatening diseases. In this chapter, the uptake and noxious effects of Pb on photosynthetic rate, germination rate, yield, nutrient uptake, accumulation, ultrastructural and oxidative damage, carbon metabolism, and alteration in enzymes activities were reported.

Keywords Lead (Pb) · Heavy metal · Crops · Soil · Human health

10.1 Introduction

Lead is recalcitrant, highly pernicious, and non-degradable heavy metal after arsenic which contributes 0.002% of Earth's crust (Zulfiqar et al. 2019). Anthropogenic activities including smelting and mining of Pb ores, automobiles, activities, etc.,

M. Rani · M. Lathwal
Department of Botany, Panjab University, Chandigarh 160014, India

Vikas (✉) · R. Kumar · A. Kamboj
Department of Soil Science, CCS Haryana Agricultural University, Hisar 125001, India
e-mail: vikastandon38288@gmail.com

N. Kumar and A. K. Jha (eds.), *Lead Toxicity Mitigation: Sustainable Nexus Approaches*, Environmental Contamination Remediation and Management,
https://doi.org/10.1007/978-3-031-46146-0_10

released excessive amounts of Pb into the environment that causes harmful effects on environment and human health and also known as protoplasmic poison (Hadi and Aziz 2015; Lathwal et al. 2023). Pb intake in human should not exceed 25 $\mu g\ kg^{-1}$ of human body per week. In soil, Pb content may occur up to 10 ppm, but for human being, acceptable permissible range is 0.003 and 0.005 mg/l (Agrawal 2009; WHO 2017; Gaur et al. 2021).

The various forms of lead are present in soil which can be differentiated by their bioavailability, mobility, and toxicity. Bioavailability can be defined as the concentration available for absorption by living organisms such as plants, humans, and animals which is strictly related to the metal chemical form. However, in soil, the chemical forms of lead is affected by various processes such as ion exchange; precipitation and dissolution; complexation; absorption and desorption; immobilization and biological mobilization by plants. Lead accumulation contaminates the food by interfering with soil and roots. Absorption of Pb may vary from species to species, but absorbed Pb is mainly accumulated in roots, and only a small fraction is transported to the aerial tissue (Collin et al. 2022). As a result, root vegetables like potato, carrot, radish, and sweet potatoes may contain highest level of Pb (Collin et al. 2022) and leafy vegetables like lettuce and Swiss chard absorbed least Pb content (Aponte et al. 2020). Pb contamination had stimulatory effect on soil enzymatic activities and microbial biomass at low concentration but inhibitory effect at higher level. High Pb content causes increase in membrane permeability, changes in the catalytic activities of various enzymes, decline in content of photosynthetic pigments, disruption of nutrition mineral balance, inhibition on biomass production and plant growth, and alters the genes. In a study, Liu et al. (2009) reported more than 1310 genes were altered in response to Pb treatment in *Arabidopsis*. Moreover, in plants, Pb-induced excess reactive oxygen species (ROS) have been detected (Reddy et al. 2005). Some plants have tolerance to the high level of Pb contamination; they complete their life cycle without showing any stress symptoms and accumulate > 1000 ppm of Pb content in their plant biomass which are known as hyper-accumulators. Over 400 hyper-accumulating plant species from all over the world can accumulate high concentrations of metals from contaminated soils.

10.2 Sources of Lead (Pb)

Pb have various sources as natural processes such as rock weathering, volcanic eruptions, forest fires, and soil-forming or can be originated from anthropogenic processes such as fertilizer applications, industrial waste, smelting, and sewage disposal (Fig. 10.1). Pb is persistent and non-biodegradable in nature and found in ionic form such as Pb (II) in soil at low pH, but at higher pH, it is found in more stable form.

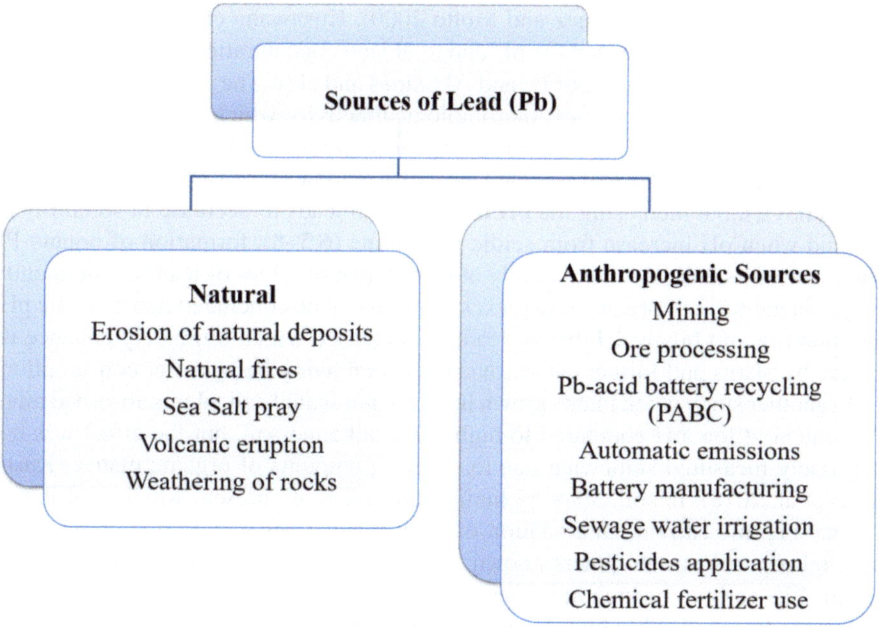

Fig. 10.1 Sources of Pb in environment

10.3 Absorption and Accumulation of Pb in Plants

Absorption of Pb depends upon many factor such as forms of Pb in soil, water and air, soil pH, organic matter, chelating agents, plant type, and concentration of Pb in soil. In soil, Pb acts as a weak Lewis acid which imparts an intense covalent character to majority of ionic bonds it forms. Owing to this intense chemical binding with colloidal and organic materials, it is suggested that only a tiny portion of the lead in soil is soluble, and consecutively available for plant uptake (Kopittke et al. 2008a, b; Punamiya et al. 2010). Solubility of Pb in soils having pH ranged from 5.5 to 7.5 is regulated by carbonate and phosphate precipitates. Pb may exist as free metal ion, organic ligands (e.g., amino acids, humic acids, and fulvic acids) and complexed with inorganic constituents (e.g., HCO^{3-}, $CO_3{}^{2-}$, $SO_4{}^{2-}$, and Cl^-). Anthropogenic sourced lead generally accumulates primarily in the surface layer of soil, and its concentration decreases with depth which makes it challenging to calculate the bioavailable amount of Pb (Cecchi et al. 2008). Availability of Pb to plants primarily depends on soil conditions. The Pb in soil is adsorbed according to Langmuir adsorption isotherm and pegged to soil pH in the range of 3.0–8.5 (Lee et al. 1998). Soil pH significantly helps in retention and uptake of lead from soil. At acidic pH, lead is present as free ionic species while lead is principally found as lead carbonates and phosphates at high pH which are insoluble in nature. When soil pH was below 5.2 ± 0.2, solubility of lead increases and significantly increased its availability to plants (Martınez and Motto 2000). Kushwaha et al. (2018) reported negative correlations between the pH and total lead concentration in all the horizons of soil except horizon deficit of Fe and Al oxides and clay. The alteration of pH may be the indirect consequence of microbiological activity which in turn controls the reduction and oxidation of iron, and manganese. Sauve et al. (1998) found that at highly acidic pH, 30–50% of soluble lead exists in ionic (Pb^{2+}) and in ion pairs form (e.g., $PbSO_4$), but increasing the pH from 3 to 6.5 leads to decrease in solubility of Pb, and when pH increase from acidic to alkaline (6.5–8), formation of organo-Pb complexes was increased, and finally at neutral pH, 80–99% of lead predominantly occurs in the form of organic complexes. Thus, it can be concluded that at acidic pH, the mobility and bioavailability of lead are increased which ultimately enhance its uptake by plants and further cause toxic effects on living beings after consumption.

In another study when plants grown in Pb contaminated soil, plants absorbed more Pb content at low pH compared to high pH or alkaline soil, but the effect was not accurately measured somewhat due to differing amounts of organic matter (Kushwaha et al. 2018). In soil, 80–90% humic substances are present which contributed as total organic carbon. Degradation of plant residue generated three-dimensional interlinked, aromatic polymers known as humic acids. These units have functional groups like carbonyl which have free electron pair and are available for the coordinative binding with Pb which completes the coordinative sphere of Pb. Thus, it is the most significant reaction for Pb adsorption by humus. Binding of Pb is found to be directly proportional to pH and inversely proportional to ionic strength (Xiong et al. 2013). Similar notion was observed by Ahmed et al. (2019) that atmospheric lead remains in the upper 2–5 cm of undisturbed soils with 5% organic matter at $pH \geq 5$ while insoluble organic lead complexes are formed in having organic content at pH 6–8. When the amount of organic matter is low in soil and pH is 6–8, it leads to precipitation of Pb with carbonate and phosphate ions or form complexes with hydrous oxide. These organic lead complexes are solublized and become available for plant uptake at pH 4–6. Pb was mainly accumulated in vascular bundles and humic acid transport the Pb content vascular bundles to shoot and in young stem (Xu et al. 2018).

Ethylenediaminetetraacetic acid (EDTA) as chelating agent is introduced to achieve the remarkable improvement in Pb concentration in shoot. It has great ability to form complex with Pb (Kroschwitz 1995). The large size of Pb particles renders their passage through casparian strip of root endodermis tissue. Formation of complex with EDTA simultaneously reduced their size and increased its solubility (Vassil et al. 1998). Chelation of Pb provides it an escape route from the precipitation with phosphates and carbonates and aids to avoid binding to cell wall in cation exchange process (Jarvis and Leung 2002). Moreover, the transport of solutes from parenchyma cells and vascular cylinder to vessels and tracheids of xylem is intensively selective active-carrier based and thus prevents the transport of charged Pb particles (Raven et al. 1999). However, Pb chelate complex has better chances to get transported through this route.

Labile forms (Pb^{2+}, $PbOH^+$ and $PbCO_3$) constituted a major portion of lead input from the washout of the atmospheric deposits, whereas particulate or bound forms of lead were dominant in urban runoff and ore mining. The abundant forms of Pb in sediments are lead sulfates, lead carbonates, and lead sulfides, and in surface waters, the concentrations of dissolved lead are low. In air, the main Pb compounds present are tetra methyl lead and tetra ethyl lead, which are used as gasoline antiknock additives. These are only present in immediate proximity of anthropogenic sources (Pattee and Pain 2003).

Type of plant or plant species plays very crucial role for the uptake of Pb from air, water, and soil. Plant should have tolerance to Pb contamination, strong and vast root system, and large biomass production. Rani et al. (2023) reported that bamboo has the potential to remediate the heavy metal contaminated soil due to their special characteristics such as large biomass production, ability to uptake large amount of heavy metal, resistant to abiotic and biotic stresses, large CO_2 sequestration, and worldwide distribution. The plants which absorbed Pb from atmosphere have different plant morphology and plant physiology. Barber et al. (2004) opined that plant factors such as leaf surface area, leaf longevity, functional type, and cuticular structure may affect the air-vegetation transfer. Little (1978) and Madany et al. (1990) demonstrated that leaves having rough and hairy surface tend to accumulate remarkably more lead (up to 10 times) compared to smooth leaves. Rao and Dubey (1992) also stressed the role of leaf morphological factors such as trichome density and length, and stomatal index on the efficiency of dust collection by plants. Downey leaves have high affinity for heavy metal from atmosphere (Godzik 1993). Lead particles from atmosphere in the form of lead sulfide (PbS) are caught in tiny folds of leaf and get deposited on the leaf surface and undergo oxidation resulting in the formation of secondary Pb-containing compounds such as PbO, $PbSO_4$, and $PbCO_3$. These secondary Pb particles penetrate inside the leaf through two possible routes. Firstly, Pb-containing nanoparticles noticed in the stomata may enter in the apoplasm as solid compounds and particles were identified beneath the cuticle membrane (Uzu et al. 2010). Secondly, lead formed from the dissolution of primary particles may diffuse through aqueous pores of stomata and cuticles along the hydrophilic pathway, causing necrosis augmented with lead.

10.3.1 Translocation of Lead: Soil to Root

Adsorption/Absorption by plant roots is the major mechanism of transfer of lead from soil to plant system. Roots of plants are actively engaged in the absorption of lead present in solution (Sharma and Dubey 2005; Uzu et al. 2009). The process of lead uptake by plants may occur in two steps. In the first step, lead present in soil is chemically adsorbed on the outer layers of radicular cortex which consists of rhizoderm and collenchyma/parenchyma tissues. This is achieved by binding of lead to polysaccharides of the rhizoderm cell surface and carboxyl groups of mucilage uronic acid as demonstrated by Kushwaha et al. (2018). Kopittke et al. (2007), Ginn et al. (2008), Meyers et al. (2008), Uzu et al. (2009), and Krzesłowska et al. (2009, 2010) also observed the same trend in *Vigna unguiculata*, *Festuca rubra*, *Brassica juncea*, *Lactuca sativa*, *Funaria hygrometrica*, respectively, under Pb which was adsorbed on the surface of root. After being adsorbed to roots surface, lead gains entry into root system through passive absorption along with water.

At molecular level, the determination of exact mechanism of lead migration from soil to plant root system demands further exploration. There is limited detail in literature regarding the mechanisms of Pb transport into root cells; however, a number of possible pathways have been purposed by several scientists. It is well established that lead is non-selectively absorbed (Hirsch et al. 1998). Transportation of adsorbed lead on the root surface has to pass through plasma membrane of root-cell. This can be achieved by the assistance of ionic channels/transporters. The most well-known transport pathway of these cationic channel is Ca-channels which is widely documented and reported by several researchers such as Marshall et al. (1994) and Huang et al. (1996) who isolated right-side-out plasma membrane vesicles from the roots of corn and wheat plants and identified a voltage gated Ca-channel in the plasma membrane of their root cells. Activity of these voltage gated Ca-channels was substantially suppressed by Pb in the plasma membrane of wheat crop either by blocking them or by migrating preferentially with respect to Ca^{2+} through them as established by Monferrán and Wunderlin (2013). This conclusion was concreted with the findings of Wang et al. (2007) which observed a surge in Pb accumulation with depleting Ca content in roots of maize and accredited it to stronger interaction with the transporting proteins such as calmodulin where it can compete with Ca^{2+} and bind to Ca^{2+} binding sites on the transmembrane transporting proteins.

10.4 Accumulated Pb Distribution in Plant Parts

Pb content in different plant organs tend to accumulate in the following order: seeds < inflorescence < leaves < root. The application of Pb to the foliar in *Phaseolus vulgaris* resulted higher content of Pb in roots than the control, indicating that the Pb was absorbed by leaves and translocated within to roots of the plant (Feleafel and Mirdad 2013). In an another study, Nicklow et al. (1983) observed that vegetable crops show different Pb concentration in different parts such as root peel of beets accumulate highest (90 ppm) and lowest in the root (23 ppm) while turnip had the highest Pb concentration in leaf and lowest in root peel. Burzynski (1984) reported that Pb was accumulated in roots (93–96%) predominantly and partially in hypocotyle (4–7%) in cucumber seedlings. Spinach, coriander, cabbage, lettuce, and cauliflower had Pb concentration in the following order: leaves > stem > roots, but in reddish, the lead content followed the different trend, i.e., roots > stem > leaves (Farooq et al. 2008). Root had a significant ability for lead accumulation as compared to stem and shoot. In contrast with the control, Pb accumulation increased by 10.15–40.04, 30.21–185.16, 30.61–97.62 times in shoot, stem, and root (Yongsheng et al. 2011). In *Brassica napus*, Pb was mainly accumulated in roots at flowering and physiological maturity stage (Ferreyroa et al. 2018).

10.4.1 Intracellular Localization of Pb (in Cell Wall, Vacuole, and Cell Membrane)

It has been deduced from the ultrastructural investigations that Pb is mainly amassed in intercellular space, vacuole, and cell wall, while minor deposits have been observed in other cell organelles such as dictyosome, dictyosome derived vesicles, and endoplasmic reticulum. About 90% of the adsorbed Pb is deposited in cell wall and vacuole (Wierzbicka and Antosiewicz 1993). Plants take up free Pb ion either by capillary action or from atmospheric air through cellular respiration. A well-developed root system of the plants takes up divalent Pb ions with the nutrients from the soil and also get absorbed passively and transported through xylems and unloads in the endoderm (Engwa et al. 2019).

The atmospheric Pb was absorbed via cuticle and stomata present in the surface of leaves. Absorbed Pb causes the chlorosis in leaves reaching to the endodermis region and bound to the cell wall and plasma membrane. Endodermal cells act as a barrier for the transport of Pb such as apoplastic and symplastic pathway. Casparian strip block the apoplastic pathway and then Pb can only translocate through symplast pathway. And the role of this barrier in leaf cell vacuole is only to restrict the transport of Pb (Collin et al. 2022).

10.4.1.1 Within Cell Walls

Cell wall serves as a site of Pb accumulation in the form of insoluble lead complexes such as lead phosphate complex (Lane and Martin 1982; Zegers et al. 1976) and prevents the entry of Pb into cytoplasm. The mechanisms employed in restricting the movement of high levels of lead in the cells of the tolerant clone demands further exploration (Qureshi et al. 1986). Plant cell walls contains abundance of divalent and trivalent cation binding compounds which contain functional groups like –OH, –COOH, and –SH. Phenolics, proteins, amino acids, and polysaccharides are the most significant compounds. The ability of these compounds to bind trace metal ions such as Pb is highly dependent on the number of these functional groups present in them (Pelloux et al. 2007). The amount of polysaccharides presents in cell wall which are abundant in carboxyl groups principally determines thc binding capacity of cell wall. Homogalacturonan is one of the four major polysaccharide domains which constitute pectin. Those fractions of homogalacturonan which have low degree of methyl esterification contain free carboxyl which binds the Pb^{2+} (Dronnet et al. 1996; Fritz 2007). Binding of Pb^{2+} to pectin in cell wall renders it metabolically inactive. Gall et al. (2015) said that cell wall form the first barrier for the entry of heavy metal for the cell which plays significant role in detoxification mechanism. Under transmission electron microscopy, Islam et al. (2007) reported that in *Elsholtzia argyi* main organ of the Pb accumulation was cell wall.

Another mechanism by which cell wall protects internal cell organelles from Pb^{2+} ion exposure is by separating sequestered Pb deposits in it from plasma membrane through a callose layer which metal ions are unable to penetrate; thus, it essentially acts as a barrier against intrusion of Pb^{2+} into the protoplasm (Hall 2002; Patra et al. 2004). Thus, endocytosis of sequestered Pb returning into cell wall with compounds of cell wall provides a robust safeguard to plant cells (Krzeslowska et al. 2010). Thickening of cell walls occurs to aid in restricting the trace metal ion uptake into the plants (Probst et al. 2009). This change in cell wall morphology is induced by raise peroxidase activity and lignification of cell wall (Liu 2012). The amount of trace metals entering the protoplast is reduced chemically by thickened cell walls by binding trace metals to the negatively charged substances, synthesizing lipid compounds and callose to introduce physical barrier against immigration of ions and function as a compartment as well for the accumulation of trace metals. Xu et al. (2018) showed that *Typha orientalis* when grown under hydroponic Pb stress resisted the Pb-induced damage by the isolating the Pb content in the cell wall.

10.4.1.2 Inside Vacuoles

Vacuole is an important membrane-bound cell organelle which plays a key role in detoxification of cytoplasm by sequestrating the metal ion, thus imparting the tolerance to plant against lead (Seregin and Ivanov 2001). Sahi et al. (2002) has established the existence of this mechanism in leguminous shrub *Sesbania drummondii* which procures the globular deposits of Pb in vacuoles. First, Pb ions from the external solution may enter endoplasm reticulum closely linked to the apoplast. The Pb particles which had entered the cytoplasm is firstly gathered into membrane-bounded vesicles followed by their subsequent sequestration within the vacuole, evidently through exocytosis (Koppitke et al. 2008). Inside the vacuole, two types of compounds are present (i) Compounds of organic acids such as malate and oxalate which have high affinity for Pb ions, and (ii) the compounds that interact with heavy metals to form low-soluble complexes which causes predominate localization of Pb in this organelle (Krotz et al. 1989; Mazen and El Maghraby 1997). Moreover, the introduction of heavy metal ions inside cytoplasm triggers the synthesis of metal binding peptides through the induced expression of the gsh1, gsh2, and MT genes which forms metal peptide complex and transported to vacuole (Vögeli-Lange and Wagner 1990, 1996; Seregin and Ivanov 2001). These vacuoles are called Pb-sequestering vacuoles which specifically function to entrap cytosolic Pb which is transported to them via specific intracellular mechanisms (Meyers et al. 2008). Pb-sequestering and non-sequestering vacuoles can lie side by side. Exposure to heavy metal prompts the production of additional vacuoles particularly to store toxic metals. This is evident from the observations made by Sridhar et al. (2005) in a conventional transmission electron microscopy (TEM) study of *B. juncea* root which indicated a rise in the number of vacuoles in the cortical and epidermal cells of roots following exposure to Cd and suggested the existence of a similar mechanism active in the root tips of *B. juncea* against Pb exposure. The sequestration of Pb and Cd in vacuoles is an energy-requiring process (Salt and Rauser 1995). Jiang et al. (2019) studied the effect of lead on cell and found that Pb is strategically translocated from cell wall to vacuole to avoid the damage caused to sensitive parts of cell such as mitochondria and protoplast and provide extra space to cell wall under tolerance strategy.

10.4.1.3 Within Cell Membrane

The plasma membrane functions as a living barrier of the cell to uninterrupted influx of Pb ions across the cell membrane through diffusion (Jiang and Liu 2010). Membrane transport systems located in cell membrane are an integral component of mechanisms which involves uptake, accumulation and removal of heavy metals from the cell. These mechanisms ensure protection against heavy metal toxicity by maintaining the optimum concentration of heavy metals inside cell required to perform its normal functions (Malecka et al. 2008). Strange and Macnair (1991) suggested that cell membrane is the site of primary tolerance mechanism against toxicity of heavy metal. Transmission electron microscopy and X-ray microanalysis of root sections of hyper-accumulator shrub *Sesbania drummondii* by Sahi et al. (2002) established the localization of Pb granules in plasma membrane. This can be attributed to the large number of functional groups present in cell membrane (Gardea-Torresdey et al. 2001). Accumulation of Cu in root apoplast at cell membrane deters its entry into cytosol (Ernst et al. 1992). Wojcik and Tukiendorf (2003) observed this mechanism in *Arabidopsis thaliana* plants. Root ultrastructural studies conducted by Islam et al. (2007) detected the dispersion of fine Pb particles across cell membrane. On being exposed to high concentration of heavy metal ions, there is induction of constitutive altercations in the structure of plasma membrane (Ernst et al. 1990). The enfolding's of plasmalemma produce certain vesicles which accumulate Pb inside them and deter the dissemination of free Pb ions in cytoplasm, thereby confining them to minimal space (Jiang and Liu 2010; Clemens 2006).

10.5 Effects and Responses of Plants Under Pb Stress

Metal phytotoxicity occurs when metals take up by plants from roots and transported to various parts of shoot. Excessive Pb concentration causes deleterious effects in plants such as decline in photosynthetic rate, chlorophyll synthesis, affects the Calvin cycle, closing of stomata by creating deficiency of CO_2, growth inhibition which is connected with cell division, let down mineral nutrition and water balance and enzyme activities (Fig. 10.2). It also brings the changes in lipid composition and chlorophyll b content (Kumar and Rai 2007; Collin et al. 2022). There is some unexpected possible mechanism such as changes in the permeability of the cell membrane, reaction to sulfhydryl groups with cations, possible attraction for phosphate groups and active groups of ATP and ADP (Hadi and Aziz 2015). Spraying of various rates of Pb on tomato plant causes the leaves margins burning, bending of branches and sudden decrease in flowers (El-Shebiny 1989). Apart from these, Pb toxicity enhanced the production of reactive oxygen species (ROS) which cause the oxidative stress in the plant cell. Plants activate their antioxidant system to prevent the oxidative damage by ROS enzymatically and non-enzymatically. These antioxidants (superoxide dismutase. peroxidase, catalase, glutathione reductase, and ascorbate peroxidase) appears to play a pivotal role in combating oxidative stress. Ashraf et al. (2017) reported that under Pb conditions photosynthetic pigment destructions, induction of oxidative stress with increased production of H_2O_2, MDA, protein production, and soluble sugar. Fazeli et al. (1991) study found that the Pb only affects the growth of tomatoes, and the content of Vitamin C remains constant. Mishra et al. (2006) found decreased level of sucrose in vegetables due to inhibition of carbohydrate synthesis. Pb reduced the plant height, number of leaves, and dry matter of sunflower plants at higher concentration (Hung et al. 2014). Blackberry plant's leaves have 4.5 folds higher Pb concentration than the fruits, and blackberry contains 71% of the Pb exceeding the WHO threshold by 29 times. The consumers are at high risks who take 100 g fresh blackberries which consist 8.51 mg Pb (Collin et al. 2022). Opeolu et al. (2010) reported less number of flowers under high concentration of Pb in sunflower, and histological changes in leaves such as thin blade leaf, minified xylem and phloem in vascular bundle, and reduction in diameter of xylem vessels were observed in soybean under Pb contamination (Hadi and Aziz 2015). Table 10.1 presents the effects of different Pb concentration on crops and vegetables.

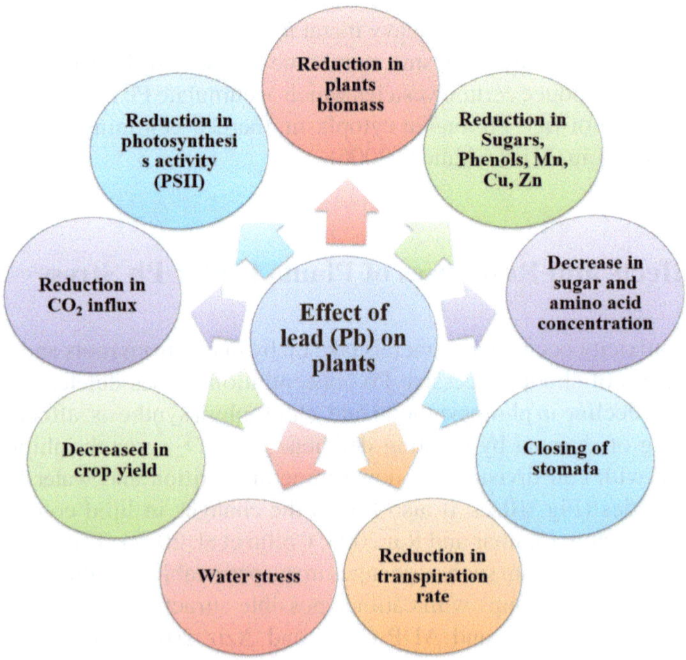

Fig. 10.2 Effects of Pb on plants

Table 10.1 Effects of Pb on some crops and vegetables

Plant species	Pb concentration	Effects on plants	References
Helianthus annuus	300, 600, and 900 mg/kg 0.5, 1, 1.5, 2, 2.5, 3, 3.5, 4, 4.5, and 5 mM	Reduction in shoot and root length, fresh weight and dry weight, Chl a and Chl b, carotenoid contents, and matter stress tolerance index were less as well	Azhar et al. (2006) and Saleem et al. (2018)
Brassica juncea	0.25, 0.50, and 0.75 mM	Lower dry weight, shoot and root length, total chlorophyll and carotenoid level. Water content and relative water content were enhanced	Kohli et al. (2017, 2018a, b)
Lactuca sativa	20 mg/l 500 μM	Delayed in germination, significant decline in dry and fresh weight of plants and reduction in carotenoids, Chl a and Chl b content	Đurđević et al. (2008) and Silva et al. (2017)
Hordeum vulgare	100 and 200 μM	Reduction in shoot length, root length, fresh weight, and dry weight	Arshad et al. (2017)
Oryza sativa	100 μM 400, 800, and 1200 ppm 500 and 1000 μM	Decrease in shoot, root lengths, fresh and dry weights, gas exchange parameters, and decrease in Chl a and Chl b, total chlorophyll, and carotenoid as compared to control	Chen et al. (2017), Ashraf et al. (2017) and Verma and Dubey (2003)
Triticum aestivum	100 μM 40 and 60 ppm 1.5, 3, and 15 mM 0.05, 0.1, 0.5, and 1 g/L	Reduction in dry weight, fresh weight, shoot length, and root length and number of tendrils. Low total chlorophyll, Chl a, and Chl b. Dose-dependent reduction in growth, fresh weight, dry weight, and germination percentage	Tripathi et al. (2016) and Lamhamdi et al. (2010)
Brassica juncea	25, 50, and 100 μM	Reduced growth in terms of percentage, germination, shoot length, root length, fresh weight, dry weight total chlorophyll, and carotenoid	Pratima and Pratima (2016)

(continued)

Table 10.1 (continued)

Plant species	Pb concentration	Effects on plants	References
Gossypium hirsutum	25, 50, and 100 μM 500 μM	Steep decline in biomass of leaf, stem and root in plants. Decline in plant height, root length, leaf area, and number of leaves per plant. Reduced content of Chl a, Chl b, total chlorophyll, and carotenoid content and gradual retardation in levels of, stomatal content, net photosynthetic rate, and transpiration rate. Changes in leaf morphology, viz. length, width, and petiole size, were reduced	Bharwana et al. (2013, 2014, 2016)
Brassica juncea L.	32, 100, 200, 400, and 800 ppm 50 mg/L	Seed germination and survival were reduced. The number of leaves, root, shoot and branches length, fresh weights, and dry weights were declined reduction in photosynthetic indices	Kaur et al. (2013) and John et al. (2012)
Brassica napus L.	33, 100, 200, and 400 μM	Lowered root length and tolerance index	Mosavian and Chaab (2012)
Vigna mungo	25, 50, 75, 100, and 150 ppm 9, 10, and 11 mg/L	Reduction in percentage germination, lengths of root and shoot, and the number of leaves. Similarly, reduction in dry weight, fresh weight, leaf area, and number of nodules were also observed Plant height, fresh and dry weights, chlorophyll, and carotenoid content also get reduced	Kumar and Jayaraman (2014) and Gupta et al. (2006)
Arachis hypogea (cultivar K6 and K9)	100, 200, 400, and 800 ppm	Pb-induced reduction in biomass, and growth in term of shoot and root length	Nareshkumar et al. (2015)
Vigna unguiculata	200 ppm 0.025, 0.050, 0.1, 0.15, 0.2, 0.3, 0.4, 0.5, 1, 1.5, and 2.50 mM	Growth was drastically lowered. Reduction in shoot and root growth, fresh mass of roots, and shoot was also low	Ojwang et al. (2015) and Kopittke et al. (2007)
Pisum sativum	0.25 mg/L	Reduction was observed in number of tendrils, plant height, and leaf length, whereas in the number of leaves and leaf width got increased	Ghani et al. (2015)
Vigna radiata	25, 0.05 and 0.3 mM	Seed length and percentage germination were reduced	Hassan and Mansoor (2014)

(continued)

Table 10.1 (continued)

Plant species	Pb concentration	Effects on plants	References
Zea mays L.	1, 25, 50, 100, 200, and 500 mM 10, 20, and 30 ppm	Germination percentage and seedling growth in terms of root and shoot length were reduced significantly. Similar reduction in fresh and dry weight of seedlings, root and shoot growth got inhibited. Decline in level of total chlorophyll	Hussain et al. (2013) and Ghani (2010)
Raphanus sativus	2.5 mM	Decline in fresh weight, dry weight, and plant height reduced net photosynthetic rate and total chlorophyll content	Anuradha et al. (2011)

10.5.1 Germination

Germination of the seed marks the beginning of plant life. Many researchers have reported the adverse effect of Pb on germination such as such as rice, *Pinus helipensis*, *Phaseolus vulgaris*, and *Pisum sativum* (Mukherji and Maitra 1976; Nakos 1979; Wierzbicka and Obidzińska 1998). Mukherji and Maitra (1976) reported that 60 mM lead acetate lowered the activity of protease and amylase by about 50% in rice endosperm. The lead which enters through symplastic pathways leads penetration into the seed embryos and delays the germination by disrupting the activity of protease and α-amylase enzymes. These enzymes are released from aleuronic layer into the endosperm and scutellum epithelial cells, and its function is to hydrolyze the storage protein into metabolizable sugars to nourish the germinating plant (Tan-Wilson and Wilson 2012). Lead disrupts their functioning by binding at their active site and leading to inhibition of seed germination and growth (Sengar et al. 2008).

Seed coats exhibit selective permeability to Pb ions, but during imbibition of seeds, their permeability varies and become highly permeable to lead. The variation in selective permeability is a consequence of chemical and physical processes due to hydration of seed coats. Selective permeability is the function of both living and dead cells containing substances like lipids and tenins and increases with the reduction in water uptake as the final stages of imbibitions approach. Mrozek and Funicelli (1982) observed the inhibitory action of Pb in seed germination of *Spartina alterniflora* by altering their selective permeability and suggested the alteration in dormancy mechanism in halophyte seeds which aid these seeds to avoid adverse salinity conditions. In rice seedlings, 14–30% of the germination and 13–45% development reduced which adversely affects the length of radical and hypocotyls under high Pb concentration (1 mM) (Gidlow 2015). Other scientists also reported the similar result regarding the inhibition of germination by the Pb in *Hordeum vulgare*, *Oryza sativa*, *Zea mays*, and *Pinus halepensis* (Collin et al. 2022). Pb interferes with metabolic process like protease and amylase enzyme activity that ultimately leads to decrease in seed germination. Inhibition of ATPase/ATP synthase enzyme results into decreased ATP production which is necessary for the growth of seedlings (Mench et al. 1987). When the seeds were incubated in Pb salts, their level of saturated fatty acid decreased and level of unsaturated fatty acid increased (18:3). Synthesis of DNA, RNA, and protein was decreased with increased concentration of Pb in the embryo axis and endodermis of the germinating rice seedlings (Wierzbicka 1987).

10.5.2 Photosynthetic Indices

The toxicity of Pb ions has a negative impact on photosynthetic rate. Reduction in photosynthetic rate may be due to the distortion in structure of chloroplast, reduction in the production of chlorophyll and carotenoids, impeded functioning of the enzymes involved in the Calvin cycle, crippled transport of electrons, and dearth of CO_2 which surfaced from the closing of stomata. Rebechini and Hanzely (1974) experimented with the *Ceratophyllum* by *demersem* plants by growing them in an aqueous solution containing lead nitrate and observed well-defined alterations in the fine structure of chloroplast. A substantial fall in grana stacks and stroma content was observed along with elimination of starch grains. Stefanov et al. (1995) also noticed a modification in the lipid composition of thylakoid membranes as a result of Pb treatment. Uptake of elements like Fe and Mg by plants is critical for chlorophyll synthesis and thus adversely affects the formation of chlorophyll pigment when disturbed by Pb (Akinci et al. 2010). Owing to its intense affinity toward protein N- and S-ligands, it heavily impairs the photosynthetic machinery (Ahmed and Tajmir-Riahi 1993). Acceleration of chlorophyllase activity causes enhancement in chlorophyll degradation in Pb-treated plants. The adverse effect of Pb treatment on Chlorophyll-b is more pronounced compared to chlorophyll-a (Vodnik et al. 1999). Donor and acceptor sites of PSII, PSI, and cytochrome b/f complex are affected by Pb. PS I electron transport is less prone to inhibition by Pb than PS II (Mohanty et al. 1989; Šeršeň et al. 1998). It also leads to the dismantling of oxygen releasing extrinsic polypeptide of PS II and removal of Cl, Ca, and Mn from oxygen-evolving complex (Rashid et al. 1991). The application of Pb reduces photosynthesis indirectly by prompting stomata closure (Bazzaz et al. 1975). This reduction in carotenoid and chlorophyll content of seedlings and cuttings under high concentration of Pb stress can be regarded as a response of the plants specifically to the metal stress, which resulted in photosynthesis inhibition and chlorophyll degradation (Gajewska et al. 2006). In ryegrass, the content of total Chl, Chl a, Chl b, and Car markedly decreased under the Pb treatment by 40.50, 44.16, 28.25, and 51.11%, respectively, in compared with control (Bai et al. 2015). In shoots, the Pb exposure decreased the K, Mg, Fe, Zn, and Cu content by 59.50, 18.51, 43.07, 28.73, and 46.85%, whereas enhancement is observed in Ca content by 92.78% in comparison with control (Bai et al. 2015). Pb at lower concentration enhances the anthocyanin content in *Brassica napus* but decreased at higher level of Pb concentration (Fatemi et al. 2021). Several study revealed that Pb toxicity has decreased the photosynthetic pigment in many plant species such as *Brassica napus* (Kanwal et al. 2014). In *Pisum sativum*, Rodriguez et al. (2015) reported that chlorophyll A and chlorophyll B concentration increased with elevated level of Pb and reach maximum at 1000 mg kg^{-1}. *Hordeum vulgare* under hydroponic Pb condition showed toxic symptoms such as change in chloroplast morphology and decreased number of thylakoids damage to chloroplast directly inhibits photosynthesis and adversely affects the growth and development of the plants (Legocka et al. 2015). Pb reduces the photosynthetic pigments (Chl a and Chl b) of bean and pea seedlings (Hameed et al. 2010). It can be caused due to the impaired uptake of essential elements such as Mg and Fe or due to reduced leaf size (Feleafel and Mirdad 2013). Chlorophylase (Chlase) and Mg-Dechelatase (MDCase) involved mainly in the destruction of chlorophyll, and their activity was stimulated under Pb stress. The net photosynthetic rate declines significantly in *Davidia involucrata* with increasing Pb concentration and maximum at Pb 200 mg/kg (Yang et al. 2020).

10.5.3 Growth

Contamination of lead in soil impairs the early plant growth. Hadi and Aziz (2015) attributed the poor germination of seeds and retarded growth in seedlings due to toxic effects of Pb on chlorophyll synthesis, cell division, root growth, and transpiration. Pb has harmful ramifications on growth of radish plants as observed by Tomulescu et al. (2004). When Jiang and Liu (2010) investigated Pb-induced changes in cell after 2–3 days of Pb exposure, they found that Pb caused the loss of endoplasmic reticulum, dictyosome and cristae, and the mitochondrial structure of root meristematic cells, and damaged biological membranes. The excess migration of Pb in roots declined root growth and facilitated the loss of apical dominance leading to a decline of 10% in the fresh biomass of plants as a result of Pb activity in roots and shoot, respectively, at 0.3 and 0.07 μM concentration (Kopittke et al. 2007). Pb toxicity leads to expansion of interphase stage of mitosis which reduced cell division leading to decreased plant growth (Patra et al. 2004). Pb caused a remarkable decline in the sprouting, seedling development, and growth in wheat, and inhibitory effect was also observed in *Jatropha curcas* (Dey et al. 2007; Shu et al. 2012). Even low concentration of Pb suppresses the growth of roots as well as aerial parts of the plants and detrimental effects on growth of the roots are observed than other plant parts (Kopittke et al. 2007; Liu et al. 2008).

Toxicity of lead leads to abnormalities in root morphology and structure as swollen, short, and stubby roots that exhibit a rise in the number of secondary roots and their length (Kopittke et al. 2007). As a result of Pb contamination, a reduction in the elongation of Mesquite (*Prosopis* sp.) roots was observed by Arias et al. (2010). The most obvious symptoms of growth retardation such as fewer, smaller, and brittle leaves having dark purple dorsal surfaces are clearly visible at extreme level of Pb toxicity (Islam et al. 2007). Overall, compromised nutrient metabolism, photosynthesis, and plant water relations leading to inhibition of plant growth occur as a result of Pb toxicity (Kopittke et al. 2007; Alsokari and Aldesuquy 2011). Toxic effects of Pb are different on the basis of its concentration, duration of exposure, affected species, and stage of growth in plants (Gupta et al. 2009; Gul et al. 2019). High application of Pb (1000 and 5000 μg/g) in soil causes the reduction in root and shoot growth, ceased seedling growth, very thin stems, and small leaves in radish plant (Khan and Frankland 1983). Pb concentration significantly undermined starch solubility in endosperm and α-amylase activity in seeds in rice (Gautam et al. 2010). In seedlings of wheat, Pb causes reduction in seed germination, reduction in macro (Na, Ca, Mg, K, and O) and micronutrients (Fe, Cu, and Zn) biomass shoot and root elongation in comparison of control (Lamhamdi et al. 2011). Decrease in Cu and K concentration in maize cultivar was observed under higher level of lead contamination (Rizwan et al. 2018).

10.5.4 Crop Productivity

The most widely notable manifestation of Pb toxicity is detention of the photosynthetic carbon fixation which leads to declined crop productivity (Singh et al. 2010). Lead interferes with the synthesis of plastoquinone, carotenoids, and functioning of electron transport chain and retardation of enzymatic activities vital for CO_2 fixation as stomatal and non-stomatal constraints are accountable for carbon fixation (Mishra et al. 2006; Chen et al. 2007; Qufei and Fashui 2009). Lead stress curtailed the photosynthetic activity of sunflower plants as a consequencely affect the biosynthesis of chlorophyll which ultimately leads to reduction in plants biomass production (Mukhtar et al. 2010). Oxidative stress, synthesis and activity of chlorophyllase enzyme got promoted due to Pb toxicity which ultimately leads to decline in the rate of photosynthesis as a result of chlorophyll degradation (Liu et al. 2008). Furthermore, under Pb stress, there is also a significant reduction in the activities of delta-aminolevulinic acid dehydratase (ALAD) and ferredoxin $NADP^+$ reductase, which impedes chlorophyll synthesis (Gupta et al. 2009). Dissociation of chlorophyll is a four step reaction, and the final products include phytol, Mg, and a primary product of porphyrin rings. Although, level of toxicity differs among plant species, and usually it is more intimately related to Chl b than Chl a: however, decline in photosynthetic activity is more vulnerable to Pb stress than content of photosynthetic pigments (Xiong et al. 2006). Kosobrukhov et al. (2004) analyzed the extent of structural changes and photosynthetic activity of plants grown in soil contaminated with Pb and observed a decline of 40–50% in stomatal conductance. Romanowska et al. (2006) ascribed disrupted photosynthesis under Pb stress to depletion in leaf area, total chlorophyll contents, vascular bundles, and CO_2 influx due to ill-function of stomatal closure. Qufei and Fashui (2009) stated that addition of Pb in leaves of duckweed destroyed secondary structure of photosystem II and constrained the absorption and transfer of energy among numerous enzymes. It alters the actions of photosystem I and photosystem II in *Pisum sativum*. The rate of electron transport during hill reaction and halted cyclic and non-cyclic photophosphorylation got slow down due to Pb exposure (Romanowska et al. 2008). The catalysis of Melvin Calvin cycle enzymes is also affected by Pb toxicity (Chen et al. 2007). Exposure to Pb also significantly affects ATP content and respiration of plants. Pb exposure chiefly disturbs the activity of ribulose-bisphosphate carboxylase responsible for regulating the assimilation of CO_2 in C_3 plants, devoid of affecting the oxygenase activity (Assche and Clijsters 1990). Parys et al. (1998) documented a considerable rise in CO_2 concentration in leaves of pea on being exposed to $Pb(NO_3)_2$ accrediting it to increased respiration with a simultaneous fall in photosynthesis. It was divulged by Romanowska et al. (2002) that photorespiration in this case remains constant and elevated respiration under Pb exposure is related to mitochondrial respiration (dark) only. The dark respiration instigated by Pb was observed in protoplast of barley and pea leaves (Romanowska et al. 2002, 2005, 2006). Furthermore, invigoration of respiration was linked with high ATP generation in mitochondria, raising the demand for more energy in order to cope up with the Pb toxic effects.

The decline in grain yield of crops under Pb stress can be accredited to poor nutrient uptake, incomplete carbon fixation as a result of stomatal and non-stomatal constraints, plant water relations, and escalated oxidative damage. Gu et al. (1989) reported a significant impact of Pb contamination in soils on productivity (grain and biological yield) of rice. Rehman et al. (2017) also documented 25–30% decline in the grain yield of wheat due to lead toxicity. Misra et al. (2010) perceived a marked decline in the economic yield of sugarcane crop grown under Pb stress. Decline in productivity of various crops is determined by Pb concentration in soil as argued by Codling et al. (2016) and Hussain et al. (2006) noticed a reduction of 28–32% and 24% in the economic yields of potato and mash bean, respectively. In maize seedlings, Pb stress causes a general reduction of macro- and micronutrient contents especially of K, Ca, and Mn which was observed (Wang et al. 2007). In seedlings and cuttings of ryegrass, photosynthetic rate (Pn) manifests a related trend of abrupt decrease as noted for conductance (Cond) and transpiration (Tr). Pn, Cond, and Tr were found to be comparatively less in leaves of seedlings than leaves of cuttings under lower Pb concentration, but an opposing trend is seen under high Pb concentration. The severely damaged thylakoid membrane of stroma and grana caused sharp reduction in rate of photosynthesis. In a study, Bai et al. (2015) observed the parallel change of Cond, and Pn in peanut leaves that reinforced the changes in Pn could be allocated to the changes in Cond.

Chatterjee et al. (2004) planted three variety of rice (Xinagyaxiangezhan, Meixiangezhan-2, Basmati-385) in Pb contaminated soil and found reduction in yield in the following order 69.12%, 58.05%, and 46.27% respectively. Reduction in grain yield may be due to the decrease in chlorophyll in leaves, carotene, sugars and Fe, Mn, Cu, and Zn. Moreover, the effects of Pb toxicity on germination, yield, and growth of different crops depend on time and concentration and also fluctuate with prevailing growth conditions and plant species. Ma et al. (2021) reported inhibition of grain yield of fragrant rice under soil Pb stress. Xian (1989) reported decrease in yield of kidney beans at high concentration of Pb. With the increasing concentration of Pb, pods per plant, seeds per pod, as well as total protein content of pea plant decreased in Sorial and Abd El-Fattah (2001) study. In spinach, ribulose-bisphosphate carboxylase/oxygenase activity was inhibited even at low concentration of 5 μM Pb concentration (Vallee and Ulmer 1972). Pb is known as most potent metal ion for the inhibition of chloroplastic ATP synthetase/ATPase activity.

10.5.5 Biomass

Plants grown in contaminated soil, especially in high concentrations of Pb (1500 mg/kg), had dramatically lower biomass, almost 60% compared with the control plant (Fatemi et al. 2021). The plants heave lower biomass when grown in Pb contaminated soil which is about 60% of control plant biomass. The Pb exposure inhibited the growth of ryegrass seedlings compared with control, and the reductions of plant height, fresh mass, dry mass, root volume, and root/shoot ratio were 35.77, 18.28, 55.18, 54.12, and 34.62%, respectively (Bai et al. 2015). Pb toxicity causes decline in biomass in *B. napus* and coriander (Fatemi et al. 2020a, b). Growth of coriander was decreased under Pb toxicity (Fatemi et al. 2021). Sidhu et al. (2016) reported that root and shoot biomass increased under Pb stress but declined at 29 mg kg^{-1} Pb concentration in *Cronopus didymus.* Xu et al. (2018) observed that under Pb stress tea plants show poor biomass production loss of photosynthetic pigment reduction in total caffeine, free amino acid but increased catechin concentration. Compared to the control in tea plant under Pb stress, 17–65% root biomass, 3–50% stem, 20–73% leaves biomass were reduced that significantly affects the yield of the tea plants. *Chrysanthemum indicium* has caused significant reduction in root (32.7 mg kg^{-1}) and shoots (41.3 mg kg^{-1}) biomass with 50 mg kg^{-1} Pb concentration. However, minimum application of Pb can promote the growth of the plant for some extent (Mani et al. 2015).

10.5.6 Antioxidant Enzymes

The production of hydrogen peroxide, superoxide, and hydroxyl radical increased when plants are exposed to heavy metals stress. The possible reason behind this phenomenon can be that the heavy metal stress reduces the capability of plants to assimilate carbon and escalate influx of photosynthetic electrons to molecular oxygen. Antioxidants rapidly scavenge reactive oxygen species that damaged the proteins, lipids, and pigments (Bhaduri and Fulekar 2012). ROS is collective term used for hydrogen peroxide, superoxide radical, hydroxyl radical, and singlet oxygen generated at the time of heavy metal stress (Devi and Prasad 1998). Lipids, nucleic acids, amino acids, and proteins got damaged which leads to irreversible metabolic dysfunction and cell death (Luna et al. 1994). The antioxidant system got activated to cope and repair the damage (Shu et al. 2012). A variety of mechanisms to deal with ROS effects in cellular compartments has been evolved by the plants. This is generally coped up through the production of various anti-oxidative enzymes, such as superoxide dismutase (SOD), catalase (CAT), and peroxidase (POD) (Wang et al. 2007). Pb exposer shows simulative effects on electrolytes leakage H_2O_2 level MDA content and activity of antioxidant (SOD, CAT, APX, GPX, GR), but at higher concentration SOD, CAT, H_2O_2, and MDA declined in *Cronopus didymus* (Sidhu et al. 2016). SOD, POD, and CAT were up-regulated compared to the control in maize seedlings (Wang et al. 2007). The activity of SOD and POD decreased in *Elsholzia argyi* after addition of Pb (Islam et al. 2007). SOD activities increased, but POD activities decreased under Pb stress in *S. japonica* which indicated that antioxidant system behaved differently in different plant species (Zhong et al. 2017). Ascorbate peroxidase participates in the detoxification of H_2O_2 into water and oxygen with the consumption of ascorbic acid. Glutathione reductase catalyzes the GSSH to GSH, and both these helps the plant cell to increase the antioxidants level under the metal stress (Qureshi et al. 2007). Proline not only regulates the osmotic potential of the cell but also plays a great role in the removal of reactive oxygen species and protects the cell membrane by maintaining structural stability and proton pumps from Pb toxicity (Cai et al. 2022).

10.5.6.1 Superoxide Dismutase (SOD)

SOD, a metallo-enzyme present in different cellular compartments, can catalyze the dismutation of O_2^- into H_2O_2 and O_2, and subsequently H_2O_2 can be effectively scavenged by CAT and POD. The first step of ROS generation is superoxide formation, superoxide radicals (precursor of the other ROS) (Bhaduri and Fulekar 2012). Ashraf et al. (2017) found that SOD was the initial of scavenger of reactive oxygen species in rice under soil Pb stress. Heavy metal ions can increase the activity of SOD in oat and rice (Luna et al. 1994; Verma and Dubey 2003). Heavy metals may also decrease or not affect at all the SOD activity (Reddy et al. 2005). SOD activity of seedlings culminated at higher metal concentrations than those of cuttings of ryegrass, suggesting that SOD has better protection against oxidant damage. The decrease by 46.74% in shoots and by 55.99% in roots of SOD activity in ryegrass plant after Pb treatment has been observed (Bai et al. 2015). Multiple SOD genes encoding at least three Fe-SODs, three Cu–Zn-SODs, and one Mn-SOD had been reported in *Arabidopsis* (Bhaduri and Fulekar 2012). SOD increased but not dramatically in *B. napus* under various concentration of Pb (Fatemi et al. 2021).

10.5.6.2 Peroxidase (POD)

H_2O_2 is utilized in the oxidation of various organic and inorganic substrates by peroxidase. Guaiacol acts as electron donor when utilized by peroxidase in vitro known as guaiacol peroxidases. A strong increase in POD activity has been observed in response to Pb was reported in *rice, A. thaliana* and *Zea mays* (Verma and Dubey 2003; Wang et al. 2007). However, POD activity has been inhibited due to heavy metals which has also been observed in oat leaves (Luna et al. 1994). POD activity increased by approximately 2.26 times in comparison with controls at 3 mM of Pb concentration in seedlings. The activity of POD is high at lower concentration in comparison with higher concentration, in seedlings and cuttings of *Jatropha curcas*. POD participating in lignin biosynthesis can build up a physical barrier against toxic heavy metals. A physical barrier can be build up against toxic heavy metals by participation of POD in lipid biosynthesis indication that seedlings are more efficient in avoiding damage than cuttings (Shu et al. 2012). POD activity significantly increased in *Brassica napus* by 47% at the Pb level 1500 mg kg^{-1} (Fatemi et al. 2021).

10.5.6.3 Catalase (CAT)

CAT is a universal heme-containing and oxidoreductase enzyme that decomposes H_2O_2 to water and molecular oxygen, and acts as one of the key enzymes implicated in the removal of toxic peroxides. Generally, CAT activity gets stimulated under heavy metal stress. Shu et al. (2012) reported that at higher concentration, CAT activity is less in comparison with lower concentration. When seedlings had exposed to high Pb stress, CAT activity in leaves was quite high. CAT activity in both roots and shoots of the Pb-treated ryegrass plants increased significantly compared to control (Bai et al. 2015). CAT activity get stimulated up to 500 mg kg^{-1} of Pb while show decline trend at higher concentration in *B. napus* (Fatemi et al. 2021).

10.5.6.4 Ascorbate Peroxidase (APX)

Ascorbate peroxidase is an important peroxidase which is ubiquitously present in plants. APX is universal housekeeping protein in the chloroplasts and cytosol of plant cells. Ascorbate work as a substrate and believed to scavenge excess H_2O_2 formed in plant cells under both stress and normal conditions (Bhaduri and Fulekar 2012). The ascorbate-free radical is the product of oxidation of ascorbate which got reduced back to dehydroascorbate with NADPH as the electron donor by the enzyme mono-hydroascorbate reductase (Asada et al. 1996). Several scientists reported increase in ascorbate peroxidase activity in response to air pollutants specially with O_3 in several species such as in wheat spinach, pumpkin, and *Picea abies* (Tanaka et al. 1985; Bender et al. 1994; Ranieri et al. 1996; Sehmer et al. 1998). The Pb stress also increased the H_2O_2 content in comparison of control by 181.86% in leaves and by 235.95% in roots of ryegrass plant (Bai et al. 2015).

10.5.7 Malondialdehyde (MDA)

When plants are subjected to oxidative stress, malondialdehyde is the term used to measure lipid peroxidation because it is a final product of the peroxidation of membrane lipid. Lipid peroxidation enhancement indicates that Pb and/or IAA caused oxidative stressin maize seedlings (Wang et al. 2007). Shu et al. (2012)

reported that the MDA contents of cuttings in rye grass increased about 100.91%, while in seedlings the increment is about 108.81% for seedlings at a highly toxic Pb level compared to the control. In *B. napus*, MDA content increased up to 1000 mg kg^{-1} Pb concentration after that downward trend was observed (Fatemi et al. 2021). Excessive proline in the body may also participate in the clearance of reactive oxygen species and effectively keeping the MDA content low in aromatic rice and sunflowers under soil Pb stress (Liao et al. 2021). At different Pb concentration (500–2500 μM) in culture medium, there is an increase in MDA and H_2O_2 content in roots of wheat (Kaur et al. 2012). Similarly, in Maize and rice, the MDA content increased in respect of duration of exposure and dose (Thakur et al. 2017).

10.5.8 Protein

A large number of enzymes having sulfhydryl groups get inhibited at different sites when exposed to metal stress resulting in deleterious effects in the normal protein formation and folding pathways. According to Shu et al. (2012), the increase in protein content has been observed while the protein content decreased in leaves of seedlings in ryegrass plants. Soluble proteins can decrease the osmotic potential of the cell to ensure extracellular turbulence and stability (Jiang et al. 2019).

10.6 Conclusion

Pb pollution is a leading and common cause for stress in plants. The plants are adversely affected in terms of growth and physiological activities. To overcome the stress, plants evolutionally develop antioxidants system. When plants are under stress, a surge in free radical species observed which enhanced the activity of enzymatic and non-enzymatic antioxidants. The antioxidant system helps in maintaining the cell components structural integrity. This chapter detailed the Pb sources, impact and distribution in different parts, overall growth of the plants, and antioxidants activity in plants under Pb stress. Recently, world population and industrialization are growing exponentially which causes increase in food demand and at the same time reduction in cultivable land. Future agriculture requires stress-tolerant varieties, and to develop these varieties, the comprehensive knowledge of antioxidant system in plants is obligatory. Despite the advances in understanding of synthesis of antioxidants under Pb pollution, the detailed study on biochemical interaction of antioxidants and Pb is required.

References

Agrawal R (2009) Study of physico-chemical parameters of groundwater quality of Dudu town in Rajasthan. Rasayan J Chem 2(4):969–971

Ahmed A, Tajmir-Riahi HA (1993) Interaction of toxic metal ions Cd^{2+}, Hg^{2+}, and Pb^{2+} with light-harvesting proteins of chloroplast thylakoid membranes. An FTIR spectroscopic study. J Inorg Biochem 50(4):235–243

Ahmed IM, Helal AA, El Aziz NA, Gamal R, Shaker NO, Helal AA (2019) Influence of some organic ligands on the adsorption of lead by agricultural soil. Arab J Chem 12(8):2540–2547

Akinci IE, Akinci S, Yilmaz K (2010) Response of tomato (*Solanum lycopersicum* L.) to lead toxicity: growth, element uptake, chlorophyll and water content. Afr J Agric Res 5(6):416–423

Alsokari SS, Aldesuquy HS (2011) Synergistic effect of polyamines and waste water on leaf turgidity, heavy metals accumulation in relation to grain yield. J Appl Sci Res 7(3):376–384

Anuradha S, Rao R, Eeta S (2011) Amelioration of lead toxicity in radish (*Raphanus sativus* L.) plants by brassinolide. J Appl Biol Sci 5(3):43–48

Aponte H, Meli P, Butler B, Paolini J, Matus F, Merino C, Cornejo P, Kuzyakov Y (2020) Meta-analysis of heavy metal effects on soil enzyme activities. Sci Total Environ 737:139744

Arias JA, Peralta-Videa JR, Ellzey JT, Ren M, Viveros MN, Gardea-Torresdey JL (2010) Effects of *Glomus deserticola* inoculation on Prosopis: enhancing chromium and lead uptake and translocation as confirmed by X-ray mapping, ICP-OES and TEM techniques. Environ Exp Bot 68(2):139–148

Arshad T, Maqbool N, Javed F, Arshad MU (2017) Enhancing the defensive mechanism of lead affected barley (*Hordeum vulgare* L.) genotypes by exogenously applied salicylic acid. J Agric Sci 9(2):139–146

Asada H, Kawamura Y, Maruyama K, Kume H, Ding RG, Ji FY, Kanbara N, Kuzume H, Sanbo M, Yagi T, Obata K (1996) Mice lacking the 65 kDa isoform of glutamic acid decarboxylase (GAD65) maintain normal levels of GAD67 and GABA in their brains but are susceptible to seizures. Biochem Biophys Res Commun 229(3):891–895

Ashraf U, Kanu AS, Deng Q, Mo Z, Pan S, Tian H, Tang X (2017) Lead (Pb) toxicity; physio-biochemical mechanisms, grain yield, quality, and Pb distribution proportions in scented rice. Front Plant Sci 8:259

Assche F, Clijsters H (1990) Effects of metals on enzyme activity in plants. Plant Cell Environ 13(3):195–206

Azhar NAZILA, Ashraf MY, Hussain M, Hussain F (2006) Phytoextraction of lead (Pb) by EDTA application through sunflower (*Helianthus annuus* L.) cultivation: seedling growth studies. Pak J Bot 38(5):1551–1560

Bai XY, Dong YJ, Wang QH, Xu LL, Kong J, Liu S (2015) Effects of lead and nitric oxide on photosynthesis, antioxidative ability, and mineral element content of perennial ryegrass. Biol Plant 59:163–170

Barber JL, Thomas GO, Kerstiens G, Jones KC (2004) Current issues and uncertainties in the measurement and modelling of air–vegetation exchange and within-plant processing of POPs. Environ Pollut 128(1–2):99–138

Bazzaz FA, Carlson RW, Rolfe GL (1975) Inhibition of corn and sunflower photosynthesis by lead. Physiol Plant 34(4):326–329

Bender J, Weigel HJ, Wegner U, Jäger HJ (1994) Response of cellular antioxidants to ozone in wheat flag leaves at different stages of plant development. Environ Pollut 84(1):15–21

Bhaduri AM, Fulekar MH (2012) Antioxidant enzyme responses of plants to heavy metal stress. Rev Environ Sci Bio/Technol 11:55–69

Bharwana SA, Ali S, Farooq MA, Iqbal N, Abbas F, Ahmad MSA (2013) Alleviation of lead toxicity by silicon is related to elevated photosynthesis, antioxidant enzymes suppressed lead uptake and oxidative stress in cotton. J Bioremed Biodegr 4(4):4153–4172

Bharwana SA, Ali S, Farooq MA, Iqbal N, Hameed A, Abbas F, Ahmad MSA (2014) Glycine betaine-induced lead toxicity tolerance related to elevated photosynthesis, antioxidant enzymes suppressed lead uptake and oxidative stress in cotton. Turk J Bot 38(2):281–292

Bharwana SA, Shafaqat A, Farooq MA, Mujahid F, Nusrat B, Rehan A (2016) Physiological and biochemical changes induced by lead stress in cotton (*Gossypium hirsutum* L.) seedlings. Acad J Agric Res 4(4):160–167

Burzynski M (1984) Influence of lead on the chlorophyll content and on initial steps of its synthesis in greening cucumber seedlings. Acta Soc Bot Pol 54:95–105

Cai X, Fu J, Li X, Peng L, Yang L, Liang Y, Jiang M, Ma J, Sun L, Guo B, Yu X (2022) Low-molecular-weight organic acid-mediated tolerance and Pb accumulation in centipedegrass under Pb stress. Ecotoxicol Environ Saf 241:113755

Cecchi M, Dumat C, Alric A, Felix-Faure B, Pradère P, Guiresse M (2008) Multi-metal contamination of a calcic cambisol by fallout from a lead-recycling plant. Geoderma 144(1–2):287–298

Chatterjee C, Dube BK, Sinha P, Srivastava P (2004) Detrimental effects of lead phytotoxicity on growth, yield, and metabolism of rice. Commun Soil Sci Plant Anal 35:255–265

Chen Q, Zhang X, Liu Y, Wei J, Shen W, Shen Z, Cui J (2017) Hemin-mediated alleviation of zinc, lead and chromium toxicity is associated with elevated photosynthesis, antioxidative capacity; suppressed metal uptake and oxidative stress in rice seedlings. Plant Growth Regul 81(2):253–264

Chen J, Zhu C, Li L, Sun Z, Pan X (2007) Effects of exogenous salicylic acid on growth and H_2O_2-metabolizing enzymes in rice seedlings under lead stress. J Environ Sci 19(1):44–49

Clemens S (2006) Toxic metal accumulation, responses to exposure and mechanisms of tolerance in plants. Biochimie 88(11):1707–1719

Codling EE, Chaney RL, Green CE (2016) Accumulation of lead and arsenic by potato grown on lead–arsenate-contaminated orchard soils. Commun Soil Sci Plant Anal 47(6):799–807

Collin MS, Kumar Venkataraman S, Vijayakumar N, Kanimozhi V, Arbaaz SM, Stacey RS, Anusha J, Choudhary R, Lvov V, Tovar GI, Senatov F (2022) Bioaccumulation of lead (Pb) and its effects on human: a review. J Hazard Mater Adv 100094

Devi SR, Prasad MNV (1998) Copper toxicity in *Ceratophyllum demersum* L. (Coontail), a free floating macrophyte: response of antioxidant enzymes and antioxidants. Plant Sci 138(2):157–165

Dey SK, Dey J, Patra S, Pothal D (2007) Changes in the antioxidative enzyme activities and lipid peroxidation in wheat seedlings exposed to cadmium and lead stress. Braz J Plant Physiol 19:53–60

Dronnet VM, Renard CMGC, Axelos MAV, Thibault JF (1996) Heavy metals binding by pectins: selectivity, quantification and characterisation. In: Progress in biotechnology, vol 14. Elsevier, pp 535–540

Đurđević B, Lisjak M, Stošić M, Engler M, Popović B (2008) Influence of Pb and Cu toxicity on lettuce photosynthetic pigments and dry matter accumulation. Cereal Res Commun 36:1951–1954

El-Shebiny GM (1989) Lead relationships with soil and plant. Ph.D. thesis, Faculty of Agriculture, Alexandria University

Engwa GA, Ferdinand PU, Nwalo FN, Unachukwu MN (2019) Mechanism and health effects of heavy metal toxicity in humans. In: Poisoning in the modern world—new tricks for an old dog, vol 10, pp 70–90

Ernst WHO, Schat H, Verkleij JAC (1990) Evolutionary biology of metal resistance in *Silene vulgaris*. Evol Trends Plants 4(1):45–51

Ernst WHO, Verkleij JAC, Schat H (1992) Metal tolerance in plants. Acta Bot Neerl 41(3):229–248

Farooq M, Anwar F, Rashid U (2008) Appraisal of heavy metal contents in different vegetables grown in the vicinity of an industrial area. Pak J Bot 40:2099–2106

Fatemi H, Pour BE, Rizwan M (2020a) Isolation and characterization of lead (Pb) resistant microbes and their combined use with silicon nanoparticles improved the growth, photosynthesis and antioxidant capacity of coriander (*Coriandrum sativum* L.) under Pb stress. Environ Pollut 114982

Fatemi H, Esmaielpour B, Sefidkon F, Soltani A-A, Nematollahzadeh A (2020b) How mycorrhiza symbiosis help coriander (*Coriandrum sativum* L.) plants grow better under contaminated soil? J Plant Nutr 1–14

Fatemi H, Esmaiel Pour B, Rizwan M (2021) Foliar application of silicon nanoparticles affected the growth, vitamin C, flavonoid, and antioxidant enzyme activities of coriander (*Coriandrum sativum* L.) plants grown in lead (Pb)-spiked soil. Environ Sci Pollut Res 28:1417–1425

Fazeli MS, Sathyanarayan S, Satish PN, Muthanna L (1991) Effect of paper mill effluents on accumulation of heavy metals in coconut trees near Nanjangud, Mysore District, Karnataka, India. Environ Geol Water Sci 17:47–50

Feleafel MN, Mirdad ZM (2013) Hazard and effects of pollution by lead on vegetable crops. J. Agric Environ Ethics 26:547–567

Ferreyroa GV, Gelma J, Sosa MD, Orellana Benitez MA, Tudino MB, Lavado RS, Molina FV (2018) *Brassica napus* growth in lead-polluted soil: bioaccumulation in plant organs at different ontogenetic stages and lead fractionation in soil. Water Air Soil Pollut 229:1–11

Fritz E (2007) Measurement of cation exchange capacity (CEC) of plant cell walls by X-ray microanalysis (EDX) in the transmission electron microscope. Microsc Microanal 13(4):233–244

Gajewska E, Skłodowska M, Słaba M, Mazur J (2006) Effect of nickel on antioxidative enzyme activities, proline and chlorophyll contents in wheat shoots. Biol Plant 50:653–659

Gall HL, Philippe F, Domon JM, Gillet F, Pelloux J, Rayon C (2015) Cell wall metabolism in response to abiotic stress. Plants 4(1):112–166

Gardea-Torresdey JL, Arteaga S, Tiemann KJ, Chianelli R, Pingitore N, Mackay W (2001) Absorption of copper (II) by creosote bush (*Larrea tridentata*): use of atomic and X-ray absorption spectroscopy. Environ Toxicol Chem Int J 20(11):2572–2579

Gaur VK, Sharma P, Gaur P, Varjani S, Ngo HH, Guo W, Chaturvedi P, Singhania RR (2021) Sustainable mitigation of heavy metals from effluents: toxicity and fate with recent technological advancements. Bioengineered 12(1):7297–7313

Gautam M, Sengar RS, Chaudhary R, Sengar K, Garg S (2010) Possible cause of inhibition of seed germination in two rice cultivars by heavy metals Pb^{2+} and Hg^{2+}. Toxicol Environ Chem 92(6):1111–1119

Ghani A (2010) Toxic effects of heavy metals on plant growth and metal accumulation in maize (*Zea mays* L.). Iran J Toxicol 4(3):325–334

Ghani A, Khan I, Ahmed I, Mustafa I, Abd-Ur R, Muhammad N (2015) Amelioration of lead toxicity in *Pisum sativum* (L.) by foliar application of salicylic acid. J Environ Anal Toxicol 5:292

Gidlow DA (2015) Lead toxicity. Occup Med 65(5):348–356

Ginn BR, Szymanowski JS, Fein JB (2008) Metal and proton binding onto the roots of *Fescue rubra*. Chem Geol 253(3–4):130–135

Godzik B (1993) Heavy metals content in plants from zinc dumps and reference areas. Pol Bot Stud 5

Gu SH, Zhu JZ, Gu ZL (1989) Study on the critical lead content of red paddy soil. Agro-Environ Prot 8:17–22

Gul I, Manzoor M, Silvestre J, Rizwan M, Hina K, Kallerhoff J, Arshad M (2019) EDTA-assisted phytoextraction of lead and cadmium by Pelargonium cultivars grown on spiked soil. Int J Phytorem 21(2):101–110

Gupta DK, Srivastava A, Singh VP (2006) Phytoremediation of induced lead toxicity in *Vigna mungo* (L.) hepper by vetiver grass. Rohilkhand University, Bareilly

Gupta DK, Nicoloso FT, Schetinger MRC, Rossato LV, Pereira LB, Castro GY, Srivastava V, Tripathi RD (2009) Antioxidant defense mechanism in hydroponically grown *Zea mays* seedlings under moderate lead stress. J Hazard Mater 172(1):479–484

Hadi F, Aziz T (2015) A mini review on lead (Pb) toxicity in plants. J Biol Life Sci 6(2):91–101

Hall JÁ (2002) Cellular mechanisms for heavy metal detoxification and tolerance. J Exp Bot 53(366):1–11

Hameed A, Afzal I, Iqbal N (2010) Seed priming and salinity induced variations in wheat (*Triticum aestivum* L.) leaf protein profile. Seed Sci Technol 38(1):236–241

Hassan M, Mansoor S (2014) Oxidative stress and antioxidant defense mechanism in mung bean seedlings after lead and cadmium treatments. Turk J Agric For 38(1):55–61

Hirsch RE, Lewis BD, Spalding EP, Sussman MR (1998) A role for the AKT1 potassium channel in plant nutrition. Science 280(5365):918–921

Huang JW, Pellet DM, Papernik LA, Kochian LV (1996) Aluminum interactions with voltage-dependent calcium transport in plasma membrane vesicles isolated from roots of aluminum-sensitive and-resistant wheat cultivars. Plant Physiol 110(2):561–569

Hung NM, Hiep NV, Dung BN, Hai NX (2014) Lead accumulation in different parts of okra plant (*Abelmoschus esculentus*). ARPN J Agric Biol Sci 9(6):190–194

Hussain I, Chaudhry Z, Muhammad A, Asghar R, Naqvi SS, Rashid H (2006) Effect of chlorocholine chloride, sucrose and BAP on in vitro tuberization in potato (*Solanum tuberosum* L. cv. Cardinal). Pak J Bot 38(2):275

Hussain A, Abbas N, Arshad F, Akram M, Khan ZI, Ahmad K, Mansha M, Mirzaei F (2013) Effects of diverse doses of lead (Pb) on different growth attributes of *Zea mays* L.

Islam E, Yang X, Li T, Liu D, Jin X, Meng F (2007) Effect of Pb toxicity on root morphology, physiology and ultrastructure in the two ecotypes of *Elsholtzia argyi*. J Hazard Mater 147(3):806–816

Jarvis MD, Leung DWM (2002) Chelated lead transport in *Pinus radiata*: an ultrastructural study. Environ Exp Bot 48(1):21–32

Jiang W, Liu D (2010) Pb-induced cellular defense system in the root meristematic cells of *Allium sativum* L. BMC Plant Biol 10:1–8

Jiang M, Liu S, Li Y, Li X, Luo Z, Song H, Chen Q (2019) EDTA-facilitated toxic tolerance, absorption and translocation and phytoremediation of lead by dwarf bamboos. Ecotoxicol Environ Saf 170:502–512

John R, Ahmad P, Gadgil K, Sharma S (2012) Heavy metal toxicity: effect on plant growth, biochemical parameters and metal accumulation by *Brassica juncea* L. Int J Plant Prod 3(3):65–76

Kanwal U, Ali S, Shakoor MB, Farid M, Hussain S, Yasmeen T, Adrees M, Bharwana SA, Abbas F (2014) EDTA ameliorates phytoextraction of lead and plant growth by reducing morphological and biochemical injuries in *Brassica napus* L. under lead stress. Environ Sci Pollut Res 21:9899–9910

Khan DH, Frankland B (1983) Effects of cadmium and lead on radish plants with particular reference to movement of metals through soil profile and plant. Plant Soil 70:335–345

Kaur G, Singh HP, Batish DR, Kumar RK (2012) Growth, photosynthetic activity and oxidative stress in wheat (*Triticum aestivum*) after exposure of lead to soil. J Environ

Kaur L, Gadgil K, Sharma S (2013) Comparative study of natural phyto-extraction and induced phyto-extraction of lead using mustard plant (*Brassica juncea* arawali). Int J Bioassays 2(1):352–357

Kohli SK, Handa N, Bali S, Arora S, Sharma A, Kaur R, Bhardwaj R (2018a) Modulation of antioxidative defense expression and osmolyte content by co-application of 24-epibrassinolide and salicylic acid in Pb exposed Indian mustard plants. Ecotoxicol Environ Saf 147:382–393

Kohli SK, Handa N, Sharma A, Gautam V, Arora S, Bhardwaj R, Wijaya L, Alyemeni MN, Ahmad P (2018b) Interaction of 24-epibrassinolide and salicylic acid regulates pigment contents, antioxidative defense responses, and gene expression in *Brassica juncea* L. seedlings under Pb stress. Environ Sci Pollut Res 25(15):15159–15173

Kohli SK, Handa N, Sharma A, Kumar V, Kaur P, Bhardwaj R (2017) Synergistic effect of 24-epibrassinolide and salicylic acid on photosynthetic efficiency and gene expression in *Brassica juncea* L. under Pb stress. Turk J Biol 41(6):943–953

Kopittke PM, Asher CJ, Kopittke RA, Menzies NW (2007) Toxic effects of Pb^{2+} on growth of cowpea (*Vigna unguiculata*). Environ Pollut 150(2):280–287

Kopittke PM, Asher CJ, Blamey FPC, Auchterlonie GJ, Guo YN, Menzies NW (2008a) Localization and chemical speciation of Pb in roots of signal grass (*Brachiaria decumbens*) and Rhodes grass (*Chloris gayana*). Environ Sci Technol 42:4595–4599

Kopittke PM, Asher CJ, Menzies NW (2008b) Prediction of Pb speciation in concentrated and dilute nutrient solutions. Environmen Pollut 153(3):548–554

Kosobrukhov A, Knyazeva I, Mudrik V (2004) Plantago major plants responses to increase content of lead in soil: growth and photosynthesis. Plant Growth Regul 42:145–151

Kroschwitz JI (1995) Iron compounds. In: Kirk RE, Othmer DF (eds) Encyclopedia of chemical technology, vol 4, pp 887–888

Krotz RM, Evangelou BP, Wagner GJ (1989) Relationships between cadmium, zinc, Cd-peptide, and organic acid in tobacco suspension cells. Plant Physiol 91(2):780–787

Krzesłowska M, Lenartowska M, Mellerowicz EJ, Samardakiewicz S, Woźny A (2009) Pectinous cell wall thickenings formation—a response of moss protonemata cells to lead. Environ Exp Bot 65(1):119–131

Krzesłowska M, Lenartowska M, Samardakiewicz S, Bilski H, Woźny A (2010) Lead deposited in the cell wall of *Funaria hygrometrica* protonemata is not stable—a remobilization can occur. Environ Pollut 158(1):325–338

Kumar M, Jayaraman P (2014) Toxic effect of lead nitrate [$Pb(NO_3)_2$] on the black gram seedlings (*Vigna mungo* L. Hepper). Int J Adv Res Biol Sci 1(9):209–213

Kumar G, Rai PK (2007) EMS induced karyomorphological variations in maize (*Zea mays* L.) inbreds. Turk J Biol 31(4):187–195

Kushwaha A, Hans N, Kumar S, Rani R (2018) A critical review on speciation, mobilization and toxicity of lead in soil-microbe-plant system and bioremediation strategies. Ecotoxicol Environ Saf 147:1035–1045

Lamhamdi M, Bakrim A, Aarab A, Lafont R, Sayah F (2010) A comparison of lead toxicity using physiological and enzymatic parameters on spinach (*Spinacia oleracea*) and wheat (*Triticum aestivum*) growth. Moroccan J Biol 6(7):64–73

Lamhamdi M, Bakrim A, Aarab A, Lafont R, Sayah F (2011) Lead phytotoxicity on wheat (*Triticum aestivum L.*) seed germination and seedlings growth. C R Biol 334(2):118–126

Lane SD, Martin ES (1982) An ultrastructural examination of lead localisation in germinating seeds of *Raphanus sativus*. Z Pflanzenphysiol 107(1):33–40

Lathwal M, Rani M, Indira A, Chongtham N (2023) Bamboo: a sustainable alternative for biochar production. In: Bamboo science and technology. Springer Nature Singapore, Singapore, pp 265–295

Lee SZ, Chang L, Yang HH, Chen CM, Liu MC (1998) Adsorption characteristics of lead onto soils. J Hazard Mater 63(1):37–49

Legocka J, Sobieszczuk-Nowicka E, Wojtyla Ł, Samardakiewicz S (2015) Lead-stress induced changes in the content of free, thylakoid- and chromatin-bound polyamines, photosynthetic parameters and ultrastructure in greening barley leaves. J Plant Physiol 186:15–24

Liao J, Cai X, Yang Y, Chen Q, Gao S, Liu G, Sun L, Luo Z, Lei T, Jiang M (2021) Dynamic study of the lead (Pb) tolerance and accumulation characteristics of new dwarf bamboo in Pb-contaminated soil. Chemosphere 282:131089

Little PE (1978) Deposition of exhaust lead and its impact on plants. In: Symposium "the impact of road traffic on plants", Sept 1978, vol 513, pp 49–54

Liu CJ (2012) Deciphering the enigma of lignification: precursor transport, oxidation, and the topochemistry of lignin assembly. Mol Plant 5(2):304–317

Liu D, Li TQ, Jin XF, Yang XE, Islam E, Mahmood Q (2008) Lead induced changes in the growth and antioxidant metabolism of the lead accumulating and non-accumulating ecotypes of *Sedum alfredii*. J Integr Plant Biol 50(2):129–140

Liu T, Liu S, Guan H, Ma L, Chen Z, Gu H, Qu LJ (2009) Transcriptional profiling of Arabidopsis seedlings in response to heavy metal lead (Pb). Environ Exp Bot 67(2):377–386

Luna CM, González CA, Trippi VS (1994) Oxidative damage caused by an excess of copper in oat leaves. Plant Cell Physiol 35(1):11–15
Ma L, Kong L, Gui R, Yang X, Zhang J, Gong Q, Qin D, Zhuang M, Ashraf U, Mo Z (2021) Application of hydrogen-rich water modulates physio-biochemical functions and early growth of fragrant rice under Cd and Pb stress. Environ Sci Pollut Res 28:58558–58569
Madany IM, Ali SM, Akhter MS (1990) Assessment of lead in roadside vegetation in Bahrain. Environ Int 16(2):123–126
Małecka A, Piechalak A, Morkunas I, Tomaszewska B (2008) Accumulation of lead in root cells of *Pisum sativum.* Acta Physiol Plant 30:629–637
Mani D, Kumar C, Patel NK, Sivakumar D (2015) Enhanced clean-up of lead-contaminated alluvial soil through *Chrysanthemum indicum* L. Int J Environ Sci Technol 12:1211–1222
Marshall J, Corzo A, Leigh RA, Sanders D (1994) Membrane potential-dependent calcium transport in right-side-out plasma membrane vesicles from *Zea mays* L. roots. Plant J 5(5):683–694
Martınez CE, Motto HL (2000) Solubility of lead, zinc and copper added to mineral soils. Environ Pollut 107(1):153–158
Mazen AMA, El Maghraby OMO (1997) Accumulation of cadmium, lead and strontium, and a role of calcium oxalate in water hyacinth tolerance. Biol Plant 40:411–417
Mench M, Morel JL, Guckert A (1987) Metal binding properties of high molecular weight soluble exudates from maize (*Zea mays* L.) roots. Biol Fert Soils 3:165–169
Meyers DE, Auchterlonie GJ, Webb RI, Wood B (2008) Uptake and localisation of lead in the root system of *Brassica juncea.* Environ Pollut 153(2):323–332
Mishra S, Srivastava S, Tripathi RD, Kumar R, Seth CS, Gupta DK (2006) Lead detoxification by coontail (*Ceratophyllum demersum* L.) involves induction of phytochelatins and antioxidant system in response to its accumulation. Chemosphere 65(6):1027–1039
Misra P, Nath K, Tandon PK (2010) Effect of heavy metals (Ni and Pb) stress on sugarcane (*Saccharum officinarum* L.). Res Environ Life Sci 3(4):183–188
Mohanty N, Vass I, Demeter S (1989) Impairment of photosystem 2 activity at the level of secondary quinone electron acceptor in chloroplasts treated with cobalt, nickel and zinc ions. Physiol Plant 76(3):386–390
Monferrán MV, Wunderlin DA (2013) Biochemistry of metals/metalloids toward remediation process. In: Heavy metal stress in plants, pp 43–71
Mosavian SN, Chaab A (2012) Response of morphologic and physiologic two variety canola to stress concentration of lead. Adv Environ Biol 6:1503–1509
Mrozek E Jr, Funicelli NA (1982) Effect of zinc and lead on germination of *Spartina alterniflora* Loisel seeds at various salinities. Environ Exp Bot 22(1):23–32
Mukherji S, Maitra P (1976) Toxic effects of lead on growth and metabolism of germinating rice (*Oryza sativa* L.) seeds on mitosis of onion (*Allium cepa*) root tip cells. Indian J Exp Biol 14:519–521
Mukhtar SAIMA, Bhatti HN, Khalid M, Haq MAU, Shahzad SM (2010) Potential of sunflower (*Helianthus annuus* L.) for phytoremediation of nickel (Ni) and lead (Pb) contaminated water. Pak J Bot 42(6):4017–4026
Nakos G (1979) Fate of lead in the soil and its effects on *Pinus halepensis.* Plant Soil 53:427–443
Nareshkumar A, Veeranagamallaiah G, Pandurangaiah M, Kiranmai K, Amaranathareddy V, Lokesh U, Venkatesh B, Sudhakar C (2015) Pb-stress induced oxidative stress caused alterations in antioxidant efficacy in two groundnut (*Arachis hypogaea* L.) cultivars. Agric Sci 6(10):1283
Nicklow CW, Comas-Haezebrouck PH, Feder WA (1983) Influence of varying soil lead levels on lead uptake of leafy and root vegetables. J Am Soc Hort Sci 108:193–195
Ojwang LO, Banerjee N, Noratto GD, Angel-Morales G, Hachibamba T, Awika JM, Mertens-Talcott SU (2015) Polyphenolic extracts from cowpea (*Vigna unguiculata*) protect colonic myofibroblasts (CCD18Co cells) from lipopolysaccharide (LPS)-induced inflammation–modulation of microRNA 126. Food Funct 6(1):145–153

Opeolu BO, Adenuga OO, Ndakidemi PA, Olujimi OO (2010) Assessment of phyto-toxicity potential of lead on tomato (*Lycopersicon esculentum* L) planted on contaminated soils. Int J Phys Sci

Parys E, Romanowska E, Siedlecka M, Poskuta JW (1998) The effect of lead on photosynthesis and respiration in detached leaves and in mesophyll protoplasts of *Pisum sativum*. Acta Physiol Plant 20:313–322

Patra M, Bhowmik N, Bandopadhyay B, Sharma A (2004) Comparison of mercury, lead and arsenic with respect to genotoxic effects on plant systems and the development of genetic tolerance. Environ Exp Bot 52(3):199–223

Pattee OH, Pain DJ (2003) Lead in the environment. In: Handbook of ecotoxicology, vol 2, pp 373–399

Pelloux J, Rusterucci C, Mellerowicz EJ (2007) New insights into pectin methylesterase structure and function. Trends Plant Sci 12(6):267–277

Pratima M, Pratima H (2016) Copper toxicity causes oxidative stress in *Brassica juncea* L. seedlings. Indian J Plant Physiol 14(4):397–401

Probst A, Liu H, Fanjul M, Liao B, Hollande E (2009) Response of *Vicia faba* L. to metal toxicity on mine tailing substrate: geochemical and morphological changes in leaf and root. Environ Exp Bot 66(2):297–308

Punamiya P, Datta R, Sarkar D, Barber S, Patel M, Das P (2010) Symbiotic role of *Glomus mosseae* in phytoextraction of lead in vetiver grass [*Chrysopogon zizanioides* (L.)]. J Hazard Mater 177:465–474

Qufei L, Fashui H (2009) Effects of Pb^{2+} on the structure and function of photosystem II of *Spirodela polyrrhiza*. Biol Trace Elem Res 129:251–260

Qureshi JA, Hardwick K, Collin HA (1986) Intracellular localization of lead in a lead tolerant and sensitive clone of *Anthoxanthum odoratum*. J Plant Physiol 122(4):357–364

Qureshi MI, Abdin MZ, Qadir S, Iqbal M (2007) Lead-induced oxidative stress and metabolic alterations in *Cassia angustifolia* Vahl. Biol Plant 51:121–128

Rani M, Lathwal M, Singh AN, Chongtham N (2023) Bamboo act as a phytoremediation candidate for heavy metal contaminated soil: a synthesis. In: Bamboo science and technology. Springer Nature Singapore, Singapore, pp 125–161

Ranieri A, D'Urso G, Nali C, Lorenzini G, Soldatini GF (1996) Ozone stimulates apoplastic antioxidant systems in pumpkin leaves. Physiol Plant 97(2):381–387

Rao MV, Dubey PS (1992) Occurrence of heavy metals in air and their accumulation by tropical plants growing around an industrial area. Sci Total Environ 126(1–2):1–16

Rashid A, Bernier M, Pazdernick L, Carpentier R (1991) Interaction of Zn^{2+} with the donor side of Photosystem II. Photosynth Res 30:123–130

Raven JA, Evans MC, Korb RE (1999) The role of trace metals in photosynthetic electron transport in O_2-evolving organisms. Photosynth Res 60:111–150

Rebechini HM, Hanzely L (1974) Lead-induced ultrastructural changes in chloroplasts of the hydrophyte *Ceratophyllum demersum*. Z Pflanzenphysiol 73(5):377–386

Reddy AM, Kumar SG, Jyothsnakumari G, Thimmanaik S, Sudhakar C (2005) Lead induced changes in antioxidant metabolism of horsegram (*Macrotyloma uniflorum* (Lam.) Verdc.) and bengalgram (*Cicer arietinum* L.). Chemosphere 60(1):97–104

Rehman MZU, Rizwan M, Ali S, Sabir M, Sohail MI (2017) Contrasting effects of organic and inorganic amendments on reducing lead toxicity in wheat. Bull Environ Contam Toxicol 99:642–647

Rizwan M, Ali S, Rehman MZU, Javed MR, Bashir A (2018) Lead toxicity in cereals and its management strategies: a critical review. Water Air Soil Pollut 229:1–16

Rodriguez E, da Conceição Santos M, Azevedo R, Correia C, Moutinho-Pereira J, Ferreira de Oliveira JMP, Dias MC (2015) Photosynthesis light-independent reactions are sensitive biomarkers to monitor lead phytotoxicity in a Pb-tolerant *Pisum sativum* cultivar. Environ Sci Pollut Res 22:574–585

Romanowska E, Igamberdiev AU, Parys E, Gardeström P (2002) Stimulation of respiration by Pb^{2+} in detached leaves and mitochondria of C3 and C4 plants. Physiol Plant 116(2):148–154

Romanowska E, Pokorska B, Siedlecka M (2005) The effects of oligomycin on content of adenylates in mesophyll protoplasts, chloroplasts and mitochondria from Pb2+ treated pea and barley leaves. Acta Physiologiae Plant 27:29–36

Romanowska E, Wróblewska B, Drozak A, Siedlecka M (2006) High light intensity protects photosynthetic apparatus of pea plants against exposure to lead. Plant Physiol Biochem 44(5–6):387–394

Romanowska E, Wróblewska B, Drożak A, Zienkiewicz M, Siedlecka M (2008) Effect of Pb ions on superoxide dismutase and catalase activities in leaves of pea plants grown in high and low irradiance. Biol Plant 52:80–86

Sahi SV, Bryant NL, Sharma NC, Singh SR (2002) Characterization of a lead hyperaccumulator shrub, *Sesbania drummondii*. Environ Sci Technol 36(21):4676–4680

Saleem M, Asghar HN, Zahir ZA, Shahid M (2018) Impact of lead tolerant plant growth promoting rhizobacteria on growth, physiology, antioxidant activities, yield and lead content in sunflower in lead contaminated soil. Chemosphere 195:606–614

Salt DE, Rauser WE (1995) MgATP-dependent transport of phytochelatins across the tonoplast of oat roots. Plant Physiol 107(4):1293–1301

Sauve S, McBride M, Hendershot W (1998) Soil solution speciation of lead (II): effects of organic matter and pH. Soil Sci Soc Am J 62(3):618–621

Sehmer L, Fontaine V, Antoni F, Dizengremel P (1998) Effects of ozone and elevated atmospheric carbon dioxide on carbohydrate metabolism of spruce needles. Catabolic and detoxification pathways. Physiol Plant 102(4):605–611

Sengar RS, Gautam M, Sengar RS, Sengar RS, Garg SK, Sengar K, Chaudhary R (2008) Lead stress effects on physiobiochemical activities of higher plants. Rev Environ Contam Toxicol 196:73–93

Seregin IV, Ivanov VB (2001) Physiological aspects of cadmium and lead toxic effects on higher plants. Russ J Plant Physiol 48:523–544

Šeršeň F, Král'ová K, Bumbalova A (1998) Action of mercury on the photosynthetic apparatus of spinach chloroplasts. Photosynthetica 35:551–559

Sharma P, Dubey RS (2005) Lead toxicity in plants. Braz J Plant Physiol 17:35–52

Shu X, Yin L, Zhang Q, Wang W (2012) Effect of Pb toxicity on leaf growth, antioxidant enzyme activities, and photosynthesis in cuttings and seedlings of *Jatropha curcas* L. Environ Sci Pollut Res 19:893–902

Sidhu GPS, Singh HP, Batish DR, Kohli RK (2016) Effect of lead on oxidative status, antioxidative response and metal accumulation in *Coronopus didymus*. Plant Physiol Biochem 105:290–296

Silva S, Pinto G, Santos C (2017) Low doses of Pb affected *Lactuca sativa* photosynthetic performance. Photosynthetica 55(1):50–57

Singh R, Tripathi RD, Dwivedi S, Kumar A, Trivedi PK, Chakrabarty D (2010) Lead bioaccumulation potential of an aquatic macrophyte *Najas indica* are related to antioxidant system. Bioresour Technol 101(9):3025–3032

Sorial ME, Abd El-Fattah MA (2001) Alleviation of the adverse effects of lead and cadmium on growth and chemical changes of pea plant by phosphorus and calcium treatments. J Pest Control Environ Sci 9:161–194

Sridhar BM, Diehl SV, Han FX, Monts DL, Su Y (2005) Anatomical changes due to uptake and accumulation of Zn and Cd in Indian mustard (*Brassica juncea*). Environ Exp Bot 54(2):131–141

Stefanov K, Seizova K, Popova I, Petkov V, Kimenov G, Popov S (1995) Effect of lead ions on the phospholipid composition in leaves of *Zea mays* and *Phaseolus vulgaris*. J Plant Physiol 147(2):243–246

Strange J, Macnair MR (1991) Evidence for a role for the cell membrane in copper tolerance of *Mimulus guttatus* Fischer ex DC. New Phytol 119(3):383–388

Tanaka K, Suda Y, Kondo N, Sugahara K (1985) O_3 tolerance and the ascorbate-dependent H_2O_2 decomposing system in chloroplasts. Plant Cell Physiol 26(7):1425–1431

Tan-Wilson AL, Wilson KA (2012) Mobilization of seed protein reserves. Physiol Plant 145(1):140–153
Thakur S, Singh L, Zularisam AW, Sakinah M, Din MFM (2017) Lead induced oxidative stress and alteration in the activities of antioxidative enzymes in rice shoots. Biol Plant 61:595–598. Biology 33(2):265
Tomulescu IM, Radoviciu EM, Merca VV, Tuduce AD (2004) Effect of copper, zinc and lead and their combinations on the germination capacity of two cereals. Acta Agrar Debr 15:39–42
Tripathi DK, Singh VP, Prasad SM, Dubey NK, Chauhan DK, Rai AK (2016) LIB spectroscopic and biochemical analysis to characterize lead toxicity alleviative nature of silicon in wheat (*Triticum aestivum* L.) seedlings. J Photochem Photobiol B Biol 154:89–98
Uzu G, Sobanska S, Aliouane Y, Pradere P, Dumat C (2009) Study of lead phytoavailability for atmospheric industrial micronic and sub-micronic particles in relation with lead speciation. Environ Pollut 157(4):1178–1185
Uzu G, Sobanska S, Sarret G, Munoz M, Dumat C (2010) Foliar lead uptake by lettuce exposed to atmospheric fallouts. Environ Sci Technol 44(3):1036–1042
Vallee BL, Ulmer DD (1972) Biochemical effects of mercury, cadmium and lead. Annu Rev Biochem 41:91–128
Vassil AD, Kapulnik Y, Raskin I, Salt DE (1998) The role of EDTA in lead transport and accumulation by Indian mustard. Plant Physiol 117(2):447–453
Verma S, Dubey RS (2003) Lead toxicity induces lipid peroxidation and alters the activities of antioxidant enzymes in growing rice plants. Plant Sci 164(4):645–655
Vodnik D, Jentschke G, Fritz E, Gogala N, Godbold DL (1999) Root-applied cytokinin reduces lead uptake and affects its distribution in Norway spruce seedlings. Physiol Plant 106(1):75
Vögeli-Lange R, Wagner GJ (1990) Subcellular localization of cadmium and cadmium-binding peptides in tobacco leaves: implication of a transport function for cadmium-binding peptides. Plant Physiol 92(4):1086–1093
Vögeli-Lange R, Wagner GJ (1996) Relationship between cadmium, glutathione and cadmium-binding peptides (phytochelatins) in leaves of intact tobacco seedlings. Plant Sci 114(1):11–18
Wang HH, Shan XQ, Wen B, Owens G, Fang J, Zhang SZ (2007) Effect of indole-3-acetic acid on lead accumulation in maize (*Zea mays* L.) seedlings and the relevant antioxidant response. Environ Exp Bot 61(3):246–253
Wierzbicka M (1987) Lead accumulation and its translocation barriers in roots of *Allium cepa* L.—autoradiographic and ultrastructural studies. Plant Cell Environ 10(1):17–26
Wierzbicka M, Antosiewicz D (1993) How lead can easily enter the food chain—a study of plant roots. Sci Total Environ 134:423–429
Wierzbicka M, Obidzińska J (1998) The effect of lead on seed imbibition and germination in different plant species. Plant Sci 137(2):155–171
Wójcik M, Tukiendorf A (2003) Response of wild type of *Arabidopsis thaliana* to copper stress. Biol Plant 46:79–84
World Health Organization (2017) Lead poisoning and health fact sheet
Xian X (1989) Response of kidney bean to concentration and chemical form of cadmium, zinc, and lead in polluted soils. Environ Pollut 57(2):127–137
Xiong ZT, Zhao F, Li MJ (2006) Lead toxicity in *Brassica pekinensis* Rupr.: effect on nitrate assimilation and growth. Environ Toxicol Int J 21(2):147–153
Xiong J, Koopal LK, Tan W, Fang L, Wang M, Zhao W, Liu F, Zhang J, Weng L (2013) Lead binding to soil fulvic and humic acids: NICA-Donnan modeling and XAFS spectroscopy. Environ Sci Technol 47(20):11634–11642
Xu Q, Duan D, Cai Q, Shi J (2018) Influence of humic acid on Pb uptake and accumulation in tea plants. J Agric Food Chem 66(46):12327–12334
Yang Y, Zhang L, Huang X, Zhou Y, Quan Q, Li Y, Zhu X (2020) Response of photosynthesis to different concentrations of heavy metals in *Davidia involucrata*. PLoS ONE 15(3):e0228563
Yongsheng W, Qihui L, Qian T (2011) Effect of Pb on growth, accumulation and quality component of tea plant. Procedia Eng 18:214–219

Zegers PV, Harmet KH, Hanzely L (1976) Inhibition of IAA-induced elongation in Avena coleoptile segments by lead: a physiological and electron microscopic study. Cytobios 15:23–35

Zhong B, Chen J, Shafi M, Guo J, Wang Y, Wu J, Ye Z, He L, Liu D (2017) Effect of lead (Pb) on antioxidation system and accumulation ability of Moso bamboo (*Phyllostachys pubescens*). Ecotoxicol Environ Saf 138:71–77

Zulfiqar U, Farooq M, Hussain S, Maqsood M, Hussain M, Ishfaq M, Ahmad M, Anjum MZ (2019) Lead toxicity in plants: impacts and remediation. J Environ Manage 250:109557

Chapter 11
Physico Chemical and Biological Treatment Techniques for Lead Removal from Wastewater: A Review

Simmi Goel

Abstract The release of untreated industrial effluent loaded with heavy metals especially lead ions seems to have adverse effects on various components of ecosystem. A variety of physicochemical methods for the treatment of lead-contaminated water have been used on a commercial level. But these methods are having a lot of limitations like reduced efficiency, costly due to input of a load of chemicals and not environment friendly. In lieu of this, major focus is given on the use of biological methods especially biosorption for treatment of industrial effluent. In this review, various sources and health effects of lead-contaminated water has been given. A variety of physiochemical methods for the treatment of industrial wastewater and also limitations have been discussed. Significance of various biological methods over other conventional treatments including the use of agricultural waste, plant- or animal-based sorbents and enlisting the list of microbes like bacteria, fungi and algae as sorbents for treatment of lead waste water including their mechanism of action have been reviewed. The factors which need to be optimized for maximum removal of lead ions to increase the efficiency of treatment have also been discussed.

Keywords Lead contamination · Biosorption · Bioremediation · Biomass · Industrial effluent · Sorbents · Optimized parameters · Wastewater treatment

11.1 Introduction

A large number of electroplating industries discharge untreated wastewater into the ecosystem. The easy absorption of heavy metals in humans resulted due to their high dissolution capability in water. These heavy metals once released into environment their concentration increases with each trophic level resulted in bioaccumulation. Consumption of heavy metal–contaminated water beyond the permitted

---

S. Goel (✉)
Department of Biotechnology, Mata Gujri College, Fatehgarh Sahib, Punjab, India
e-mail: simmibiotech@matagujricollege.org

N. Kumar and A. K. Jha (eds.), *Lead Toxicity Mitigation: Sustainable Nexus Approaches*, Environmental Contamination Remediation and Management,
https://doi.org/10.1007/978-3-031-46146-0_11

level results in serious health disorders. The heavy metal–contaminated wastewater when discharged untreated in the environment resulted in the ecosystem degradation. The various manmade activities like batteries manufacturing, automobile industries, mining activities, leaching of ores, agricultural inputs in the form of chemicals are the major contributors of heavy metals into the ecosystem. These are difficult to degrade under natural conditions, altering the bio-geochemical cycles, disturbance of ecological diversity and ultimately destroying the whole ecosystem (Chug et al. 2022). Once these metals enter into the food chain, there occurs a huge concentration rise with each successive trophic level. The consumption of heavy metal–contaminated water causes kidney infections, respiratory dysfunction and cardiovascular imbalance and carcinogenic in acute cases. The industries release a large amount of heavy metal–loaded wastewater, among which lead (Pb) seems to be the major toxicity contributor released from metal finishing, electroplating and paint manufacturing, leaching, battery operations, etc. Lead once discharged persists into environment for a long period due to its non-biodegradable nature (Raut et al. 2015). Lead is discharged naturally from leaching of ores and also due to major human source like exhaust fumes of automobiles from where lead gets accumulated on the surface of road side vegetation or deposited within soil particles and ultimately through rain water enters into the rivers and ground water. Lead enters in human body through contaminated drinking water and accumulated in bones causing neurotoxicity and carcinogenicity (Biela and Sopikova 2017). Once lead is ingested in human body, it acts as metabolic poison resulted in inhibiting various enzymatic actions (Lo et al. 1999). In addition, the intake of lead-contaminated water can cause infections in kidneys, nervous breakdown and dysfunction of reproductive organs, heart, liver and brain (Naiyaa et al. 2009). In adults, lead poisoning can cause neurological disorders resulting in weakness, irritation, poor attention span, headaches, muscle cramps, memory loss and hallucinations and ultimately death. Lead ions (Pb^{2+}) can cause mental retardation, kidney damage, anaemia, central nervous system dysfunction, alters haemoglobin production, reproductive failure and gastrointestinal tract infections. The lead particles released in atmosphere get deposited on the surface of fruits, vegetables, soil and groundwater adversely affecting the health of especially the pregnant women and young children (Ghosal et al. 2021). In lieu of all these facts, there seems to be an urgent need to develop effective treatment technologies for the removal of Pb from water and wastewater to keep the safety and protection of human health and environment.

11.2 Sources of Lead

Major sources of lead are acid battery manufacturing, metal plating and finishing, mining operations, battery services, ammunition, tanneries, petroleum refining, tetraethyl lead manufacturing, paint industries, use of heavy metal–loaded chemicals in agricultural fields, pigment manufacture, coal combustion power plants, ceramic and glass usage, printing and photographic activities, lead water pipes (Goel et al.

2005; Momcilovic et al. 2011). Other contributors of lead used to be the exhaust gases of automobiles, which ultimately accumulate on the surface of crops growing alongside roads, entering the food chain, retained in water droplets in atmosphere and released in form of rains leading to the contamination of surface water and groundwater bodies. The surface and groundwater lead contamination also occurs due to the accidental leakages from rusting of the water pipes and waste water from the ferrous metallurgy operations, batteries discharge and glass manufacturing industry (Pitter 2009; Biela and Sopikova 2017).

11.3 Effects of Lead Exposure

The consumption of lead-contaminated water can severely affect the functioning of kidneys, liver, brain, nervous disorders, reproductive failure, hypertension ultimately leading to illness or death of individuals. Fatal exposure to lead-loaded wastewater can cause reproductive failures and foetus loss resulting in abortions. Exposure to lead ions (Pb^{2+}) can also cause mental retardation, kidney failure, anaemia, retards the functioning of central nervous system, haemoglobin production, reproductive failure and gastrointestinal infections. Lead particles released in air contaminate the agricultural food by depositing on surface of fruits, vegetables, soil and water. Other health effects of lead exposure are uneasiness, irritable behaviour, loss of focus/ concentration, migraines, muscular tremors, abdominal cramps, infections in kidney, hallucinations and memory loss. (WHO 2011; Biela and Sopikova 2017). Some of essential ions can be displaced or substituted from their cellular locations resulting in retardation of functioning of enzyme, polynucleotides and essential nutrient transport systems due to consumption of mercury and lead-contaminated water. Additionally, these also resulted in the denaturation and inactivation of enzymes and alters the integrity of cellular and organelle membranes (Zhang et al. 2016). In lieu of all these adverse effects, a very minute concentration of lead in drinking water seems to be highly toxic and so there is a strong need for developing an efficient removal, environment friendly and economical treatment technology (Mahmoud et al. 2012; Ghosal et al. 2021).

Atmospheric lead particles get deposited on vegetation and absorbed within aquatic organisms (Jamali et al. 2009), finds its way to enter the human body through the food chain. The lead contaminated soil alters the physiological, morphological and biochemical characteristics of plants by making their availability in the roots, stems, leaves and fruits through contaminated soil. Other severe threats of exposure to lead resulted in carcinogenesis, teratogenesis and gene mutations in humans (Wendt and Lee 2010). All these threats diverted the attention of scientists and researchers to explore effective and economical methods to remediate lead (Pb) pollution from wastewater (Jing et al. 2021a, b).

The intake of lead (Pb) through contaminated fruits, vegetables, water etc., causes anaemia, renal infections, nervous breakdown and ultimately death. Excessive intake leads to hepatic and renal failure, respiratory infections, reproductive disorders, capillary damage, gastrointestinal irritation and central nervous system irritation, nausea, encephalopathy, severe migraine and vomiting, learning disability, mental disorders, hyperactive behaviour, muscular tremors, liver cirrhosis, dysfunction of thyroid gland, sleep disorders, weakness, schizophrenia (Kale et al. 2018). Heavy metal intake can cause allergies, carcinogenic, organ damage, impairs growth and development due to their persistent or non-biodegradable nature. Overall lead poisoning can cause anaemic behaviour, kidney disorder, brain disorder and even death in extreme poisoning situation (Acharya et al. 2013; Singha and Das 2015). According to WHO, the acceptable safe limit of lead in drinking water is 0.01 g/L.

11.4 Methods for the Removal of Lead from Wastewater

The common modes for the removal of lead (Pb) from contaminated wastewater include physical methods such as ultra-filtration, coagulation, flocculation, adsorption, membrane filtration, floatation, reverse osmosis and ion exchange and chemical methods include neutralization, solvent extraction, chemical precipitation and electrochemical treatment.

Among these methods, adsorption seems to be highly effective and economical mode. The commonly used adsorbents for wastewater treatment are silica gels, activated alumina, metal oxides and hydroxides, zeolites, clay minerals, synthetic polymers, and carbonaceous materials, such as activated carbon and molecular carbon sieves.

11.4.1 Adsorption

The adsorption capacity of hydroxyapatite nanorods and chitosan nanocomposite was assessed for the removal of lead ions from aqueous solution in a batch mode of experimentation (Mohammad et al. 2015).

11.4.2 Chemical Precipitation

Chemical precipitation is based on the reaction of heavy metals with added chemicals resulting in the formation of insoluble precipitates in the form of hydroxides, sulphides, carbonates which can further separated by filtration. Addition of caustic soda increases the precipitation of dissolved lead (Pb) from wastewater in the form of solid metal hydroxide particles. Various coagulants and flocculants are added to increase their particle size so that these can be easily removed in the form of sludge. The precipitates can be further separated using sedimentation or filtration techniques.

Various parameters which need to be optimized for increasing the efficacy of lead removal are low pH, temperature, content of lead ions, contact time, presence of other ions and charge of ions (Ahluwalia and Goyal 2007).

11.4.3 Ion Exchange

This method is based on the ion exchange capacity of various cations or anions like zeolites or resins to remove the metal ions in the solution. The method is based on the concept of attracting soluble ions from the liquid phase and their transition to the solid phase, commonly used method in water treatment industry.

Ion exchange resins have the property of absorbing cations or anions from an electrolyte solution and release other ions with the same charges into the solution in an equivalent amount. Strong acid cation or weak acid cation or anion resins are used for lead removal. For example, the positively charged ions in cationic resins such as hydrogen and sodium ions are exchanged with positively charged ions, such as nickel, copper, zinc, copper, silver, cadmium, gold, mercury, lead, chromium, iron, tin, arsenic, selenium, molybdenum, cobalt, manganese and aluminium ions in the solutions. The negative ions in the resins such as hydroxyl and chloride ions can be replaced by the negatively charged ions such as chromate, sulphate, nitrate, cyanide and dissolved organic carbon.

11.4.4 Coagulation–Flocculation

This process is based on the capacity of the electrostatic bonding between heavy metal and coagulant–flocculants agents to form multi-charged polynuclear complexes. The commonly used metal coagulants to hydrolyse the metal ions are aluminium sulphate, aluminium chloride, ferric sulphate, ferrous sulphate, ferric chloride, hydrated lime and magnesium carbonate. Various flocculating agents are added to the wastewater to increase the particle size. These flocculated large sized particles can be easily removed by filtration, straining or floatation. Naturally occurring cactus juice can be effectively used as a bio-flocculant to reduce chromium concentration in wastewater.

11.4.5 Membrane Separation

The method depends on the efficacy of permeable barriers to remove contaminants by passing wastewater through porous membranes under pressure. The variations in the pore size of membranes allow certain particles to easily pass through while retaining others based on principle of size exclusion. The commonly used membrane separation techniques like ultra-filtration, nano-filtration and reverse osmosis can be utilized for heavy metal removal from wastewater (Chaemiso and Nefo 2019).

11.4.6 Ultra-Filtration

This method is based on the efficiency of membrane permeability to separate heavy metals, macromolecules and suspended solids from wastewater on the basis of the pore dimensions and molecular weight of the impurities. Various water-soluble polymers can be added which can bind metal ions resulting in the formation of macromolecular complexes by producing metal ions free effluent. An integrated and hybrid approach of using metal-binding polymers in combination with ultra-filtration to remove heavy metals from aqueous solution was investigated by various researchers (Qasem et al. 2021).

11.4.7 Reverse Osmosis

The method of reverse osmosis is based on the concept of applying forced pressure to solution resulted in the retaining of the solute particles and allows the pure solvent to pass through the membrane. The type of membrane in reverse osmosis is semipermeable in behaviour. It allows the selective passage of pure solvent but not for metals, i.e. solute particles. The semipermeable membranes used for reverse osmosis have a dense barrier layer in the polymer matrix where most of the separation occurs.

11.4.8 Electrodialysis

The method of electrodialysis involves the passage of various ionic species through ion exchange semipermeable membrane under the action of electric potential. The membranes are made of thin sheets of plastic and selective for either positively charged or negatively charged ions. Anions move towards anode and cations move towards cathode. One of the modifications of this is cation-selective membranes with negatively charged matter, which rejects negatively charged ions and allows positively charged ions to flow through. Selective membranes are fitted between the electrodes in electrolytic cells and under continuous electrical current, the associated ion migrates, allowing the recovery of lead ions (Qasem et al. 2021). Various factors like flow rate of ions, voltage and temperature affect the efficiency of removal.

11.4.9 Electro-Coagulation

This method is based on the use of electrical current to remove suspended solids, tannins, dyes and dissolved metals especially lead ions from wastewater. When these ions and other charged particles are neutralized with ions of opposite electrical charges provided by electro-coagulation system, they become destabilized and precipitated in a stable form (Arbabi et al. 2015; Hunsom et al. 2005; Rahman et al. 2015).

Factors need to be controlled for maximum removal of lead ions:

i. content of lead in effluent, acidic or alkaline conditions, temperature, flow rate.
ii. Organic and inorganic load of effluent.
iii. Cost investment and the maximum permissible limits as set by government agencies.

Limitations: These physicochemical methods are not cost-effective and possess certain disadvantages like sludge production, low metal ions removal efficiency, energy consumption and low selectivity which limits their usage especially in small-scale industrial treatment plants. Moreover, these techniques are too expensive on large scale or commercial level and also dangerous for constant monitoring as these cannot completely treat the wastewater (Siddiquee et al. 2015; Tarekegn et al. 2020).

Physicochemical methods used for the removal of heavy metal ions from wastewater are often ineffective if the concentration of heavy metals is very low. The heavy metals present in dissolved form in wastewater cannot be separated using physical methods. The methods for heavy metal removal like chemical precipitation, chemical oxidation, ion exchange, membrane separation, reverse osmosis, electro dialysis are not very effective, non-economical and require high energy input. These are also associated with generation of toxic sludge, disposal of which makes it expensive and non-eco-friendly in nature (Kale et al. 2018).

11.5 Biological Methods for Lead Removal from Wastewater

In view of all these facts, sorption seems to be an effective and economical method for the removal of heavy metals especially lead from wastewater (Kale et al. 2018). The sorbents used for the removal of pollutants can be of chemical or biological origin. Out of these biological sorbents like viable or non-viable microbes or plant- or animal-based products proved to be an attractive alternative over the chemical sorbents due to their better efficacy, easy availability, economical and pollution-free approach (Table 11.1). The technique of using biological originated sorbents for the removal of pollutants is known to be biosorption or bioremediation (Fig. 11.1).

The various features of biosorption including low capital investment, metal specificity, increased efficiency, no sludge generation, no need of chemical additives,

Table 11.1 List of various biosorbents for lead removal from wastewater

List of biological sorbents	Examples
Agricultural sorbents	Rice husk, saw dust, peanut husk, wheat bran, groundnut husk, banana pith, cork powder, corncob, coir pith, sugar beet pulp, hazelnut shell, jackfruit, maize cob or husk, rice straw, coconut shell, sawdust of walnut tree, almond hulls, sugarcane bagasse, maize husks, shea butter seed husks, coconut fibre, sugar beet pulp, nut shells
Sludge sorbent	Activated sludge, sewage sludge, alum sludge
Plant-based sorbent	Stem, leaves, roots, vegetable and fruit peels, Pomegranate peel, cork, bark, sunflower stalk, tree sawdust, seaweeds, lichen, pine barks, tea leaves, plant tissues, date stones, grape fruit peel, peat and nut shells, coconut shells, rice husk, tea waste, peanut hulls, almond shells, peach stones, citrus peels
Animal-based sorbent	Egg shells, shells from aquatic animals

Source (Shartooh et al. 2014; Reddad et al. 2002; Saeed et al. 2005; Chockalingam and Subramanian 2006; Montanher et al. 2005; Khan et al. 2004; Lu et al. 2008; Husoon 2011; Kale et al. 2018; Chowdhary et al. 2022)

recovery of biosorbent and metal make it highly reliable and effective mode of treatment (Volesky 1994). The usage of crop and forest waste including agricultural wastes as sorbents not only allows for sustainable waste utilization, but also helps to remove toxic heavy metal ions from wastewater (Liu et al. 2018; Jin et al. 2020; Zhang et al. 2021). The use of waste biomass materials, including cotton stalks and grapefruit peels (Trakal et al. 2014; Fu et al. 2021; Shartooh et al. 2014), as precursors to process various activated carbon adsorbents to remove toxic heavy metals from wastewater is an active area of research.

11.5.1 Mechanism of Biosorption

The mechanism of sorption of heavy metal ions using microbes includes two pathways. First is the initial passive and rapid uptake which occurs via surface adsorption on the cell wall components and polysaccharides. Second is the further active and slow uptake which occurs through the membrane transporting metal ions within the cells.

Various mechanisms involved in the process of biosorption are as follows:

- Toxic states of heavy metals can be transformed into non-toxic states by alkylation or various redox reactions. The availability of metals depends upon the dissolution capacity and movement which further depends on their valency and anionic or cationic form of metals. For example, hexavalent form of chromium is more toxic and hazardous than its trivalent form.

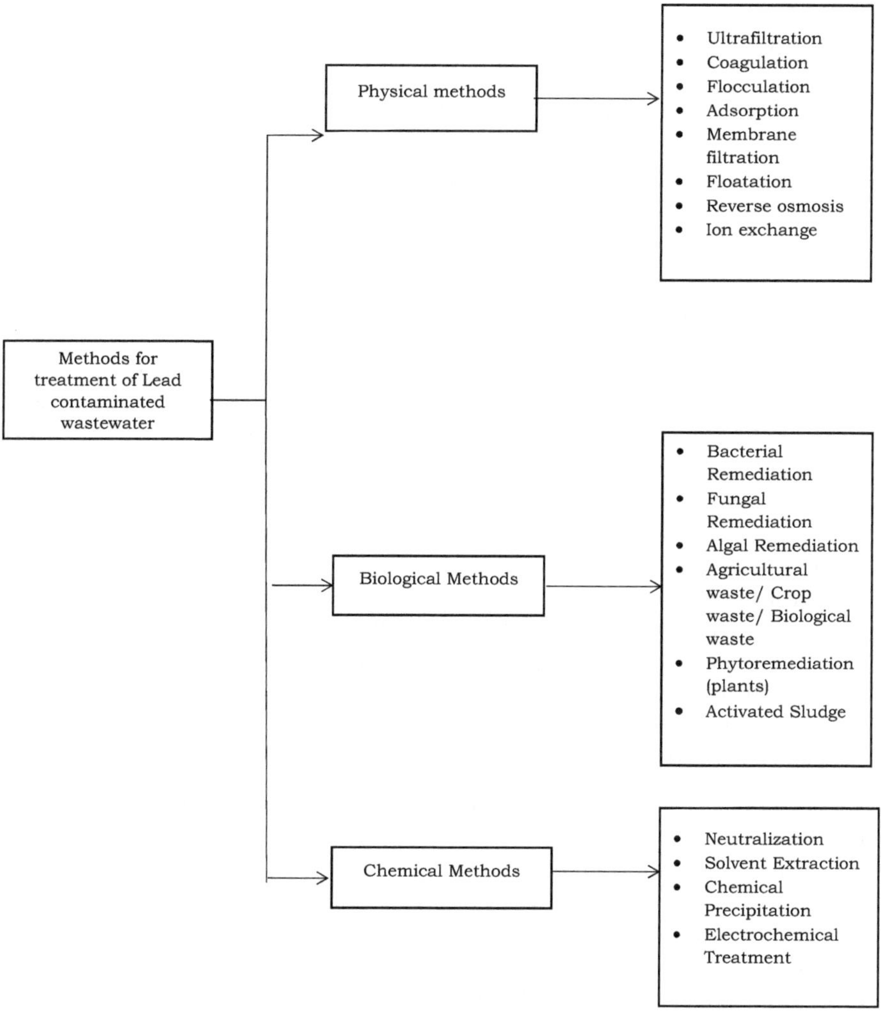

Fig. 11.1 Flowchart enlisting various methods for the removal of lead from wastewater

- Passive sorption is metabolism-free process, in which metals bind to functional groups present on the cell surface through electrostatic attraction, precipitation, surface complexation, ion exchange and physical adsorption. The factors including temperature, ionic strength, concentration and type of the sorbate and sorbent, state of biomass: suspension or immobilized and the presence of other anion and cations in the growth medium controls the efficacy of the removal of metal ions from wastewater.

- Active sorption is the metabolism-dependent intracellular uptake of heavy metals within the living cells within cytoplasm. Heavy metals are removed from wastewater by binding with metal-binding proteins or metallothioneins as low molecular mass cysteine-rich proteins, and metallochaperones present within the bacterial cell.
- Various parameters which need to be optimized are pH, temperature, salinity, media composition, biochemical and physiological features or genetic variability of biomass and toxicity of metals towards biosorbents. A variety of microbial strains like *Cyanobacteria*, *Pseudomonads* and M*ycobacteria* have the ability to synthesize metal-binding proteins which can be further used for the removal of zinc, copper, cadmium, mercury and lead (Thi Pham et al. 2022).
- The removal of metal ions can also be carried out by a complex mechanism of releasing EPS like proteins, DNA, RNA and polysaccharides resulting in the slippery layer on the outside of the cell wall. These further retard the penetration of metals within the intracellular environment. *Stenotrophomonas maltophilia*, *Azotobacter chroococcum* and *Bacillus cereus* possess the ability to secrete EPS. Bioremediation efficiency by this mechanism relies on the type and amount of carbon source available and other abiotic stress factors like pH, temperature and the growth phase of each bacterium.
- The functional groups like carboxyl, phosphonate, amine and hydroxyl groups present in the cell walls of bacteria are able to bind heavy metal ions present in the wastewater and their further removal. The efficacy of biosorption depends on the diversity of cell wall structures. Gram-positive bacteria have been shown to contain a high sorption capacity because of their thicker peptidoglycan layer.

Biosorption is commonly used for the removal of lead and chromium from industrial effluent. It involves the use of viable or non-viable microbes for pollutants removal from aqueous solutions and industrial wastewater. Most commonly used biosorbents for the removal of heavy metals from industrial effluent are microbial biomass (Volesky and Holan 1995), and biological wastes like peat and nut shells, coconut shells, rice husk, tea waste, peanut hulls, almond shells, peach stones, citrus peels, (Reddad et al. 2022; Saeed et al. 2005; Khan et al. 2004; Chockalingam and Subramanian, 2013). These biosorbent materials are economical, high sorption efficacy, metal specificity, no sludge generation, regeneration and metal ion recovery (Tunali et al. 2006) and environmentally safer to use (Husoon et al. 2013). The presence of various functional groups like carboxyl, hydroxyl, sulphate, phosphate and amino groups in biological sorbents further assist in binding of metal contaminants from wastewater (Fig. 11.2).

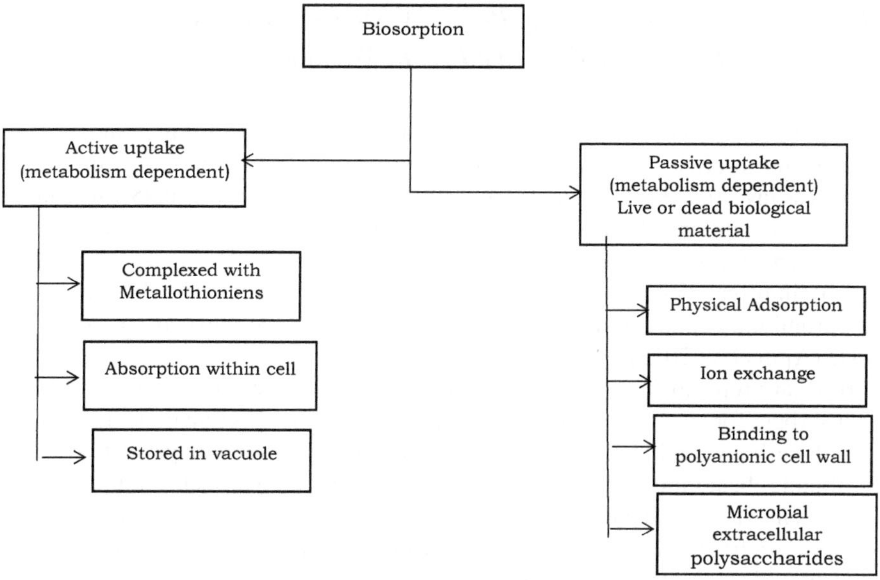

Fig. 11.2 Flowchart showing the mechanism of biosorption by microbial biomass

11.5.2 Factors Affecting the Efficacy of Sorption of Metals Contaminants: (Shartooh et al. 2014; Yarkandi et al. 2014; Dehagni et al. 2023)

- pH, Temperature, moisture,
- shaking speed, incubation time, aeration,
- initial concentration of metal and chemical nature of each contaminant, chemical state of the site or availability and affinity between site and metal or binding strength,
- chemical characteristics of metal like ionic potential, ionic radius, ionic stability limit,
- amount of biosorbent, size of the biomass, number of sites of biosorbent material, the accessibility of sites,
- interaction between different metallic ions and ionic strength,
- toxicity of the pollutants to viable microbial cell (Regine and Volesky, 2000),
- shaking or stationary conditions.

11.5.3 Biological or Agricultural Waste as Biosorbent for the Removal of Lead from Wastewater

Banana peel was successfully used as biosorbent for the lead ions removal from an aqueous solution, and the impact of varying operational conditions in a batch mode was investigated. The maximum lead removal was observed at 98.146% under optimized conditions of lead concentration 100 mg/L, pH 5, amount of sorbent 0.55 g and sorbent size 75 μm (Afolabi et al. 2021).

Certain researchers showed the maximum adsorption efficiency of banana peels for chromium, cadmium and lead and considered it as the inexpensive household waste (Ajmi et al. 2018).

Biochar prepared from agricultural, animal and wood residues proved to be efficient sorbent for lead removal owing to the binding of various functional groups like phenol, carboxyl and hydroxyl to metal ions. While modified clays such as montmorillonite, bentonite, kaolinite, vermiculite, polymeric hydrogels are mostly used for the removal of lead and mercury from wastewater (Aranda and Rivas 2022).

Various agricultural wastes like stems, roots, fruit peels, rinds, saw dust, husks, hulls, dried leaves, fruit shells and seeds have been efficiently used to remove metal contaminants from synthetic solutions (Sun and Shi 1998; Al-Asheh and Duvnjak 1998; Meunier et al. 2003; Sekhar et al. 2003; Wang and Qin 2005).

The peels of grapes were also used to remove lead, copper and zinc from factory wastewater. The fresh fruit peels, dried small pieces and powdered peels were tested for treatment of heavy metal–loaded wastewater by optimization of various parameters like pH, temperature and exposure time, concentration of metal contaminants. The highest removal efficiency was for lead metal as compared to copper and zinc. The Fourier transform infrared spectroscopy analysis (FTIR) studies illustrated that hydroxyl, carboxyl and carbonyl groups were the major binding sites for Pb, Cu and Zn ions removal using grape fruit peels (Shartooh 2012).

Maize cob was used as sorbent for lead ions removal from synthetic solution. Batch mode of sorption studies were performed under well-optimized experimental conditions of 500 ppm concentration, 2.5 g dosage, 400 min exposure time, 400 rpm agitation speed and 5 pH. The efficiency of removal for Pb (II) ions was 95% using maize cob as adsorbent, and studies can be further extended for the treatment of metal-contaminated wastewater (Muthusamy and Murugan 2016).

Activated carbon prepared from leaves of medicinal plant *Militia ferruginea* was utilized efficiently for the removal of Pb (II) ions from wastewater. The maximum adsorption of lead was more than 97% from industrial effluent at 3 h of contact time for 4.0 g of adsorbent and at pH of 4.0. The amount of lead ion adsorbed per gram of the adsorbent increased with decreasing concentration of Pb^{2+} ions. The percentage of adsorption had increased with the increasing temperature (Mengistie et al. 2008).

Undaria pinnatifida was immobilized in sodium alginate beads and further utilized efficiently for the removal of Pb(II) ions from wastewater. To understand the mechanism of sorption, the resulting biosorbent was characterized by Fourier transform infrared spectroscopy (FTIR) and scanning electron microscopy coupled with energy-dispersive X-ray spectroscopy (SEM–EDS). The effect of various conditions on Pb (II) ion removal efficiency such as temperature, pH, ionic strength, time and underlying biosorption mechanisms was also observed (Namkoong et al. 2022).

11.5.4 Bacteria as Biosorbent for the Removal of Lead from Wastewater

Positively charged metal ions can be easily removed by Gram-positive bacteria having higher electronegative charge density which might be due to the presence of teichoic and teichuronic acids linked by phosphodiester bonds which are further attached to peptidoglycan layer in the cell wall. The presence of various functional groups having oxygen, nitrogen, sulphur or phosphorus in bacterial cells are responsible for metal ion removal from wastewater (Redha 2020) (Table 11.2).

Bacterial proteins were utilized for the treatment of lead-contaminated wastewater. Lead oxide nanoplates were synthesized by interactions, and the removal of lead ions was confirmed using various advance techniques of inductively coupled plasma analysis, X-ray spectroscopy and X-ray diffraction (Ghosal et al. 2021).

Table 11.2 Examples of microbial biosorbents for lead removal from wastewater

Microbial biosorbents	Examples
Bacteria	*Bacillus* sp., *Pseudomonas* sp., *Arthrobacter* sp., *Alcaligenes sp., Azotobacter* sp., *Rhodococcus sp.*, Acinetobacter *sp.* and methanogens *Pseudomonas putida, Pseudomonas aeruginosa* and *Escherichia coli, Bacillus thuringiensis, Streptomyces sp., Cellulosimicrobium funkei, Lactiplantibacillus plantarum, Cellulosimicrobium sp., Methylobacterium sp., Aerobacillus pallidus, Arthrobacter viscosus, Klebsiella pnuemoniae,, Rhodotorula* sp., *Bacillus megaterium, Vibrio parahaemolyticus, Klebsiella* sp, *Staphylococcus epidermidis, Oceanbacillus profundus, Micrococcus luteus, Flavobacterium, Enterobacter, Acinetobacter sp., Micrococcus luteus, Bacillus subtilis, Aspergillus niger* and *Penicillium sp*
Fungi	*Aspergillus* sp., *Penicillium* sp., *Rhizopus* sp., *Mucor* sp., *Alternaria* sp., and *Cladosporium* sp, *A. ferrooxidans* and *A. thiooxidans, Desulfovibrio desulfuricans, Coprinopsis atramentaria, A. niger, Rhizopus oryzae, Saccharomyces cerevisiae, Penicillium chrysogenum, Candida sphaerica*
Yeast	*Hansenula polymorpha, S. cerevisiae, Yarrowia lipolytica, Rhodotorula pilimanae, Pichia guilliermondii*, and *Rhodotorula mucilage*
Algae	*C. vulgaris, Gelidium amansii, Phormidium ambiguum, Porphyra leucosticte, Spirogyra sp., Sargassum muticum, Chlorella miniate* and *Spirulina platensis*

Source (Thi Pham et al. 2022; Kumar and Goyal 2009; Kareem and Anwar 2020; Amasha and Aly 2019; Olusola and Aransiola 2015; Rao and Bhargavi 2013; Sheba and Nandini 2016; Villegas et al. 2018; Iram and Abrar 2015; Rastogi et al. 2019; Aracagok 2022)

A continuous column treatment setup was developed in up-flow anaerobic sludge blanket reactor (UASB) loaded with anaerobic sulphate-reducing bacteria and used for the continuous removal of lead and mercury ions from wet flue gas desulfurization (FGD) wastewater. Lead and mercury were removed in the form of sulphides and gets accumulated in sludge. The reactor was operated under various optimized experimental conditions at metal loading rates of 9.2 g/m^3d Pb (II) and 2.6 g/m^3d Hg (II) for retention time of 50 days. The UASB reactor removed $72.5 \pm 7\%$ of sulphite and more than 99.5% of both Hg(II) and Pb(II) and found to be very efficient for the treatment of metal-contaminated wastewater (Zhang et al. 2016). In another similar study, lead was removed from sulphide-rich effluent using sulphate-reducing bacteria in the form of lead sulphide precipitate. The whole treatment was performed in UASB reactor in a continuous mode with initial feeding load of effluent containing 45–50 mg/L concentration of lead ions. The maximum lead removal 85–90% was achieved in UASB reactor (Hoa et al. 2007).

The efficiency of metal tolerance and accumulation of about 164 isolated heterotrophic bacterial strains was studied especially for lead, cadmium and zinc. The metal tolerance studies of all the isolated bacterial isolates showed that about 45% of the total isolates showed very high tolerance of greater than 6000 μg/ml towards lead ions as compared to cadmium and zinc towards which bacterial strains have comparatively low tolerance. Further, one of screened bacterial strain *Bacillus sp.* was found to be more efficient in the bioaccumulation of lead ions (Varghese et al. 2012).

11.5.5 Fungi as Biosorbent for the Removal of Lead from Wastewater

Many metal-tolerant fungal strains were isolated from sewage sludge and industrial wastewater especially tolerant towards lead, cadmium, chromium and nickel. These isolated fungal strains were characterized through various morphological, biochemical and genetic identification tests. The identified and screened fungal strains were *Aspergillus foetidus, Phanerochaete chrysosporium, Aspegillus awamori, Rhizopus sp., Aspergillus flavus, Trichoderma viridae* which were further used for the removal of different metals from wastewater. The screened fungal strains were found to tolerate and ability to grow up to 400 ppm concentration of cobalt, lead, cadmium, chromium, copper and nickel metal ions. All the above-mentioned fungal strains have the remarkable efficacy to be used as biosorbent for the removal of cobalt, lead, cadmium, chromium, copper and nickel metal ions from wastewater and industrial effluents (Dwivedi et al. 2012).

The efficacy of non-viable biomass of *Penicilluim sp.* was estimated for the removal of lead ions from synthetic solution. All the treatments were done in the batch mode under optimized conditions of 10 mg/l lead ion concentration, 1 g/l biomass dosage, 2 h of exposure time and found to achieve 78.03% removal of lead ions. The mechanism of sorption found out the involvement of carbonyl, methylene, phosphate, carbonate and phenolic groups in removal of lead ions from industrial effluent (Rastogi et al. 2019).

The efficiency of fungal biomass *Aspergillus neoalliaceus* for sorption of lead ions as a function of pH, biomass dosage, contact time and initial lead concentration was studied. The removal of lead ions with *Aspergillus neoalliaceus* followed Langmuir isotherm and pseudo-second-order kinetic models compared to other used models (Aracagok 2022).

The pre-treated biomass of *Aspergillus niger* was found as an efficient sorbent for the removal of heavy metals especially lead and nickel from wastewater. The various optimized experimental conditions for maximum removal of lead and nickel were pH of 7 and 6 and equilibration time for maximum biosorption at 5 h and 8 h, respectively. In the presence of co-ion lead, the percentage removal of nickel was 92% which was greater than using the single metal system removal (Rao and Bhargavi 2013).

The agricultural waste edible fungi residue was found to adsorb 76.34% of Pb (II) ions from wastewater. All the treatments were performed under optimized experimental conditions of 483.83 mg/L of lead ion concentration, 4.99 g/L of fungi residue at pH of 5.89. The FTIR characterization of fungi residue both before and after treatment confirmed the involvement of various functional groups which controlled the sorption of heavy metals (Jing et al. 2021a, b).

11.5.6 Algae as Biosorbent for the Removal of Lead from Wastewater

The various factors like the presence of functional groups, high surface area and high binding capacity in algae make it an efficient sorbent for the removal of heavy metal ions especially lead ions from wastewater. This might be due to the presence of chitin, polysaccharides, proteins and lipids in cell wall of algae resulting in higher biosorption ability (Davis et al. 2003). In the first rapid extracellular passive sorption, heavy metals are adsorbed over the cell surface by not involving cellular metabolism. The various factors like bioavailability of metals, availability of metal-binding groups on the algal cell surface, metal uptake and storage efficiency of algal cells determine the efficacy of sorption. Biosorption can be carried out by both viable and non-viable biomass. Algae can either exchange metal ions with calcium, magnesium, sodium or potassium ions or form complex with the functional groups on the surface of algal cell. Contaminants can easily bind to the surface of algae due to the presence of polysaccharides, lipids and proteins. The presence of sulphate, carboxyl, amino and hydroxyl groups in the cell wall of microalgae makes it suitable as the binding site for the pollutants. Heavy metals can also be transported across the cell membrane within the cytoplasm or organelles through an active uptake and can be carried out by viable biomass only and also dependent on cellular metabolism. This process is also known as slow intracellular active accumulation and requires energy for the accumulation of heavy metals inside the microalgal cells.

Various self-defence mechanisms of gene regulation, complexation, ion exchange, chelation, produce reducing agents or anti-oxidants and cause heavy metal immobilization enhances the efficiency of algae to fight against the toxicity of heavy metal ions (Chugh et al. 2022).

The algae *Chlorella vulgaris* has the ability to remove various heavy metals, especially lead and cadmium in one metal solution system (Moustafa and Idris 2003). The mechanism of sorption revealed that the removal of lead occurs in two consecutive steps, the first is the adsorption on its surface followed by fixation. The algae was able to remove 60% lead and 65% cadmium efficiently from synthetic solution (Dhokpande and Kaware 2013; Sonali et al. 2013).

Sludge-based adsorbent was prepared by ferric activation through pyrolysis and further used for the sorption of lead ions from aqueous solution. The ferric-activated sludge-based adsorbent showed a favourable porous structure development and lead ions removal with the maximum sorption capacity of 42.96 mg/g (Yang et al. 2019).

11.6 Conclusions

This review article summarizes the various sources of heavy metal contamination along with the hazards of especially lead-contaminated wastewater, their current physiochemical treatments and their limitations. This review also highlighted the biosorption technology used for the treatment of lead-contaminated wastewater along with the in-depth knowledge of major mechanisms involved through which biosorbents remove metals from wastewater. Furthermore, a brief discussion on the effect of lead contamination on various components of ecosystem and remediation of heavy metals using an elaborated list of biosorbents has been reviewed. To make this bioremediation technique more efficient and successful, recent advancements, challenges and strategies to carry out in the future have been explored. The need of biosorption, factors affecting sorption, utilization of biological waste as biosorbents has been detailed. It has been concluded from detailed literature survey that the use of biosorbents seems to be a more promising alternative for the removal of lead ions from wastewater as compared to the other conventional methods of treatment. These biosorbents can be either plant- or animal-originated or microbial biomass. Microbial sorbents can be used in form of live or dead biomass, more economical, pollution free, easily availability and regeneration ability, effectively utilized for continuous treatment in columns for industrial waste treatment.

References

Acharya J, Kumar U, Meikap B (2013) Thermodynamic characterization of adsorption of lead (II) ions on activated carbon developed from tamarind wood from aqueous solution. S Afr J Chem Eng 18(1):70–76

Afolabi FO, Musonge P, Bakare BF (2021) Evaluation of Lead (II) removal from wastewater using banana peels: optimization study. Pol J Environ Stud 30(2):1487–1496

Ahluwalia SS, Goyal D (2007) Microbial and plant derived biomass for removal of heavy metals from wastewater. Bioresour Technol 98(12):2243–2257

Ajmi RN, Sultan M, Hanno SH (2018) Bioabsorbent of chromium, cadmium and lead from industrial waste water by waste plant. J Pharm Sci Res 10(3):672–674

Al-Asheh S, Duvnjak Z (1998) Binary metal sorption by pine bark: study of equilibrium and mechanisms. Sep Sci Technol 33(9):1303–1329

Amasha RH, Aly MM (2019) Removal of dangerous heavy metal and some human pathogens by dried green algae collected from jeddah coast. Pharmacophore 10(3):5–13

Aracagök YD (2022) Biosorption of lead by a soil isolate *Aspergillus neoalliaceus*. In: Archives of microbiology

Aranda FL, Rivas BL (2022) Removal of mercury and lead by bioadsorbents: an overview. J Chil Chem Soc 67(4):5682–5691

Arbabi M, Hemati S, Amiri M (2015) Removal of lead ions from industrial wastewater: removal methods review. Int J Epidemiol Res 2(2):105–109

Biela R, Sopikova L (2017) Efficiency of sorption materials on the removal of lead from water. Appl Ecol Environ Res 15(3):1527–1536

Chaemiso TD, Nefo T (2019) Removal methods of heavy metals from laboratory wastewater. J Nat Sci Res 9(2):36

Chockalingam E, Subramanian S (2013) Studies on removal of metal ions and sulphate reduction using rice husk and *Desulfotomaculum nigrificans* with reference to remediation of acid mine drainage. Chemosphere 62(5):699–708

Chowdhary IR, Chowdhary S, Mazumder MAJ, Al-Ahmed A (2022) Removal of lead ions (Pb^{2+}) from water and wastewater: a review on the low-cost adsorbents. Appl Water Sci 12:185

Chugh M, Kumar L, Shah MP, Bharadvaja N (2022) Algal Bioremediation of heavy metals: an insight into removal mechanisms, recovery of by-products, challenges, and future opportunities. Energy Nexus 7:100129

Davis TA, Volesky B, Mucci A (2003) A review of the biochemistry of heavy metal biosorption by brown algae. Water Res 37(18):4311–4330

Dehghani MH, Sardari SA, Afsharina M, Qasemi M, Shams M (2023) Removal of toxic lead from aqueous solution using a low-cost adsorbent. Sci Rep 13:3278

Dhokpande SR, Kaware JP (2013) Biological methods for heavy metal removal: a review. Int J Eng Sci Innov Technol (IJESIT) 2(5)

Dwivedi SN, Mishra RP, Alava S (2012) Phytochemistry, pharmacological studies and traditional benefits of *Trachyspermum ammisprague*. Int J Pharm Life Sci 3:1705–1709

Fu Y, Yang Z, Xia Y, Xing Y, Gui X (2021) Adsorption of ciprofloxacin pollutants in aqueous solution using modified waste grapefruit peel. Energy Sourc A Recovery Utilization Environ Effects 43(2):225–234

Ghosal C, Alam N, Tamang A, Chattopadhyay B (2021) Removal of lead contamination through the formation of lead-nanoplates by a hot spring microbial protein. Adv Microbiol 11:681–693

Goel J, Kadirvelu K, Rajagopal C, Kumar Garg V (2005) Removal of lead (II) by adsorption using treated granular activated carbon: batch and column studies. J Hazard Mater 125(1-3):211–220

Hoa TTH, Liamleam W, Annachhatre AP (2007) Lead removal through biological sulfate reduction process. Biores Technol 98:2538–2548

Hunsom M, Pruksathorn K, Damronglerd S, Vergnes H, Duverneuil P (2005) Electrochemical treatment of heavy metals (Cu^{2+}, Cr^{6+}, Ni^{2+}) from industrial effluent and modeling of copper reduction. Water Res 39(4):610–616

Husoon A, Al-Azzawi MNA, Al-Hiyaly SAK (2013) Investigation biosorption potential of copper and lead from industrial waste water using orange and lemon peels. J Al-Nahrain Univ 16(2):713–719

Iram S, Abrar S (2015) Biosorption of copper and lead by heavy metal resistant fungal isolates. Int J Sci Res Publ 5(1)

Jamali HA, Mahvi AH, Nazmara S (2009) Removal of cadmium from aqueous solutions by hazel nut shell

Jin W, Fu Y, Hu M, Wang S, Liu Z (2020) Highly efficient SnS-decorated Bi_2O_3 nanosheets for simultaneous electrochemical detection and removal of Cd(II) and Pb(II). J Electroanal Chem 856:113744

Jing Y, Li Z, Li Y, Lei G, Li L, Yang X, Zhang Z, Yang W (2021) The ability of edible fungi residue to remove lead in wastewater, vol 9, Article 723087

Jing Y, Li Z, Li Y, Lei G, Li L, Yang X, Zhankg Z, Yang W (2021) Efficient treatment of Industrial wastewater with microbiome and synthetic biology, The ability of edible fungi residue to remove lead in wastewater. Front Environ Sci Sec Water Wastewater Manage 9

Kale A, Jadhav S, Kharade A, Khurmure C, Kumatkar D, Wankhede B (2018) Removal of heavy metals from wastewater by using low cost adsorbents. Int J Adv Res Sci Eng 07(3):1231–1237

Kareem APM, Anwer SS (2020) Biosorption of cadmium and lead using microalgae *Spirulina sp.* isolated from koya city (Iraq). Appl Ecol Environ Res 18(2):2657–2668

Khan NA, Ibrahim S, Subramanian P (2004) Elimination of heavy metals from wastewater using agricultural wastes as adsorbents. Malaysian J Sci 23:43–51

Kumar R, Goyal D (2009) Comparative biosorption of Pb^{2+} by live algal consortium and immobilized dead biomass from aqueous solution. Indian J Exp Biol 6:690–694

Liu S, Duan Z, He C, Xu X, Li T, Li Y, Li X, Wang Y, Xu L (2018) Rapid removal of Pb^{2+} from aqueous solution by phosphate modified baker's yeast. RSC Adv 8:8026–8038

Lo W, Chua H, Lam KH, Bi SP (1999) A comparative investigation on the biosorption of lead by filamentous fungal biomass. Chemosphere 39:2723–2736

Mahmoud ME, Osman MM, Ahmed SB, Abdel-Fattah TM (2012) Enhanced removal of Pb by chemically and biologically treated carbonaceous materials. Sci World J 2012:1–11

Mengistie AA, Rao TS, Prasada Rao AV, Singanan M (2008) Removal of lead(II) ions from aqueous solutions using activated carbon from *Militia ferruginea* plant leaves. Bull Chem Soc Ethiop 22(3):349–360

Meunier N, Laroulandie J, Blais JF, Tyagi RD (2003) Cocoa shells for heavy metal removal from acidic solutions. Biores Technol 90:255–263

Mohammad AM, Eldin TAS, Hassan MA, El-Anadouli BE (2015) Efficient treatment of lead-containing wastewater by hydroxyapatite/chitosan nanostructures. Arab J Chem 10:683–690

Momčilović M, Purenović M, Bojić A, Zarubica A, Ranđelović M (2011) Removal of lead (II) ions from aqueous solutions by adsorption onto pine cone activated carbon. Desalin 276(1):53–59

Moustafa M, Idris G (2003) Biological removal of heavy metals from wastewater. Alex Eng J 42:767–771

Muthusamy P, Murugan S (2016) Removal of lead ion using maize cob as a bioadsorbent. Int J Eng Res Appl 6(6)(Part-1):05–10

Naiyaa TK, Bhattacharya AK, Mandal S, Das SK (2009) The sorption of lead (II) ions on rice husk ash. J Hazard Mater 163:1254–1264

Namkoong H, Biehler E, Namkoong G, Abdel-Fattah TM (2022) Efficient removal of lead ions from aqueous media using sustainable sources based on marine algae. ACS Omega 7:39931–39937

Pitter P (2009) Hydrochemistry. University of Chemistry and Technology Prague, Prague (in Czech)

Qasem NAA, Mohammed RH, Lawal DU (2021) Removal of heavy metal ions from wastewater: a comprehensive and critical review. Clean Water Npj 4:36

Rahman SH, Yesmin RA, Islam SMN, SIiraj S, Adyel TM, Ahmed MS (2015) Removal of lead from battery effluent by electrocoagulation. J Bangladesh Acad Sci 39(2):125–134

Rao PR, Bhargavi C (2013) Studies on biosorption of heavy metals using pretreated biomass of fungal species. Int J Chem Chem Eng 3(3):171–180

Rastogi S, Kumar J, Kumar R (2019) An investigation into the efficacy of fungal biomass as a low cost bioadsorbent for removal of lead from aqueous solution. IRJET 6(3):7144–7149

Raut PA, Chahande A, Moharkar Y (2015) Various techniques for removal of Cr and Pb from wastewater: a review. Int J Emerging Trends Eng Basic Sci 2(2):64–67

Reddad Z, Gerente C, Andres Y, Le Cloirec P (2002) Adsorption of several metal ions onto a low cost biosorbent: kinetic and equilibrium studies. Env Sci Technol 36(9):2067–2073

Redha AA (2020) Removal of heavy metals from aqueous media by biosorption. Arab J Basic Appl Sci 27(1):183–193

Regine HSF, Volesky B (2000) Biosorption: a solution to pollution. Int Microbiol 3:17–24

Saeed A, Iqbal M, Akhtar MW (2005) Removal and recovery of lead(II) from single and multimetal (Cd, Cu, Ni, Zn) solutions by crop milling waste (black gram husk). J Hazard Mater 117(1):65–73

Sekhar KC, Kamala CT, Chary NS, Anjaneyulu Y (2003) Removal of heavy metals using a plant biomass with reference to environmental control. Int J Miner Process 68:37–45

Shartooh SM (2012) Comparative study on the biological treatment of some heavy metals in industrial wastewater of State Electrical Manufacturing Company in Al-Waziriya/Baghdad- Iraq. Ph.D. thesis, University of Baghdad, College of Science

Shartooh SM, Gabur HS, Mahmood OS (2014) The removal of lead, copper and zinc from industrial wastewater using grape fruit peels. Raf J Sci 25(2):22–31

Sheba MC, Nandini GKM (2016) Heavy metal lead removal by biosorption—a review. Int J Eng Res Technol (IJERT) 5(11):519–524

Siddiquee S, Rovina K, Al Azad S, Naher L, Suryani S, Chaikaer P (2015) Heavy metal contaminants removal from wastewater using the potential filamentous fungi biomass: a review. J Microb Biochem Technol 7(6):384–393

Singha B, Das SK (2015) Removal of Pb (II) ions from aqueous solution and industrial. Int J Epidemiol Res 2(2):105–109, 109

Sonali R, Dhokpande SR, Kaware JP (2013) Biological methods for heavy metal removal: a review. Int J Eng Sci Innov Technol 2(5):304–309

Sun G, Shi W (1998) Sunflower stalks as adsorbents for the removal on metal ions from wastewater. Ind Eng Chem Res 37:1324–1328

Tarekegn MM, Salilih FZ, Ishetu AI, Yildiz F (2020) Microbes used as a tool for bioremediation of heavy metal from the environment. Cogent Food Agric 6(1)

Thi Pham VH, Kim J, Chang S, Chung S (2022) Review bacterial biosorbents, an efficient heavy metals green clean-up strategy: prospects, challenges, and opportunities. Microorganisms 10:610

Trakal P, Sigut R, Sillerova H, Faturikova D, Komarek M (2014) Copper removal from aqueous solution using biochar: effect of chemical activation. Arab J Chem 7(1):43–52

Tunali S, Akar T, Özcan AS, Kiran I, Özcan A (2006) Equilibrium and kinetics of biosorption of lead(II) from aqueous solutions by *Cephalosporium aphidicola*. Sep Purif Technol 47:105–112

Varghese R, Krishna MP, Arun Babu V, Mohamed Hatha AA (2012) Biological removal of lead by *Bacillus sp.* obtained from metal contaminated industrial area. J Microbiol Biotechnol Food Sci 2(2):756–770

Villegas LC, Llamado AL, Catsao KV, Raymundo AK (2018) Removal of heavy metals from aqueous solution by biofilm-forming bacteria isolated from mined-out soil in Mogpog, Marinduque, Philippines. Philippine Sci Lett 11:18–27

Volesky B (1994) Advances in biosorption of metals: selection of biomass types. FEMS Microbiol Rev 14:291–302

Volesky B, Holan ZR (1995) Biosorption of heavy metals. Biotechnol Progr 11:235–250

Wang XS, Qin Y (2005) Study on equilibrium sorption isotherm for metal ion of Cu^{2+} on rice bran. Process Biochem 40(2):677–680

Wendt JOL, Lee SJ (2010) High-temperature sorbents for Hg, Cd, Pb, and other trace metals: mechanisms and applications. Fuel 89(4):894–903

World Health Organization (2011) Lead in drinking water. Report No. WHO/SDE/WSH/03.04/09/Rev/1, World Health Organization, Geneva

Yang X, Xu G, Yu H (2019) Removal of lead from aqueous solutions by ferric activated sludge-based adsorbent derived from biological sludge. Arab J Chem 12:4142–4149

Yarkandi NH (2014) Removal of lead (II) from waste water by adsorption. Int J Curr Microbiol App Sci 3(4):207–228

Zhang L, Lin X, Wang J, Jiang F, Wei L, Chen G, Hao X (2016) Effects of lead and mercury on sulphate-reducing bacterial activity in a biological process for flue gas desulfurization wastewater treatment. Sci Rep 6:30455. https://doi.org/10.1038/srep30455

Zhang Y et al (2021) Insight into the efficient co-removal of Cr (VI) and Cr(III) by positively charged UiO-66-NH_2 decorated ultrafiltration membrane. Chem Eng J 404:126546

Chapter 12
Antioxidant Defense: Key Mechanism of Lead Intolerance

Manish Kumar Verma, Amitava Paul, and Moon Roy

Abstract Lead (Pb) is a toxic heavy metal that can have significant adverse effects on human and animal health, especially when exposure occurs at high levels or over prolonged periods. The harmful health effects of lead are well-documented, and even low-level exposure can be dangerous. Its contaminated natural sources are mining and smelting operations, paints containing lead, paper and pulp, petrol and explosives, as well as the dumping of sewage sludge that has been enriched with lead from municipal sewage systems. The reproductive, neurological, immunological, cardiovascular, and other systems as well as developmental processes are negatively impacted by its interactions with biomolecules. Pb reacts with the sulfhydryl groups on enzymes to decrease their activity at the cellular level. Reactive oxygen species (ROS) are known to be produced more frequently as a result of oxidative stress caused by lead. Antioxidants defenses to lead toxicity may constitute different strategies. The first step is to either prevent lead from entering the cell by excluding it or to bind lead to the cell wall and other ligands, such as organic acids, amino acids, and glutathione, to render them harmless at the initial stage of lead entrance. One of the main detoxifying pathways for Pb is cell wall binding. Secondary defense system constitutes several antioxidants to fight against increased ROS production caused by lead.

Keywords Lead · Antioxidants · Cell damage · Free radicals · Reactive oxygen species

M. K. Verma
Department of Veterinary Pharmacology and Toxicology, International Institute of Veterinary Education and Research, Rohtak, Haryana, India

A. Paul (✉)
Department of Veterinary Pathology, IIVER, Bahu-Akbarpur, Rohtak, Haryana 124001, India
e-mail: amit01paul@gmail.com

M. Roy
Department of Pediatric and Preventive Dentistry, TIDSHRC, Bilaspur, Chhattisgarh 495220, India

N. Kumar and A. K. Jha (eds.), *Lead Toxicity Mitigation: Sustainable Nexus Approaches*, Environmental Contamination Remediation and Management,
https://doi.org/10.1007/978-3-031-46146-0_12

12.1 Introduction

An antioxidant is a chemical that effectively inhibits the activity of a pro-oxidant while simultaneously generating low- or non-toxic by-products. Indeed, a broader definition of antioxidant was suggested by Halliwell et al. (1995), as "any substance that when present at low concentrations, compared to those of an oxidizable substrate significantly delays or prevents oxidation of that substrate". Thus, by this theory, only those chemical agents that are capable of defending the biological target from damage are referred to as antioxidants (Galano and Reiter 2018). This defense may be based on a number of mechanisms, such as prevention of production and scavenging activity against reactive oxygen species (ROS)/reactive nitrogen species (RNS), metal chelation, reducing power, antioxidative enzyme action, and the suppression of oxidative enzymes (Lü et al. 2010; Truong et al. 2018; Pisoschi et al. 2021). Human and animal body cell naturally maintains a balance between the quantity of free radicals it produces and the antioxidants that shield it from harmful consequences. Amounts of antioxidant components are frequently found under typical physiological settings may not be enough to completely neutralize free radicals produced (Bhattacharya 2015). It follows that adding antioxidants to our diet will help us to prevent from hazardous illnesses. Therefore, a rise in concern regarding "Natural antioxidants" made from herbal materials in the food sector and in preventative medicine increased drastically. Because of this, antioxidants are growing in popularity across the globe (Khan et al. 2019). Antioxidants are natural or synthetic substances that protect or delayed cell death. Antioxidants are naturally found in many foods, including fruits, vegetables, and medicinal herbs. They are also available as dietary supplements. Antioxidants examples include lutein, lycopene, selenium, beta-carotene, vitamin (A, E, and C), etc. (Yadav et al. 2016). Antioxidants had an increasing interest due to their role in protecting food and medicinal products against oxidative decaying. It also protects against pathological alteration related to oxidative stress in the cell or body. Antioxidant is also referred to as "a substance that either directly scavenges ROS, indirectly up-regulates antioxidant defenses, or inhibits ROS production." By delaying the lipid peroxidation process, which is the major cause of cell toxicities and the degradation of pharmaceutical and food items during manufacturing and storage, antioxidant chemicals can scavenge free radicals and also extend shelf life (Li et al. 2016; Ali et al. 2020). Nutritional antioxidant molecules can be used in food systems to delay lipid peroxidation and secondary lipid peroxidation products formation, helping to preserve the texture, flavor, and color of the food items during storage (Domínguez et al. 2019). The amino acids reduction, interaction of lipid-derived carbonyls with proteins and oxidation of protein that cause change in protein function are further processes that antioxidants help with (Hu et al. 2017).

Lead is the most important harmful heavy metal in the environment. Its use dates back to historic times due to its important physico-chemical properties. It is a broadly dispersed, serious, and hazardous chemical that are harmful to the environment worldwide (Khan et al. 2021a, b). It is hard to give up because of its important characters like corrosion resistance, malleability, ductility, softness, and low conductivity. Its non-biodegradable nature and continuous use increases the risks associated with its presence in the environment (Velvizhi et al. 2020). Human and animal exposure to lead and its derivatives primarily appear in lead-related profession with a various sources such as lead-containing petrol, industrial activities such as lead smelting and its combustion, boat construction, pottery, lead-based pigments, book printing, painting, battery recycling, grids, arm industry, and lead-containing pipes. Lead affects almost all of the body's organs (Nath et al. 2019). Chronic exposure of lead in low concentration induces a various behavioral, physiological, and biochemical dysfunctions. Its harmful effect is not well understood, and several mechanisms have been suggested to explain it. These theories of mechanism of toxicity include disturbances in the, demyelization of nerve tissues, inactivation of several enzymes, and mineral metabolism. Nervous system is commonly affected by lead exposure in both man and animals. However, the effect of the toxicity level is more in children than adults. This is due to softer tissues of organ than adults. Both human and animal adults exposed for a long time may perform poorly on a number of cognitive tests that measure how well their nervous systems are functioning (Boskabady et al. 2018). Due to their increased susceptibility to Pb, new born and young children may suffer with decreased IQ, behavioral issues, and learning difficulties. Prolonged Pb exposure has been linked to high rise of blood pressure and anemia, primarily in both young and old people. Exposure to high levels of lead, which resulted in mortality, was found to be associated with serious brain and kidney damage in man and animals (Reuben 2018). Greater level of chronic Pb exposure in male decreased fertility and in females during pregnancy may causes miscarriage. Lead intoxication frequently causes blood disorders and nervous system damage. It is due to the disturbance of antioxidant and pro-oxidant balance by generation of ROS. It is likely that a number of radical scavenging enzymes maintain a threshold level of ROS in the cell in order to maintain proper cell signaling. However, if the level of ROS increases above the specified threshold level, a rise in the formation of ROS resulting excessive signaling to the cell and directly damage components of essential signaling pathway (Adwas et al. 2019). Additionally, ROS can permanently harm important macromolecules. The main cytosolic low molecular weight sulfhydryl molecule, protein-bound thiol and non-protein thiol, works as a cellular reducing and protecting reagent against a variety of hazardous chemicals, including the majority of inorganic pollutants, through the –SH group (Briffa et al. 2020; Elsayed et al. 2016). Thereforc, the first line of defense against oxidative stress is often thiol. Antioxidants have been found to play major role in the detoxification of Pb against oxidative stress.

12.2 Mechanism of Lead Toxicity

Since ancient times, metallic Pb and its organic and inorganic salts have been known and used extensively. Pb concentrations greater than 0.6 μM are related with health issues. Lead can alter proteins, interfering with their capacity to carry out enzymatic processes or change functioning by binding with cellular components. Nucleic acid

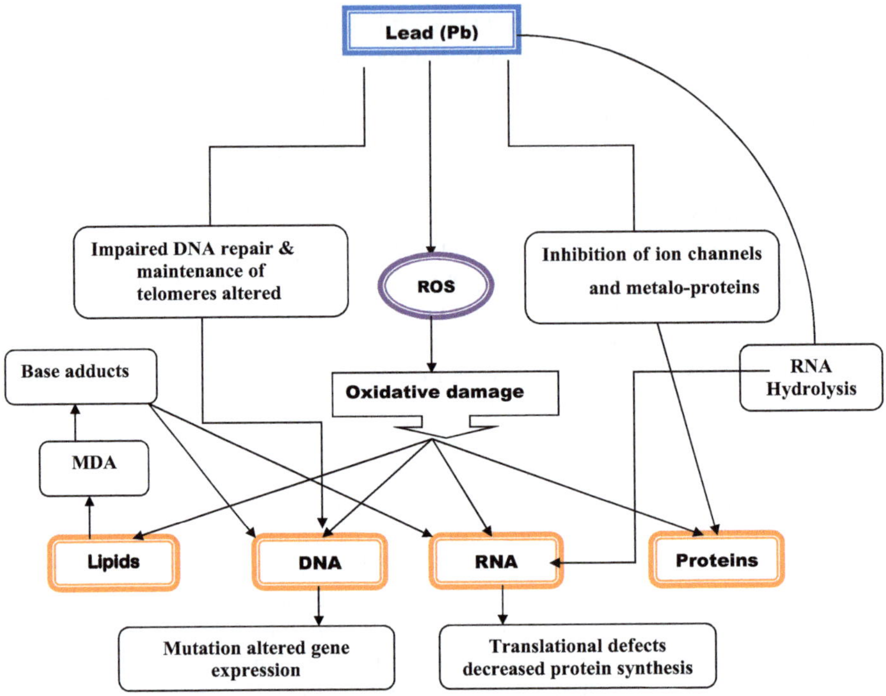

Fig. 12.1 Molecular targets of lead and their mechanism affected by lead exposure

deterioration and causing of oxidative stress are two more processes (Szymanski 2014). Numerous cellular functions are impacted by lead exposure, and each one of these effects may, in part, be responsible for the clinical signs of lead poisoning. Each mechanism's involvement may vary with degree of exposure and involvement kind of cell (Fig. 12.1).

12.2.1 Oxidative Stress

Oxidative stress is two sided, and the maintaining of a physiological amount of oxidant challenge, also known as oxidative stress, is essential for regulating essential life process through redox signaling, whereas excess oxidative stress damages or alter biomolecules (Sies et al. 2017). The ROS are continuously produced in aerobic organisms' cells as by-products of a variety of metabolic processes, such as respiration and the body's own anabolic and catabolic pathways. However, under typical circumstances, the antioxidant enzymes functioning and the creation of low molecular weight antioxidant molecules maintain the quantities of different ROS at safe levels (Tretter et al. 2021). Imbalance in between generation and scavenging of ROS leads to oxidative stress. Basically, either increased ROS generation or a compromised ROS detoxification system can cause such an imbalance. Overproduction of free radicals under oxidative stress causes inflammation, apoptosis, and damage components of cells and tissue. ROS targets proteins, nucleic acids, and lipids which can result in changes, damage, and deactivation (Bhatti et al. 2022). Free radicals are short lived, neutral, and unstable chemical species normally associated with an odd/unpaired electron. They are highly reactive if paired up with odd electron and form stable configuration. They are able to damage the healthy cells causing them to lose their structure and normal physiological functions (Taghavi and Moosavi-Movahedi 2019). Increased ROS generation in brain cells is thought to be one of the main contributors to the emergence of degenerative nervous disorders like Alzheimer's and Parkinson's. Its interaction with oxyhemoglobin has caused the generation of ROS in red blood cells (Wojtunik-Kulesza et al. 2019).

12.2.2 Interactions with Proteins

Previous researches confirm that lead can directly interfere with proteins functions by ionic displacement or by binding cysteines thiol groups, competing for binding with ion transporters or metal-dependent enzymes of natural ligands (Bridges and Zalups 2005).

12.2.2.1 δ Aminolevulinic Acid Dehydratase (ALAD)

δ Aminolevulinic acid dehydratase (ALAD) enzyme is a main target of lead poisoning in human, animals, and plants. It is also known as porphobilinogen synthase or ALA dehydratase, or aminolevulinate dehydratase encoded by the *ALAD* gene in humans (Qader et al. 2021). ALAD catalyzes two molecules of δ-aminolevulinic acid by asymmetric condensation to form the porphobilinogen in the presence of Porphobilinogen synthase. Porphobilinogen is precursor of cobalamins, plant pigment chlorophylls, cytochromes, and blood-heme (Dailey et al. 2017). The ALAD function depends on the binding of divalent zinc cation (Zn_2^+) in human, animal, fungal, and bacterial ALADs as well as magnesium (Mg_2^+) in enzymes of plant (Spencer and Jordan 2008). Basically ALAD is expressed in all the cell types in animal but is strongly expressed in liver and red blood cells (Phillips 2019). ALAD is frequently utilized as a biomarker of lead exposure due to its sensitivity to lead. Increased concentration of ALAD's substrate, -aminolevulinic acid (-ALA), is mainly because of the decrease in enzymatic activity of ALAD (Cid et al. 2018). By enolization and auto-oxidation at normal physiological pH, which leads to the creation of a superoxide anion, or by linked oxyhemoglobin oxidation, an accumulation of -ALA results in the formation of ROS (Miazek et al. 2022). The integrity of the DNA may potentially be impacted by the accumulation of -ALA. It has been demonstrated that the by-product of -ALA oxidation known as 4,5-Dioxovaleric acid which has alkylating properties capable of altering guanines (Gonçalves et al. 2020). Gamma-aminobutyric acid (GABA) and ALAD have structural resemblance due to which -ALA can bind to GABA receptors in the nervous system easily contributing to their oxidative degradation resulting in nervous system disorder (Hurkacz et al. 2021).

12.2.2.2 Calcium-Binding Proteins

Calcium acts as a second messenger and key component in the cell. It mainly depends on the large number of heterogeneous actions of calcium-binding proteins (CBP) which is capable of binding of the calcium ion in their specific domains. Calmodulin is multifunctional calcium-binding second messenger protein sensitive to alterations in cellular calcium ion concentration which is expressed in all eukaryotic cells, i.e., affected by lead exposure (Filadi et al. 2017). Lead binds to calmodulin with a higher affinity than calcium ions, although not all lead and calcium-binding sites are the same. Lead binds to the calcium-binding sites at low level, which may cause calmodulin to become activated (Kasten-Jolly and Lawrence 2018). However, greater lead levels bind with the linker region, changing its structural configuration and blocking CaM target proteins interactions, and it also displaces calcium from two of its binding sites (Garza et al. 2006). Thus, the concentration of lead may affect how lead affects calmodulin-dependent pathways. Lead may be the cause of hyperactivation at low concentrations (de Souza et al. 2018). Lead is also a potent activator of calcium/phospholipid-dependent protein kinase C and an inhibitor of the voltage-gated Ca_2^+ and K^+ channels, two proteins linked to many brain activities such as memory and learning (Reyes Gaido et al. 2023).

12.2.3 Effects of Lead on DNA

Pb has a genotoxic effect that affects the integrity of molecular proteins in addition to causing oxidative stress and interacting with protein function. It has long been known that lead and its derivatives can cause mutations. Pb can alter the chromatin by affecting DNA methylation in addition to the direct and indirect consequences (Jeena and Pandey 2021).

12.2.3.1 Oxidative Damage

The oxidative damage DNA for a very long period recognized as a serious threat for cells, due to its major toxicological effect such as aging and carcinogenesis. Every component of the DNA, including deoxyribose, nucleobases, and phosphodiester linkages, can be targeted by ROS, which can hydrolyze the phosphate-sugar backbone or modify bases (Cadet et al. 2017). 8-oxo-adenine, 5-hydroxyuracil, and thymine glycol are likewise quite carcinogenic, according to research on the effects of other oxidized DNA nucleobases. This implies how developmental disruptions brought on by lead may have consequences over time that promotes neurodegeneration (Klaunig and Kamendulis 2004).

12.2.3.2 Base Adducts

Polyunsaturated fatty acids in phospholipids are major target of oxygen radicals inside the cell (Shchepinov 2020). The oxidation of lipids results in a variety of extremely reacting chemicals, a few of which can react with the side chains of amino acids in proteins and bases in nucleic acids, in addition to having an impact on the integrity of cell membranes. According to research, endogenous DNA damage is mostly caused by lipid peroxidation (Juan et al. 2021). Malondialdehyde is most common by-products formed by lipid peroxidation and is commonly used as a sign of lead exposure (Moazamian et al. 2015). Low quantities of Malondialdehyde, a by-product of the prostaglandin production pathway, are visible in the cells when there is no stress (Wali et al. 2020). By creating exocyclic adducts with bases, elevated amounts of malondialdehyde and other aldehydes from lipid oxidation can cause DNA damage and resulting in mutations (Tudek et al. 2017).

12.2.3.3 DNA Methylation

DNA methylation is an inherited epigenetic marker in which DNA methyltransferases irreversibly transfer a methyl group at the C-5 position of the DNA cytosine ring (Mancia 2022). Gene expression is highly dependent on the epigenetic alterations of chromatin. One of the essential heritable markers is methylation of cytosines producing 5-methylcytosine (m5C) in the DNA. The epigenetic changes in chromatin have a significant impact on gene expression. Cytosines methylation forms in 5-methylcytosine which is the DNA base cytosine that regulates gene transcription. It is also essential epigenetic marker in the DNA (Breiling and Lyko 2015). DNA methylation has a role in the control of chromatin structure, genomic imprinting, and the regulation of gene expression (Szymanski 2014; Yang et al. 2022). For the creation, establishment, and maintenance of cell identity, proper methylation patterns are essential. It was established that exposure to lead linked with altered DNA methylation state (Sun et al. 2022).

12.2.3.4 Effects of Lead on RNA

The impact of Pb exposure on the integrity of RNA molecules remains a primarily unexplored territory. The understanding of the various functions carried out by both types of nucleic acids in the cell is one of the causes of this (Zoidis et al. 2018). DNA is the genetic information's storage; any damage to it must be immediately repaired to prevent mutations that could compromise the cell's viability or proper functioning

(Pedroza-Garcia et al. 2022). The effects of RNA damage do not appear to have as great an influence on cellular processes as those of DNA damage. If the DNA's information is still intact, damaged RNA can be retrieved by the proper transcripts (Herbert 2020). On the other hand, unlike DNA, RNA molecules are more vulnerable to damage molecule from ROS created at the time of oxidative stress because certain of its bases are not involved in the construction of structural configuration and are greater accessible than paired bases inside the double helix of DNA (Cannan and Pederson 2016; Juan et al. 2021).

12.3 Coping Mechanism of Antioxidants Enzymes from Lead

One of the most important substances for preventing ROS damage to cell components is glutathione (GSH), a tripeptide made up of three amino acids glutamate, histidine, and cysteine. Protein disulfide linkages are converted to cysteines by glutathione utilizing the sulfhydryl group of the cysteine as a proton donor. Glutathione undergoes this transformation into glutathione disulfide (GSSH), which has undergone oxidation. Using NADPH as a proton source, glutathione reductase transforms glutathione disulfide into glutathione (Cheng et al. 2021). Glutathione oxidation helps in detoxification of ROS. Lead has a double impact on the glutathione concentration. Its exposure causes decreased glutathione levels and elevated GSSH levels (Hasanuzzaman et al. 2018; Haridevamuthu et al. 2022). While glutathione and its glutathione reductase emerge to be the primary targets of Pb, its exposure also affects other enzymes involved in detoxification of ROS, which results in higher amounts of free radicals in the cell (Valko et al. 2016). Additionally, lead exposure has been linked in a dose–response manner to alterations in the concentrations and action of antioxidant enzymes. These enzymes include glutathione peroxidase (GPx), catalase, and superoxide dismutase (SOD), expressions of which were found to be changed in the body (Guru et al. 2021).

SOD is particularly valuable as an antioxidant because it is helpful in preventing cellular damage. The superoxide radical anion (O_2^-), which is produced by the transfer of one electron to molecular oxygen, serves as the substrate for SODs (Ifeanyi 2018). Both the direct harm to biological macromolecules and the production of additional reactive oxygen species are caused by it. SODs act as a key function in the cells defense against oxidative stress because they keep superoxide radical concentrations at low levels (Ighodaro and Akinloye 2018).

Glutathione-S-transferases (GSTs) also known as provide defense against oxidative stress. Alterations in GST activity following lead exposure are discussed. GSTs are a family of cytosolic enzymes that detoxify lead by conjugating them to glutathione. Glutathione is essential for the regular physiological activities (Awasthi et al. 2017).

Catalase is an antioxidant that decomposes hydrogen peroxide into water and oxygen. It is a common enzyme in almost all living organisms exhibited to oxygen that catalyzes the breakdown of hydrogen peroxide. It is an essential enzyme to protect the cell from oxidative harm by ROS. Hydrogen peroxide acts as a substrate for a specific reaction which creates highly free radicals. Catalase performs a major role in the defense mechanisms of cellular antioxidants by decreasing the hydrogen peroxide accumulation. The role of catalase in the protection of body cells and tissues from oxidation has been thoroughly investigated. Over-expression of catalase makes the cells more resistant to hydrogen peroxide toxicity and oxidative-mediated damage (Lobo et al. 2010; Kabel 2014).

12.4 Non-enzymatic Antioxidants Mechanisms

Antioxidant ascorbic acid is present in both animals and plants cells, but as it cannot be synthesized by humans, it must be consumed through food. ROS can be reduced as well as neutralized. By eliminating the intermediates of free radical oxidation and interacting with the lipid radicals, vitamin E has been proven to protect cell membranes against oxidation. By eliminating singlet oxygen, beta-carotene has strong antioxidant qualities and can fight from free radicals. They can be found in grains, spinach, carrots, tomatoes, milk, butter, liver, egg yolk, and egg white (Foyer 2017; Zaaboul and Liu 2022; Sarker et al. 2022).

12.5 Natural Antioxidants Protection

Natural antioxidants protective effect has received more attention from damage caused by free radicals. Flavonoids have an essential role in protecting body cells from oxidative damage. Green vegetables, fruit, red wine, tea, and cocoas are among the many foods that contain flavonoids (Khan et al. 2021a, b). Flavonoids are present in a large variety of foods and drinks and have many biological functions of antioxidation is well known. By neutralizing reactive species, natural antioxidants strengthen the body's own antioxidant defenses against ROS (Guven et al. 2019). The antioxidant activities of phenolic compounds are related to various mechanisms, like hydrogen donation, scavenging of free radical, metal ion chelating, singlet oxygen quenching, and acting as a substrate for superoxide and hydroxyl radicals (Pisoschi and Pop 2015; Behera and Senapati 2023).

12.6 Conclusion

The above literature concludes that lead causes oxidative stress resulting in irreversible damage in cellular components that responsible for propagation of many diseases such as cardiovascular problems like high blood pressure, atherosclerosis, stroke, neurological diseases, allergic diseases, hepatitis, and development of cancer. Antioxidant reduces ROS formation and scavenging free radicals.

References

Adwas AA, Elsayed A, Azab AE, Quwaydir FA (2019) Oxidative stress and antioxidant mechanisms in human body. J Appl Biotechnol Bioeng 6(1):43–47

Ali SS, Ahsan H, Zia MK, Siddiqui T, Khan FH (2020) Understanding oxidants and antioxidants: classical team with new players. J Food Biochem 44(3):e13145

Awasthi YC, Ramana KV, Chaudhary P, Srivastava SK, Awasthi S (2017) Regulatory roles of glutathione-S-transferases and 4-hydroxynonenal in stress-mediated signaling and toxicity. Free Radical Biol Med 111:235–243

Behera PC, Senapati MR (2023) Oxidative stress and its management through phytoconstituents. In: Recent frontiers of phytochemicals, pp 483–499

Bhattacharya S (2015) Reactive oxygen species and cellular defense system. In: Free radicals in human health and disease, pp 17–29

Bhatti JS, Sehrawat A, Mishra J, Sidhu IS, Navik U, Khullar N, Kumar S, Bhatti GK, Reddy PH (2022) Oxidative stress in the pathophysiology of type 2 diabetes and related complications: current therapeutics strategies and future perspectives. Free Radical Biol Med 184:114–134

Boskabady M, Marefati N, Farkhondeh T, Shakeri F, Farshbaf A, Boskabady MH (2018) The effect of environmental lead exposure on human health and the contribution of inflammatory mechanisms, a review. Environ Int 120:404–420

Breiling A, Lyko F (2015) Epigenetic regulatory functions of DNA modifications: 5-methylcytosine and beyond. Epigenetics Chromatin 8:1–9

Bridges CC, Zalups RK (2005) Molecular and ionic mimicry and the transport of toxic metals. Toxicol Appl Pharmacol 204(3):274–308

Briffa J, Sinagra E, Blundell R (2020) Heavy metal pollution in the environment and their toxicological effects on humans. Heliyon 6(9):e04691

Cadet J, Davies KJ, Medeiros MH, Di Mascio P, Wagner JR (2017) Formation and repair of oxidatively generated damage in cellular DNA. Free Radical Biol Med 107:13–34

Cannan WJ, Pederson DS (2016) Mechanisms and consequences of double-strand DNA break formation in chromatin. J Cell Physiol 231(1):3–14

Cheng X, Xu HD, Ran HH, Liang G, Wu FG (2021) Glutathione-depleting nanomedicines for synergistic cancer therapy. ACS Nano 15(5):8039–8068

Cid FD, Fernández NC, Pérez-Chaca MV, Pardo R, Caviedes-Vidal E, Chediack JG (2018) House sparrow biomarkers as lead pollution bioindicators. Evaluation of dose and exposition length on hematological and oxidative stress parameters. Ecotoxicol Environ Saf 154:154–161

Dailey HA, Dailey TA, Gerdes S, Jahn D, Jahn M, O'Brian MR, Warren MJ (2017) Prokaryotic heme biosynthesis: multiple pathways to a common essential product. Microbiol Mol Biol Rev 81(1):e00048-e116

de Souza ID, de Andrade AS, Dalmolin RJS (2018) Lead-interacting proteins and their implication in lead poisoning. Crit Rev Toxicol 48(5):375–386

Domínguez R, Pateiro M, Gagaoua M, Barba FJ, Zhang W, Lorenzo JM (2019) A comprehensive review on lipid oxidation in meat and meat products. Antioxidants 8(10):429

Elsayed ASI, Hegazi MA, Mostafa HAE, Hegazi MM (2016) The protective role of curcumin against oxidative stress caused by gasoline. Int J Appl Biol Pharm Technol 7:56–66

Filadi R, Basso E, Lefkimmiatis K, Pozzan T (2017) Beyond intracellular signaling: the ins and outs of second messengers microdomains. In: Membrane dynamics and calcium signaling, pp 279–322

Foyer CH (2017) Ascorbic acid. In: Antioxidants in higher plants. CRC Press, pp 31–58

Galano A, Reiter RJ (2018) Melatonin and its metabolites vs oxidative stress: from individual actions to collective protection. J Pineal Res 65(1):e12514

Garza A, Vega R, Soto E (2006) Cellular mechanisms of lead neurotoxicity. Med Sci Monit 12(3):RA57

Gonçalves LC, Massari J, Licciardi S, Prado FM, Linares E, Klassen A, Tavares MF, Augusto O, Di Mascio P, Bechara EJ (2020) Singlet oxygen generation by the reaction of acrolein with peroxynitrite via a 2-hydroxyvinyl radical intermediate. Free Radical Biol Med 152:83–90

Guru A, Lite C, Freddy AJ, Issac PK, Pasupuleti M, Saraswathi NT, Arasu MV, Al-Dhabi NA, Arshad A, Arockiaraj J (2021) Intracellular ROS scavenging and antioxidant regulation of WL15 from cysteine and glycine-rich protein 2 demonstrated in zebrafish in vivo model. Dev Comp Immunol 114:103863

Guven H, Arici A, Simsek O (2019) Flavonoids in our foods: a short review. J Basic Clin Health Sci 3(2):96–106

Halliwell B, Murcia MA, Chirico S, Aruoma OI (1995) Free radicals and antioxidants in food and in vivo: what they do and how they work. Crit Rev Food Sci Nutr 35(1–2):7–20. https://doi.org/10.1080/10408399509527682

Haridevamuthu B, Manjunathan T, Guru A, Kumar RS, Rajagopal R, Kuppusamy P, Juliet A, Gopinath P, Arockiaraj J (2022) Hydroxyl containing benzo [b] thiophene analogs mitigates the acrylamide induced oxidative stress in the zebrafish larvae by stabilizing the glutathione redox cycle. Life Sci 298:120507

Hasanuzzaman M, Nahar K, Rahman A, Mahmud JA, Alharby HF, Fujita M (2018) Exogenous glutathione attenuates lead-induced oxidative stress in wheat by improving antioxidant defense and physiological mechanisms. J Plant Interact 13(1):203–212

Herbert A (2020) ALU non-B-DNA conformations, flipons, binary codes and evolution. Royal Soc Open Sci 7(6):200222

Hu L, Ren S, Shen Q, Chen J, Ye X, Ling J (2017) Proteomic study of the effect of different cooking methods on protein oxidation in fish fillets. RSC Adv 7(44):27496–27505

Hurkacz M, Dobrek L, Wiela-Hojeńska A (2021) Antibiotics and the nervous system—which face of antibiotic therapy is real, Dr. Jekyll (Neurotoxicity) or Mr. Hyde (Neuroprotection)? Molecules 26(24):7456

Ifeanyi OE (2018) A review on free radicals and antioxidants. Int J Curr Res Med Sci 4(2):123–133

Ighodaro OM, Akinloye OA (2018) First line defence antioxidants-superoxide dismutase (SOD), catalase (CAT) and glutathione peroxidase (GPX): their fundamental role in the entire antioxidant defence grid. Alexandria J Med 54(4):287–293

Jeena AS, Pandey D (2021) Metal induced genotoxicity and oxidative stress in plants, assessment methods, and role of various factors in genotoxicity regulation. In: Induced genotoxicity and oxidative stress in plants, pp 133–149

Juan CA, Pérez de la Lastra JM, Plou FJ, Pérez-Lebeña E (2021) The chemistry of reactive oxygen species (ROS) revisited: outlining their role in biological macromolecules (DNA, lipids and proteins) and induced pathologies. Int J Mol Sci 22(9):4642

Kabel AM (2014) Free radicals and antioxidants: role of enzymes and nutrition. World J Nutr Health 2(3):35–38

Kasten-Jolly J, Lawrence DA (2018) The cationic (calcium and lead) and enzyme conundrum. J Toxicol Environ Health Part B 21(6–8):400–413

Khan MK, Paniwnyk L, Hassan S (2019) Polyphenols as natural antioxidants: sources, extraction and applications in food, cosmetics and drugs. In: Plant based "green chemistry 2.0" moving from evolutionary to revolutionary, pp 197–235

Khan FSA, Mubarak NM, Tan YH, Khalid M, Karri RR, Walvekar R, Abdullah EC, Nizamuddin S, Mazari SA (2021a) A comprehensive review on magnetic carbon nanotubes and carbon nanotube-based buckypaper for removal of heavy metals and dyes. J Hazard Mater 413:125375

Khan J, Deb PK, Priya S, Medina KD, Devi R, Walode SG, Rudrapal M (2021b) Dietary flavonoids: cardioprotective potential with antioxidant effects and their pharmacokinetic, toxicological and therapeutic concerns. Molecules 26(13):4021

Klaunig JE, Kamendulis LM (2004) The role of oxidative stress in carcinogenesis. Annu Rev Pharmacol Toxicol 44:239–267

Li R, Jia Z, Trush MA (2016) Defining ROS in biology and medicine. React Oxygen Species (Apex, NC) 1(1):9

Lobo V, Patil A, Phatak A, Chandra N (2010) Free radicals, antioxidants and functional foods: Impact on human health. Pharmacogn Rev 4(8):118

Lü JM, Lin PH, Yao Q, Chen C (2010) Chemical and molecular mechanisms of antioxidants: experimental approaches and model systems. J Cell Mol Med 14(4):840–860

Mancia A (2022) Genome-wide DNA methylation DNA methylation protocol for epigenetics epigenetics studies. In: Marine genomics: methods and protocols. Springer US, New York, NY, pp 19–41

Miazek K, Beton K, Śliwińska A, Brożek-Płuska B (2022) The effect of β-carotene, tocopherols and ascorbic acid as anti-oxidant molecules on human and animal in vitro/in vivo studies: a review of research design and analytical techniques used. Biomolecules 12(8):1087

Moazamian R, Polhemus A, Connaughton H, Fraser B, Whiting S, Gharagozloo P, Aitken RJ (2015) Oxidative stress and human spermatozoa: diagnostic and functional significance of aldehydes generated as a result of lipid peroxidation. MHR Basic Sci Reprod Med 21(6):502–515

Nath S, Modak BK, Das S, Podder R (2019) Comparing the susceptibility of Chironomus Striatapennis larvae exposed to heavy metals. Recent Adv Biol Res 2:48–58

Pedroza-Garcia JA, Xiang Y, De Veylder L (2022) Cell cycle checkpoint control in response to DNA damage by environmental stresses. Plant J 109(3):490–507

Phillips JD (2019) Heme biosynthesis and the porphyrias. Mol Genet Metab 128(3):164–177

Pisoschi AM, Pop A (2015) The role of antioxidants in the chemistry of oxidative stress: a review. Eur J Med Chem 97:55–74

Pisoschi AM, Pop A, Iordache F, Stanca L, Predoi G, Serban AI (2021) Oxidative stress mitigation by antioxidants-an overview on their chemistry and influences on health status. Eur J Med Chem 209:112891

Qader A, Rehman K, Akash MSH (2021) Genetic susceptibility of δ-ALAD associated with lead (Pb) intoxication: sources of exposure, preventive measures, and treatment interventions. Environ Sci Pollut Res 28:44818–44832

Reuben A (2018) Childhood lead exposure and adult neurodegenerative disease. J Alzheimers Dis 64(1):17–42

Reyes Gaido OE, Nkashama LJ, Schole KL, Wang Q, Umapathi P, Mesubi OO, Konstantinidis K, Lucza, ED, Anderson ME (2023) CaMKII as a therapeutic target in cardiovascular disease. Annu Rev Pharmacol Toxicol 63

Sarker U, Oba S, Ercisli S, Assouguem A, Alotaibi A, Ullah R (2022) Bioactive phytochemicals and quenching activity of radicals in selected drought-resistant Amaranthus tricolor vegetable amaranth. Antioxidants 11(3):578

Shchepinov MS (2020) Polyunsaturated fatty acid deuteration against neurodegeneration. Trends Pharmacol Sci 41(4):236–248

Sies H, Berndt C, Jones DP (2017) Oxidative stress. Annu Rev Biochem 86:715–748

Spencer P, Jordan PM (2008) B metal-binding sites of the Escherichia coli enzyme. In: The biosynthesis of the tetrapyrrole pigments, p 50

Sun M, Yang Z, Liu L, Duan L (2022) DNA methylation in plant responses and adaption to abiotic stresses. Int J Mol Sci 23(13):6910

Szymanski M (2014) Molecular mechanisms of lead toxicity. BioTechnologia. J Biotechnol Comput Biol Bionanotechnol 95(2)

Taghavi F, Moosavi-Movahedi AA (2019) Free radicals, diabetes, and its complexities. In: Plant and human health, volume 2: phytochemistry and molecular aspects, pp 1–41

Tretter V, Hochreiter B, Zach ML, Krenn K, Klein KU (2021) Understanding cellular redox homeostasis: a challenge for precision medicine. Int J Mol Sci 23(1):106

Truong VL, Jun M, Jeong WS (2018) Role of resveratrol in regulation of cellular defense systems against oxidative stress. BioFactors 44(1):36–49

Tudek B, Zdżalik-Bielecka D, Tudek A, Kosicki K, Fabisiewicz A, Speina E (2017) Lipid peroxidation in face of DNA damage, DNA repair and other cellular processes. Free Radical Biol Med 107:77–89

Valko M, Jomova K, Rhodes CJ, Kuča K, Musílek K (2016) Redox-and non-redox-metal-induced formation of free radicals and their role in human disease. Arch Toxicol 90:1–37

Velvizhi G, Shanthakumar S, Das B, Pugazhendhi A, Priya TS, Ashok B, Nanthagopal K, Vignesh R, Karthick C (2020) Biodegradable and non-biodegradable fraction of municipal solid waste for multifaceted applications through a closed loop integrated refinery platform: paving a path towards circular economy. Sci Total Environ 731:138049

Wali AF, Rashid S, Rashid SM, Ansari MA, Khan MR, Haq N, Alhareth DY, Ahmad A, Rehman MU (2020) Naringenin regulates doxorubicin-induced liver dysfunction: impact on oxidative stress and inflammation. Plants 9(4):550

Wojtunik-Kulesza K, Oniszczuk A, Waksmundzka-Hajnos M (2019) An attempt to elucidate the role of iron and zinc ions in development of Alzheimer's and Parkinson's diseases. Biomed Pharmacother 111:1277–1289

Yadav A, Kumari R, Yadav A, Mishra JP, Srivatva S, Prabha S (2016) Antioxidants and its functions in human body—a review. Res Environ Life Sci 9(11):1328–1331

Yang Z, Xu F, Teschendorff AE, Zhao Y, Yao L, Li J, He Y (2022) Insights into the role of long non-coding RNAs in DNA methylation mediated transcriptional regulation. Front Mol Biosci 9

Zaaboul F, Liu Y (2022) Vitamin E in foodstuff: nutritional, analytical, and food technology aspects. Compr Rev Food Sci Food Saf 21(2):964–998

Zoidis E, Seremelis I, Kontopoulos N, Danezis GP (2018) Selenium-dependent antioxidant enzymes: actions and properties of selenoproteins. Antioxidants 7(5):66

Chapter 13
Biotechnological Approaches in Remediation of Lead Toxicity

Saurabh Gupta, Amanpreet Kaur, Ravindra Kumar, Sumanveer Kaur, Sneha, Bhairav Prasad, and Vijay Singh

Abstract Rapid industrialisation, modern farming methods, and other anthropogenic activities increase the amount of toxic heavy metals in the atmosphere, leading to a profoundly toxic influence on all forms of life. The health and safety of humans are not the only things threatened by this type of heavy metal pollution. Heavy metal pollution causes neurological issues in children because it is mutagenic, endocrine, carcinogenic, and teratogenic. Because of this, addressing heavy metal pollution needs to be a priority. Cost, labour intensity, changes in soil characteristics, and disruption of natural microflora are just some of the problems with the various physical and chemical procedures employed for this goal. However, phytoremediation is an improved approach to the issue. The term "phytoremediation" refers to the practise of using plants and the microorganisms found in soil to mitigate the harmful effects of pollution. It is widely accepted because of its novelty, low cost, efficiency, little environmental impact, and use of solar energy. It is true that a lot of research is being done on phytoremediation right now. For use in phytoremediation, scientists are looking into new, highly effective metal hyperaccumulators. Plants' mechanisms of metal absorption, transport, sequestration, and tolerance are being studied with molecular techniques.

Keywords Heavy metals · Water · Contamination · Phytoremediation · Soil system · Hotspots

---

S. Gupta (✉) · A. Kaur · S. Kaur · Sneha
Department of Microbiology, Mata Gujri College, Fatehgarh Sahib, Punjab 140406, India
e-mail: sau27282@gmail.com

R. Kumar
Department of Agriculture, Mata Gujri College, Fatehgarh Sahib, Punjab 140406, India

B. Prasad
Department of Biotechnology, Chandigarh Group of Colleges, Landran, Mohali, Punjab 140307, India

V. Singh
Department of Botany, Mata Gujri College, Fatehgarh Sahib, Punjab 140406, India

N. Kumar and A. K. Jha (eds.), *Lead Toxicity Mitigation: Sustainable Nexus Approaches*, Environmental Contamination Remediation and Management,
https://doi.org/10.1007/978-3-031-46146-0_13

13.1 Introduction

Industrial wastes contain heavy metals such as Cd, Cu, Ni, Co, Zn, and Pb, which are found in drinking water, and it is well known that municipal treatment plants are not well developed to remove heavy metal residues (Malik 2003). Lead is found in aquatic and terrestrial ecosystems due to a variety of anthropogenic causes (Selvi et al. 2019). Any metallic element with a relatively high density that is harmful at low concentrations is referred to as a heavy metal. Lead is a hazardous heavy metal with an atomic number of 82, a molecular weight of 207.2, a density of 11.34 g/cm^3, and a melting point of 621.43 F (Kumar et al. 2020). Lead is expected to contribute roughly 10% of overall heavy metal pollution (Collin et al. 2022). Lichens were employed as a bio-indicator to compare different Pb isotopes across time (Gupta et al. 2019). Pb and Pb compounds, for example, have been recognised as key hazardous chemicals in 47% of the 1219 superfund sites now on the USEPA's National Priorities List (Liu and Zhao 2007). Lead has been employed throughout history because of its physio-chemical qualities such as softness, malleability, weak conductivity, and corrosion resistance (Wani et al. 2015). Lead is most commonly used in the preparation of leaded gasoline, industrial processes such as lead smelting and combustion, pottery, boat building, lead-based paintings, lead-containing pipes, battery recycling, the arm industry, book printing, and many other applications (Wani et al. 2015) (Fig. 13.1).

Due to the propensity of lead to leach into canned foods, the food canning industry is also a significant source of lead exposure (Kumar et al. 2020; Collin et al. 2022). Lead (Pb) contamination in air, soil, and water resources has been linked to natural processes, including geochemical weathering, marine spray emissions, volcanic activity, and the remobilisation of sediment, soil, and water from mining areas (Ali et al. 2022). Today, remediation of heavy metals (lead) is a global priority as they cannot be destroyed, are toxic and unsuitable for bio-treatment, their bioaccumulation and bioaugmentation by the food chain can disrupt the normal physiological functions of humans, and even trace amounts are lethal (Anju 2017). Lead is non-biodegradable by nature, meaning that it cannot be degraded, and it accumulates in the environment, causing damage to humans, plants, and animals (Ye et al. 2002). About 20–70% of the lead that is ingested is absorbed by the body (Kumar et al. 2020). As a result, ingestion is the primary route of lead exposure for humans. In addition, lead is used as a pesticide during the cultivation of vegetables and fruits (Gall et al. 2015). It affects every organ in the human body and is responsible for reproductive effects, respiratory effects, neurological effects, birth defects, cancer, and anaemia, among others. Aside from this, expectant women and children are at greater risk from lead exposure because it inhibits foetal development in the early stages (Wani et al. 2015) (Fig. 13.2).

Lead (Pb) can be harmful to both macro and microorganisms due to its ability to directly impact biochemical and physiological processes, hinder growth, degrade cell organelles, and inhibit photosynthesis (Ahmadpour et al. 2012; Din et al. 2023). According to Malik (2003), traditional physico-chemical methods for metal remediation, such as chemical oxidation, reduction, precipitation, adsorption, electrolytic

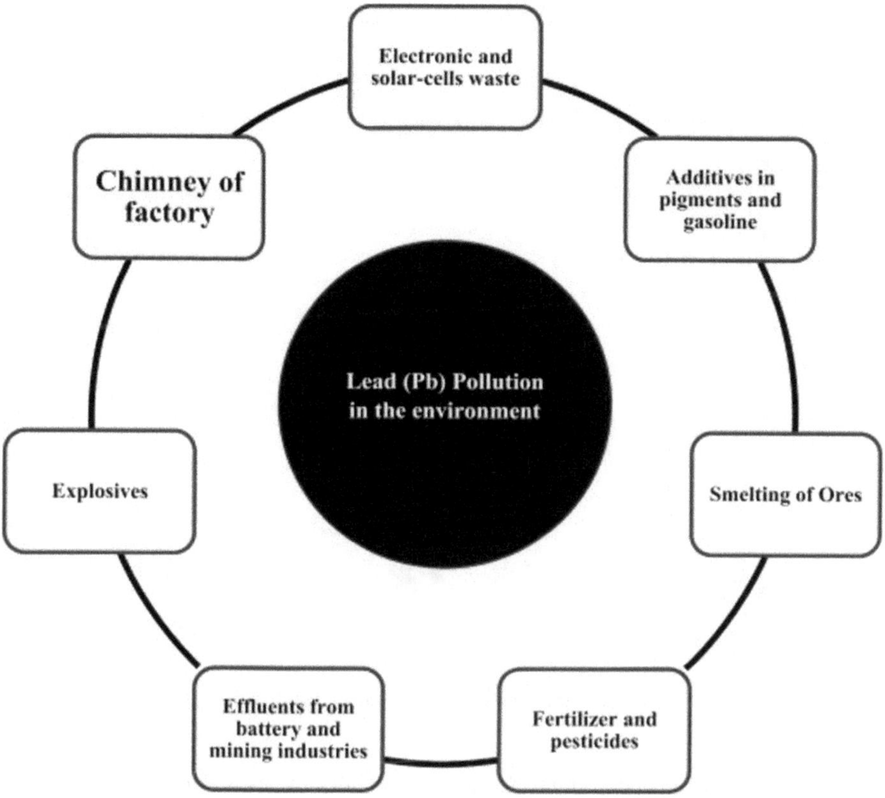

Fig. 13.1 Sources of lead pollution

recovery, and ion exchange, have been deemed costly, unsuitable, and constrained. In recent years, there has been a significant focus on biotechnological methods aimed at occupying ecological niches, as evidenced by the growing interest in this area (Malik 2003). Various methods have been devised by biotechnologists to facilitate the destruction of PB. These methods include physical, chemical, and biological approaches. Of these approaches, biological methods have been found to be the most cost-effective, efficient, and environmentally friendly (Anju 2017). The process of utilising natural and recombinant microorganisms or plants for the purpose of eliminating environmental toxic pollutants is known as bioremediation. This approach is widely acknowledged for its cost-effectiveness and eco-friendliness (Mosa et al. 2016). Various defence mechanisms have been evolved by microorganisms to cope with heavy metal stress. These mechanisms include transport across cell membranes, biosorption to cell walls, entrapment in extracellular capsules, precipitation, complexation, and oxidation–reduction reactions (Rai et al. 1981; Macaskie and Dean 1989; Huang et al. 1990; Avery and Tobin 1993; Brady and Duncan 1994; Krauter et al. 1996; Veglio et al. 1997). Their capacity to uptake heavy metals from

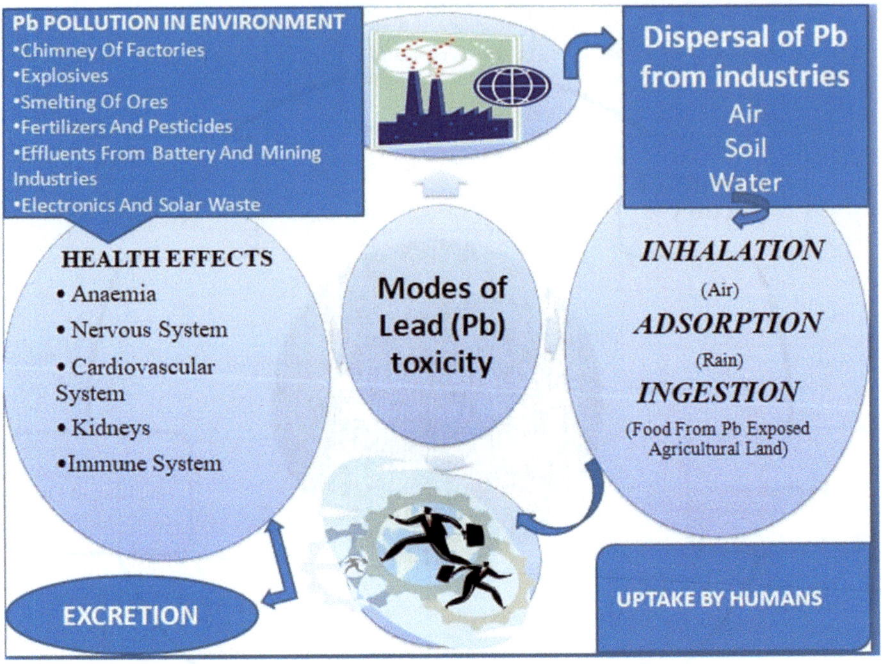

Fig. 13.2 Exposure to chemicals and their effects on human health

aqueous solutions has been demonstrated, especially in cases where the concentration of metal effluent ranges from below 1 to approximately 20 mg/l. Furthermore, biological methods for metal remediation possess the capability to manage diverse physico-chemical parameters present in effluents, exhibit selectivity towards targeted metals, and are cost-effective. These factors have stimulated a comprehensive investigation of biological approaches for the elimination of metals. Plants exhibit the requisite genetic, biochemical, and physiological attributes to position themselves as the preeminent option for the remediation of soil and water pollutants. Phytoremediation is a well-known process that involves the utilisation of plants for the purpose of extracting, sequestering, or decontaminating Pb or other heavy metal contaminants (Kavamura and Esposito 2009). The focus of this review centres on the accomplishments of biotechnological techniques and practises in the areas of environmental conservation, detoxification, and lead elimination. The present review article delves into the latest developments and potential prospects concerning bio- and phytoremediation of toxic Pb from contaminated water and soil.

13.2 Bioremediation

Bioremediation is the process of utilising natural organisms, such as bacteria, plants, or their derivatives, to break down contaminants into less toxic forms. The utilisation of bioremediation as a means of eliminating pollutants has emerged as a feasible approach that is both ecologically sustainable and economically advantageous. Bioremediation is based on the fundamental principles of altering redox reactions, modifying pH, utilising diverse complexation reactions to either increase or decrease solubility, and the process of adsorption or uptake of a substance from the surrounding environment (Smith et al. 1994). The process of utilising both recombinant and naturally occurring microorganisms to eliminate hazardous pollutants from the environment is commonly referred to as bioremediation. The microorganisms possess the ability to modify the oxidative state or organic complex of heavy metals (Xu et al. 2010). Further, the efficacy of microorganism-mediated remediation is primarily influenced by the microbe's ability to withstand heavy metal exposure, which can be induced either autonomously or in response to metal-induced stress (Naz et al. 2016). Moreover, the aforementioned approach is considered to be a financially prudent and ecologically conscientious tactic (Flathman and Lanza 2010; Mosa 2016). There exist two distinct categories of bioremediation techniques: *in-situ* bioremediation and ex-situ bioremediation, which are differentiated based on the location of the remediation process. The term "ex situ" pertains to the intentional relocation of polluted substances to an alternative site with the aim of improving biocatalysts, as stated (Prasad et al. 2010). Conversely, when the remediation process is executed at the actual location of the contamination, it is referred to as in situ (Vidali 2001; Saadoun and Al-Ghzawi 2005). The use of in-situ bioremediation techniques presents certain benefits, notably the absence of the need for excavation of the soil that has been contaminated. The aforementioned approach offers a volumetric modality of remediation that effectively addresses both dissolved and solid forms of contaminants. Accelerated in-situ bioremediation has been observed to have a shorter duration for treating sub-surface pollution in comparison with pump-and-treat processes. There exists a potential for the complete conversion of organic contaminants into benign compounds such as carbon dioxide, water, and ethane. The method can be considered cost-effective due to the limited amount of disruption to the site. Despite its advantages, this method is subject to certain limitations, which are described below. The complete transformation of certain contaminants into benign substances may not be achievable in certain locations. In cases where transformation ceases at an intermediate compound, it is possible for the intermediate to exhibit greater toxicity and/or mobility than the parent compound. Additionally, certain recalcitrant contaminants may not be susceptible to biodegradation. Improper application of injection wells can lead to blockage caused by an excessive proliferation of microorganisms, which is attributed to the introduction of nutrients, electron donors, and electron acceptors. The presence of high concentrations of heavy metals and organic compounds has been observed to impede the activity of indigenous microorganisms. The process of in-situ bioremediation typically necessitates the acclimatisation of microorganisms, which may not occur *in situ*ations involving spills and recalcitrant compounds. Further, ex-situ bioremediation offers several advantages, including its applicability to a broad spectrum of heavy metal pollutants. Its suitability can be readily assessed through data obtained from site investigations. Furthermore, biodegradation is more pronounced in a bioreactor system than in solid-phase systems due to the greater manageability and predictability of the contaminated environment.

The primary constraint associated with ex-situ bioremediation pertains to its inapplicability for heavy metals. Numerous techniques have been developed over the years, including biosorption and bioaccumulation.

13.2.1 Biosorption

Utilising natural organisms, such as bacteria, plants, or their derivatives, to transform pollutants into less hazardous forms is known as bioremediation. Utilising bioremediation to get rid of contaminants is now recognised as a workable strategy that is both environmentally responsible and profitable. The basic principles of bioremediation are changing redox reactions, modifying pH, using various complexation reactions to either increase or decrease solubility, and the process of adsorption or uptake of a substance from the environment. Bioremediation is the term used to describe the process of using recombinant and naturally occurring microbes to remove dangerous contaminants from the environment (Smith et al. 1994). According to Xu et al. (2010) research, microbes can alter the oxidative state or organic complex of heavy metals. Further, the microbe's capacity to endure heavy metal exposure, which can be induced either autonomously or in response to metal-induced stress, has a significant impact on the effectiveness of microorganism-mediated remediation (Naz et al. 2016). In-situ and ex-situ bioremediation are two distinct categories of bioremediation techniques that differ according to where the remediation process is taking place. According to Prasad et al. (2010), "ex situ" refers to the deliberate removal of contaminated materials to a different place with the objective of enhancing biocatalysts. In contrast, the remediation procedure is referred to as "in situ" when it is carried out at the actual site of the contamination (Vidali 2001; Saadoun and Al-Ghzawi 2005). Utilising in-situ bioremediation techniques has some advantages, chief among which is the avoidance of the need to remove contaminated soil from its original location. The method described above provides a volumetric remediation mechanism that successfully handles both dissolved and solid types of pollutants. It has been found that accelerated in-situ bioremediation can cure sub-surface contamination more quickly than pump-and-treat procedures. Organic pollutants have the ability to be completely transformed into harmless substances like carbon dioxide, water, and ethane. Due to the little site interruption, the procedure might be regarded as cost-effective. This approach has some drawbacks despite its benefits, which are detailed below. It might not be possible to completely turn some toxins into benign compounds in some environments. When a compound's transformation stops at an intermediate stage, it is feasible for the intermediate to be more poisonous or mobile than the parent substance. Additionally, biodegradation might not be possible with some stubborn pollutants. Due to the entry of nutrients, electron donors, and electron acceptors, improper application of injection wells might result in blockage brought on by an excessive growth of microorganisms. High quantities of organic chemicals and heavy metals have been proven to inhibit the activity of local microorganisms. In instances involving spills and resistant substances, the acclimatisation of microorganisms—which is normally required for the in-situ bioremediation process—may not take place. Additionally, because the polluted environment is easier to control and anticipate in a bioreactor system than it is in a solid-phase system, biodegradation is more pronounced in the latter. Ex-situ bioremediation's main drawback has to do with the fact that it does not work for heavy metals. Over the years, many methods have been developed, including biosorption and bioaccumulation.

13.2.2 Biosurfactants

Biologically active chemicals, also known as biosurfactants, have been shown to be effective in the removal and detoxification of heavy metals (Bachmann et al. 2014; Akbari et al. 2018). This has been accomplished by employing biosurfactants. Surfactants are typically amphiphilic chemicals that are used to solubilise, complexate, desorb, and mobilise pollutants into liquid solutions (Liu et al. 2017). This makes it possible for bacteria to recover and reuse the pollutants (Liu et al. 2017). According to research conducted by Mulligan et al. (2014) and Mao et al. (2015), biosurfactants find widespread application in a variety of industries, including the pharmaceutical industry, the petroleum industry, the cosmetics industry, the detergent industry, the paint industry, the food industry, and bioremediation. The biosurfactants are regarded as potential candidates for bioremediation because of their ionic nature, low toxicity, strong emulsifying activity, multi-functionality, surface activity, and compatibility with the environment (Das et al. 2017; Akbari et al. 2018).

13.2.3 Bioaccumulation

13.2.3.1 Metallophytes

Metallophytes are plants that can tolerate high metal concentrations due to a unique biological mechanism (Whiting et al. 2004). They are typically endemic to their native metalliferous soils and have the biological mechanisms to resist, tolerate, or even thrive there (Baker et al. 2010). The most likely developed as a result of the natural selection of metals by plants, which may have genetic roots (Whiting et al. 2002). Populations and organisms within the same population may exhibit some quantitative genetic variations in this ability (Pollard et al. 2002). The length of exposure to metals governs the degree of specialisation of the metal resistance trait (Whiting et al. 2004). The ability of metallophytes to tolerate extremely high metal concentrations makes them ideal for use in site restoration following mine closure as well as in the rehabilitation and replanting of mines and metal-contaminated sites (Anju 2017). According to the concentration of metals found in their tissues, plants can be classified as accumulators, excluders, or indicators (Baker 1981). The majority of these metallophyte taxa (referred to as "excluders" by Baker 1981) are able to tolerate particular metals in the substrate by physiologically limiting metal entry into the root and/or their transport to the shoot. However, a small number of species have highly specialised biological mechanisms that allow them to accumulate, or even hyperaccumulate, metals in their shoots to levels that can exceed 2% of their dry weight (Pollard et al. 2014). Hyperaccumulator(s) are naturally able to absorb metals at levels that are 50–500 times greater than those of typical plants (Lasat 2000). A plant that grows naturally and has leaves that contain a metallic element at a concentration higher than a certain threshold is known as a hyperaccumulator (Anju 2017). Greater potential for metal uptake, quicker root-to-shoot translocation of metals, and greater effectiveness in metal sequestration and detoxification in above-ground plant parts are the primary characteristics that set hyperaccumulators apart from non-accumulators (Rascio and Navari-Izzo 2011). The production of phytochelatins, peptides that bind metals and protect metal-sensitive enzymes by sequestering them, is thought to be the mechanism underlying plants tolerance to metals (Zenk 1996; Cobbett 2000; Hall 2002). In one growing season, the cultivation of Silene vulgaris and Armeria maritime resulted in a statistically significant decrease in Pb (Ciarkowska and Hanus-Fajerska 2008).

13.3 Phytoremediation

The term "phytoremediation" refers to both plant-based bioremediation technologies as well as the relatively new technology known as "phytotechnologies." Phytotechnologies have only been in use for a relatively short period of time (Azadpour and Matthews 1996; Garbisu et al. 2002; Singh et al. 2003; Paquin et al. 2004; Shah and Nongkynrih 2007; Padmavathiamma and Li 2007). It is a developing alternative to the restoration of contaminated sites. Plants have the potential to decrease the detrimental impacts that heavy metals will have on the environment in the future (Ahmadpour et al. 2012). They can do this by preventing pollutants from travelling via the air or poisoning groundwater. This is a reference to the practise of using healthy vegetation to cleanse polluted groundwater and soil. However, in order to choose plants that are suitable for this approach, some characteristics need to be satisfied. These requirements include metal tolerance, a high bioaccumulation factor, and a short life cycle (Kavamura and Esposito 2009). Even though the idea of employing metal-accumulating plants to remove heavy metals and other chemicals was first introduced in 1983, the practice has really been going on for the past three hundred years (Henry 2000). The phytoremediation is a viable alternative technology that is beneficial to the environment and may be used to restore degraded soils that have been contaminated with metals (Ali et al. 2013a, b; Sidhu et al. 2017a, b). In certain circumstances, mechanical conventional cleanup treatments, which usually involve significant capital expenditures, additional labour, and energy-intensive procedures (Cunningham et al. 1995), can be supplemented or replaced by phytoremediation. Phytoremediation can also be used in conjunction with traditional mechanical cleanup treatments. According to Erakhrumen (2007), phytoremediation is also referred to as green remediation, botano-remediation, agro-remediation, and vegetative remediation. All of these terms refer to the same process. According to Pivertz (2001), this approach to cleaning up polluted environments is not only friendlier to the natural world but also more financially feasible and aesthetically pleasing. The plant that is employed in the phytoremediation approach needs to have a sizeable capacity for metal absorption, accumulation, and strength in order to cut down on the length of time that the treatment takes. The plant species *T. rotundifolium* has the ability to remove 4000 kg/ha from the 8200 mg/kg concentration by the whole part of the plant (Cunningham and Ow 1996), while the plant species Euphorbiacheradenia has the ability to remove approximately 13,249 kg/ha of Pb from the 13,500 mg/kg lead concentration by the shoots of the plant (Chehregani and Malayeri 2007). There are several different ways in which plants can clean up or remediate contaminated environments. The need to wait for new plant communities to recolonise the land is eliminated thanks to phytoremediation, which, in addition to being more sustainable for the environment and economically practical, is also more environmentally benign. According to the fate of the contaminant, phytoremediation encompasses a variety of strategies. These strategies include phytoextraction/phytoaccumulation, phytodegradation, rhizodegradation/phytotransformation, and phytofiltration/phytostabilisation (Ali et al. 2013a, 2013b; Mahar et al. 2016). Each of these strategies has a unique mechanism of action, and they are utilised for the cleanup of metal ions (Anju 2017).

13.4 Phytostabilisation

Phytostabilisation, also known as in-place activation, is the process of using plants to localise toxins in order to decrease human exposure to those contaminants. In most cases, soil treatments are carried out with the intention of producing contaminant species that are insoluble. Phytoimmobilisation is an additional name for this process (Shao et al. 2022). According to Padmavathiamma and Li (2007), this mechanism does not include the removal of heavy metals; rather, the metals are simply stabilised by collecting inside root zones. The mobility of heavy metals in soils can be halted by the roots of metal-tolerant plants and the bacteria that are associated with those roots. This can be accomplished through sorption by the roots, complexation or metal valence reduction in the rhizosphere, and precipitation (Shao et al. 2022). Heavy metal-tolerant plants are used to reduce the mobility and bioavailability of heavy metals in the environment. This helps to reduce the risk of further environmental degradation (Pulford and Watson 2003; Erakhrumen 2007). Heavy metal-tolerant plants do this by immobilising or preventing the migration of heavy metals to groundwater or their entry into the food chain.

13.5 Phytoextraction

In the field of "green technologies," the method of phytoextraction, which is a form of phytoremediation, has the benefit of not being a harmful approach and is widely used (Robinson et al. 2003). Phytoextraction is a type of phytoremediation. The mobility and bioavailability of heavy metals in soil, which exist in different fractions (1), free metal ions and soluble metal complexes in the soil solution (2), metal ions occupying ion exchangeable sites and adsorbed on inorganic soil constitutes are readily available to plants, particularly in the rhizosphere. Their speciation, the characteristics of the soil, and the capacity of concentrated plant species to absorb and uptake metals are just a few of the facts that need to be considered. According to Anju (2017), hyperaccumulation is a method that can be utilised in phytoremediation for the removal of heavy metals like lead. Hyperaccumulators are specialised plant species that have a BCF value that is larger than one (Cluis 2004). These plants are also responsible for long-term, continuous phytoextraction and have the physiological potential to collect metals during the regular growth cycle (Salt and Baker 2000). According to Ernst (1996) and Schmidt (2003), two of the most important elements that determine whether or not a phytoextraction system is successful are the capacity of plants to accumulate metals in their tissues and the level of metal concentration in the soil solution. The BCF or A factor and the TF factor are the factors that determine how effective phytoextraction is. According to Zhuang et al. (2007) and Zayed et al. (1998), the bioconcentration factor, or BCF, is determined by calculating the ratio of the initial concentration of metals in the external environment to the concentration of metals in plant tissues at the time of harvest. (B.C.F. = C harvested tissue/CC soil; C harvested tissue = target metal concentration in the harvested plant tissue; C soil = target metal concentration in the soil) "B.C.F." According to research by Radulescu et al. (2013), a plant may be an accumulator if its BCF is greater than 1, free from effect if its BCF is equal to 1, or an excluder if its BCF is less than 1. The ability of a plant to transport metal from its roots to its shoots is measured by a factor called T.F., which stands for the translocation factor. According to Padmavatiamma and Li (2007), the formula for calculating it is as follows: T.F. = C shoot/CC root, where C shoot represents the target metal concentration in plant shoots, and C root represents the target metal concentration in plant roots. When these figures are multiplied by 100, one obtains the percentage of BCF and T.F. that is present. Wilson and Pyatt (2007), as well as Zacchini et al. (2009), found that, according to Ahmadpour et al. (2010), the species of plant known as J. curcas has the potential to be used for phytoextraction of lead from polluted locations (Ginochhio et al. 2004). Pollutant-accumulating plants take in significant quantities of trace elements from the substrates through their roots. These elements are then translocated to and accumulated in the plant's aerial or above-ground harvestable biomass, such as its shoots, leaves, and other parts of the plant. For phytoextraction to be successful, plants need to be able to hyperaccumulate Pb, but they also need to be fast-growing, biomass-producing, herbivore-resistant, tolerant of nutrient-poor soil, and easy to harvest (Karenlampi et al. 2000; Garbisu et al. 2002; Punshon et al. 1996). Pb tends to concentrate in the roots more than the aerial portions of the plants due to specific obstacles that restrict the flow of metals from the roots to the aerial parts of the plants (Ahmadpour et al. 2012). However, species like Thlapsi caeruslescens, which are known as hyperaccumulators, produce less above-ground biomass but accumulate Pb to a greater extent (Dickinson et al. 2009; Rozas et al. 2006). Most methods of metal phytoextraction have two key drawbacks: the bioavailability of the target metals (Pb) and the tendency of diverse plants to store heavy metals within their above-ground biomass (Raskin and Ensley 2000). Metal phytoextraction (and plant growth) may be aided by soil microorganisms that coexist with plant roots (Shilev et al. 2001). Though no plant is currently known that fully fits the criteria of an ideal plant for phytoextraction (Anju 2017), a rapidly growing non-accumulator plant could be genetically changed and/or designed to accomplish the innovative qualities of an ideal plant for phytoextraction. Numerous supplementary methods, including genetically altered plants for phytoremediation, microbe-assisted phytoremediation, and induced phytoextraction, have been studied around the world to boost this method's efficacy (Anju 2017).

13.5.1 Induced Phytoextraction

This technique is also referred to as assisted or enhanced phytoextraction in some circles. According to Prasad and Freitas (2003), chelating compounds can be used to mobilise heavy metals *in situ*ations where the availability of the metals in the substrate is insufficient for the active root to absorb and transport the metals. According to Grčman et al. (2001), ethylene diamine tetraacetic acid (also known as EDTA) was mixed with the plant species in order to increase the amount of lead that was extracted. In addition, EDTA is the most widely used chelating agent since it is effective, readily available, and relatively inexpensive (Lestan et al. 2008). According to Anju (2017), the ability of chelators to mobilise metals is contingent on a number of factors, including the metal-to-ligand ratio, the formation constants of metal-ligand complexes, the species of metals and their distribution among the soil fractions, the pH of the soil, and the presence of competing cations. Other chelating compounds that may be utilised include trans-1,2-cyclohexylenedinitrilotetraacetic acid (CDTA), diethylenetriaminepentaacetic acid (DTPA), methylglycine diacetate (MGDA), nitrilotriacetic acid (NTA), and so on (Anju 2017). The use of microbial metabolites has various advantages over the use of chelating chemicals, including the fact that they are biodegradable, less toxic, and have several other benefits (Rajkumar et al. 2012; Ma et al. 2011). Microbial metabolites can be created in situ in rhizosphere soils. Microbe-assisted phytoremediation is a technique that involves the employment of microorganisms in conjunction with plants for the purpose of phytoextraction (Anju 2017). According to Anju (2017), both microorganisms and plants have developed a wide variety of resistance mechanisms, both active and accidental, to protect themselves from being poisoned by heavy metals. According to Glick (2012), bacteria that stimulate plant development could be free-living, involved in symbiotic partnerships, or endophytic. According to Anju (2017), rhizospheric microbial activities have the potential to improve the efficacy of phytoremediation in either a direct or indirect manner. There are two ways in which rhizospheric microbial activities might boost phytoremediation, the first of which is directly and the second of which is indirectly. According to Glick (2010), direct mechanisms are directly responsible for the bioavailability of heavy metals, the final accumulation of heavy metals by plants, and the solubilisation of heavy metals. In contrast, the microbes confer plant tolerance to metal stress through an indirect method and/or boost plant development, increase plant biomass, and prevent phytopathogens from suppressing plant growth, all of which contribute to phytoremediation (Anju 2017). A wide variety of species, such as fungi, yeast, bacteria, and plants, secrete siderophores as a response to low iron levels and to assist in the assimilation of Fe (Anju 2017). Siderophores are low-molecular-mass (400–1000 Daltons) coordination molecules. Siderophores have the ability to form stable complexes with heavy metals such as lead and trivalent iron (Neubauer et al. 2000; Gadd 2010; Glick and Bashan 1997; Schalk et al. 2011); these complexes can then be taken up by cells. Rhizospheric PGPB is responsible for the majority of the production of siderophore. Under conditions of environmental stress, such as a lack of nutrients or high levels of heavy metals, rhizospheric PGPB show their best growth and siderophore production activities. According to Rajkumar et al. (2010), this fact confers on them an especially high degree of utility for phytoremediation purposes. Studies that have been conducted recently (Dimkpa et al. 2008; Braud et al. 2009; Dimkpa et al. 2009a, b; Carrillo-Castaeda et al. 2003) have focused on the effects of SPB inoculation on metal uptake by hyperaccumulators. According to Braud et al.'s 2009 research, the bacteria Pseudomonas aeruginosa, which produces rhizospheric siderophore, raised the concentration of bioavailable lead in the rhizosphere, making it easier for maize to take up the element. Siderophores are responsible for a significant portion of both the mobilisation and accumulation of metals. Because of this, it is believed that the presence of bacteria in the rhizosphere that create siderophores is necessary for heavy metal phytoextraction (Ma et al. 2011; Rajkumar et al. 2010; Dimkpa et al. 2009a, b; Braud et al. 2009).

13.6 Phytofiltration

The process of using plants to filter pollutants out of polluted water or aqueous waste streams is known as phytofiltration (Dushenkov et al. 1995; Salt et al. 1998). The various plant parts, such as roots, seeds, and plant shoots, can be used as filters that absorb, precipitate, and concentrate toxic heavy metals in order to filter polluted water containing heavy metal Pb (Anju 2017). The use of roots as a phytofilter is known as rhizofiltration; seedlings are known as blastofiltration; and excised plant shoots are termed caulofiltration (Rozas et al. 2006; Mesjasz-Przyby et al. 2004).

In the case of Pb, the rhizofitration technique is used because Pb is accumulated in roots. Garlic with Brassica oleracea has a removal efficiency of 0.02% at higher concentrations (Kumar et al. 2020).

13.7 Phytodegradation

According to Ahmadpour (2012), another name for this process is phytotransformation. Phytodegradation is the process of using plants and microorganisms to adsorb, metabolise, and breakdown an organic pollutant (Burken and Schnoor 1997). Using plant roots in conjunction with other microbes, this technique can remove lead from soil that has been contaminated with lead (Garbisu and Alkotra 2001).

13.8 Physical Remediation

According to Yao et al. (2012), physical remediation is the act of halting or reversing the harm done to soil by the use of physical technologies such as thermal treatment, replacement, isolation, and soil. Soil replacement: Soil replacement is a process that involves using a large amount of uncontaminated soil to mix with or cover the surface of contaminated soil. Some examples of soil replacement techniques include surface capping, land filling, and encapsulation. This method requires a large quantity of uncontaminated material. The removal of dirt is an option that should be reserved for places that have extremely contaminated soil and restricted space due to its high cost and the amount of labour it requires. Additionally, it is able to efficiently lower the concentration of pollutants. The installation of barrier walls will help to separate the pollutant from the surrounding area, contain it, and prevent it from leaving the site. Impermeable physical barriers comprised of steel, cement, bentonite, and grout are utilised for the purposes of capping, vertical confinement, and horizontal containment, respectively. This strategy of soil isolation or containment was adopted to decrease the migration of heavy metals into the earth (Jainkaite and Vasarevicius 2005). Although this method is not a direct remedial process, it was used to accomplish this goal.

13.9 Chemical Remediation

Chemical reagents, reactions, and principles are used to remove contaminants in a method known as chemical remediation (Song et al. 2017). Solidification/stabilisation, vitrification, soil flushing, soil washing, and electrokinetics (EK; Jankaite and Vasarevicius 2005) are the principal remediation technologies. In order to lessen the mobility of Pb contaminants, solidification or stabilisation technology is typically applied by mixing contaminated soils with reagents or materials. While stabilisation involves chemical reactions to lessen contaminant mobility, solidification involves the physical encapsulation of contaminants in a solid matrix made of cement, bitumen, asphalt, and thermoplastic binders. By forming metal phosphates, bone meal additions (finely ground, poorly crystalline apatite, $Ca_{10}(PO_4)OH_2)$) have the potential to immobilise polluting metals in soils and decrease metal bioavailability (Hodson et al. 2000). A variety of cost-effective and environmentally friendly waste resources have been reported, only to immobilise metals in contaminated soils but also to improve soil quality, such as lime-based agents (Lim et al. 2013), calcined oyster shells (Yong et al. 2010), eggshells (Soares et al. 2015), waste mussel shells (Otero et al. 2015), and calcined cockle shell (Islam et al. 2017). By combining the contaminated soil with glass-forming precursors, heating the mixtures until they liquidise, and then cooling the liquid to produce an amorphous homogeneous glass, it is possible to solidify or stabilise materials using the vitrification, or molten glass, method that requires thermal energy (1400–2000 °C) (Yao et al. 2012). Chemical bonding and encapsulation are the two main ways that lead can be immobilised in the glass matrix (Navarro 2012). The main element in the immobilisation of heavy metals in contaminated soils is the heating temperature of the vitrification process. The vitrification process's effective additives may enhance the encapsulation of contaminants and their potential for leaching (Guo et al. 2006). According to the US EPA (2006), washing and flushing the soil with water or a suitable washing solution are efficient remediation techniques for removing contaminants from the soil. To achieve optimal heavy metal removal, washing agents such as water (Dermont et al. 2008), saponin (Maity et al. 2013), chelating agents (Jiang et al. 2011), surfactants (Sun et al. 2011), and low-molecular-weight organic acids (Almaroai et al. 2012) have been reported to be effective on stimulating the desorption of contaminants in soils. According to Lestan et al. (2008), ethylene diamine tetraacetic acid (EDTA) is the most effective chelating agent for removing heavy metals from contaminated soils. Low biodegradability, high metal removal efficiency, availability of appropriate recycling techniques, and reduced effects on soil microorganisms and enzyme activity are some of the benefits of EDTA (Qiao et al. 2017).

A recently created technique called EK remediation can successfully clean up heavy metal-polluted soils. This technique uses a direct electric current to remove heavy metals from the soil's matrix through a variety of mechanisms, including electromigration, electroosmosis, electrophoresis, and electrolysis (Jankaite and Vasarevicius 2005). Chelates are also used to increase the EK's effectiveness in contaminated soils. Song et al. (2017) examined the effect of various chelating agents (EDTA, ethylenediaminedisuccinic acid, nitrilotriacetic acid, and citric acid) in boosting EK efficiency in order to examine the mobility of heavy metals (As, Cd, Cr, Cu, Ni, Pb, and Zn). To overcome the drawback of flushing technology in fine soil, some researchers have combined flushing and EK in two ways: in series and integration. According to Kim et al. (2008), the addition of a pump to EK flushing remediation increased the effectiveness of removing Co^{2+} and Cs^{+} from the contaminated soil. A two-stage EK washing technique was studied by Ng et al. (2014) for the remediation of Pb-contaminated soil. Kim et al. (2015) decontaminated uranium-contaminated soil using a full-sized serial washing-EK separation apparatus. Flushing that has been improved by electrochemistry is known as electrochemical flushing.

13.10 Genetically Engineered Plants for Enhancement Phytoremediation

In recent years, scientists have focused increasing attention on plants'natural ability to filter out harmful chemicals in the air. The ability of plants to tolerate, remove, and breakdown contaminants has been successfully enhanced by the use of genes from microorganisms, plants, and animals. Transgenic plants were able to overcome the phytotoxic effect of nitroaromatic pollutants by expressing certain bacterial genes, leading to increased elimination of these compounds. Increased metabolism and the elimination of various organic pollutants and herbicides are the results of the overexpression of the mammalian gene encoding cytochrome P450. Efforts have been made to improve phytoremediation of metal pollution by employing genes involved in absorption and detoxification. Plants engineered with DNA from bacteria were able to detoxify mercury and selenium. Endophytes, bacteria that live inside plants, have been used to promote pollution clearance and boost the plant's tolerance to ordinarily phytotoxic substances. When plants were inoculated with the herbicide-degrading bacteria, they developed resistance to the herbicide. Toluene-degrading bacteria were found to help plants endure normal phytotoxic doses of the chemical while also producing fewer emissions, according to another study. Research on transgenic plants and the application of symbiotic endophytic microbes in plant tissues have both shown promise as methods for improving phytoremediation, and both developments are discussed here.

13.11 Conclusion

The removal of heavy metals is a worldwide priority. Phytoremediation, and in particular phytoextraction, is gaining popularity as a viable green strategy for dealing with heavy metal contamination due to its many advantageous qualities. The use of plant growth promotion (PGP) bacteria, phytoremediation by acylates, and phytoremediation by microorganisms have been the main topics of research on how to clean up polluted plants and get their useful parts out. Bacteria tend to improve the solubility and bioavailability of heavy metals directly via microbial biosurfactants, organic acids, siderophores, redox processes, and biomethylation, which maximises the effect and potential of this technique. In addition, PGPB produces growth-promoting chemicals, which aid in phytoremediation in a roundabout way. Improved plant tolerance and metal accumulation can also be achieved by the application of genetic engineering, which allows for the controlled expression of genes from specific microorganisms, plants, and animals. Recognise and understand the molecular developments and triumphs that boost the efficiency of phytoremediation. Today, phytoextraction mining and phytoremediation for heavy metals are both technically feasible and commercially available. The search for genetic components responsible for the hyperaccumulation of specific heavy metals prevalent in our environment, including those in the air, water, and soil, is continuous and developing. It is time to shift from a concentration on knowledge to one on action.

References

Ahmadpour P, Ahmadpour F, Mohamed TMM, Abdu A (2012) Phytoremediation of heavy metals: a green technology. Afr J Biotechnol 11(76):14036–14043

Ahmadpour P, Nawi AM, Abdu A, Abdul-Hamid H, Singh DK, Hassan A, Majid NM, Jusop S (2010) Uptake of heavy metals by Jatropha curcas L. planted in soils containing sewage sludge. Amer J Appl Sci 7(10):1291–1299

Akbari S, Abdurahman NH, Yunus RS, Fayaz F, Alara OS (2018) Biosurfactants—a new frontier for social and environmental safety: a mini review. Biotechnol Res Inn 2:81–90

Ali B, Wang B, Ali S, Ghani MA, Hayat MT, Yang C, Xu L, Zhou WJ (2013a) 5–Aminolevulinic acid ameliorates the growth, photosynthetic gas exchange capacity, and ultra structural changes under cadmium stress in *Brassica napus* L. J Plant Growth Regul 32:604–614

Ali AS, Bayih AA, Gari SR (2022) Meta-analysis of public health risks of lead accumulation in wastewater, irrigated soil, and crops nexus. Front Public Health 10:977721

Ali H, Khan E, Sajad MA (2013) Phytoremediation of heavy metals—concepts and applications. Chemosphere 91:869–881

Almaroai YA, Usman ARA, Ahmad M, Kim KR, Moon DH, Lee SS, Ok YS (2012) Effects of synthetic chelators and low-Molecular-Weight organic acids on chromium, copper, and arsenic uptake and translocation in maize (*Zea mays* L.). Soil Sci 177:655–663

Anju M (2017) Biotechnological strategies for remediation of toxic metal(loid)s from environment. In: Gahlawat S, Salar R, Siwach P, Duhan J, Kumar S, Kaur P (eds) Plant biotechnology: recent advancements and developments. Springer, Singapore. https://doi.org/10.1007/978-981-10-4732-9_16

Avery SV, Tobin JM (1993) Mechanism of adsorption of hard and soft metal ions to *Saccharomyces cerevisiae* and influence of hard and soft anions. Appl Environ Microbiol 59:2851–2856

Azadpour A, Matthews JE (1996) Remediation of metal-contaminated sites using plants. Remediation 6:1–18

Bachmann RT, Johnson AC, Edyvean RGJ (2014) Biotechnology in the petroleum industry: an overview. Int Biodeter Biodegr 86:225–237

Baker AJM (1981) Accumulators and excluders—strategies in the response of plants to heavy metals. J Plant Nutr 3:643–654

Baker AJM, Ernst WHO, van der Ent A, Malaisse F, Ginocchio R (2010) Metallophytes: the unique biological resource, its ecology and conservational status in Europe, central Africa and Latin 16 Biotechnological Strategies for Remediation of Toxic Metal(loid)s… 348 America. In: Batty L, Hallberg KB (eds) Ecology of industrial pollution. Cambridge University Press/British Ecological Society, Cambridge, UK, pp 7–40

Brady D, Duncan JR (1994) Chemical and enzymatic extraction of heavy metal binding polymers from isolated cell walls of *Saccharomyces cerevisiae*. Biotechnol Bioeng 44:297–302

Braud A, Jézéquel K, Bazot S, Lebeau T (2009) Enhanced phytoextraction of an agricultural Cr, Hg and Pb contaminated soil by bioaugmentation with siderophore producing bacteria. Chemosphere 74:280–286

Burken JG, Schnoor JL (1997) Uptake and metabolism of atrazine by poplar trees. Environ Sci Technol 31:1399–1406

Carrillo-Castañeda G, Muñoz JJ, Peralta-Videa JR, Gomez E, Gardea-Torresdey JL (2003) Plant growth-promoting bacteria promote copper and iron translocation from root to shoot in Alfalfa seedlings. J Plant Nutr 26:1801–1814

Chehregani A, Malayeri BE (2007) Removal of heavy metals by native accumulator plants. Int J Agric Biol 9(3):462–465

Ciarkowska K, Hanus-Fajerska E (2008) Remediation of soil-free grounds contaminated by zinc, lead and cadmium with the use of metallophytes. Pol J Environ Stud 17(5):707–712

Cluis C (2004) Junk-greedy greens: phytoremediation as a new option for soil decontamination. BioTeach J 2:61–67

Cobbett CS (2000) Phytochelatins and their roles in heavy metal detoxification. Plant Physiol 123:825–832

Collin F, Baskar A, Geevarghese DM, Ali MNVS, Bahubali P, Choudhary R, Vladislav Lvov V, Tovar GI, Senatov F, Koppala S, Swamiappan S (2022) Bioaccumulation of lead (Pb) and its effects in plants: a review. J Hazard Mater Lett 3(2022):100064. https://doi.org/10.1016/j.hazl.2022.100064

Cunningham SD, Ow DW (1996) Promises and prospects of phytoremediation. Plant Physiol 110(3):715–719

Cunningham SD, Berti WR, Huang JW (1995) Phytoremediation of contaminated soils. Trends Biotechnol 13:393–397

Das S, Dhanarajan G, RangarajanV BC, Dixit A, Ale K, Sen R (2017) Biosurfactant-biopolymer driven microbial enhanced oil recovery (MEOR) and its optimization by an ANN-GA hybrid technique. J Biotechnol 256:46–56

Dermont G, Bergeron M, Mercier G, Richer–Lafleche M (2008) Soil washing for metal removal: a review of physical/chemical technologies and field applications. J Hazard Mater 152:1–31. https://doi.org/10.1016/j.jhazmat.2007.10.043

Dickinson NM, Baker AJM, Doronila A, Laidlaw S, Reeves RD (2009) Phytoremediation of inorganics: realism and synergies. Int J Phytoremed 11:97–114

Dimkpa CO, Svatos A, Merten D, Büchel G, Kothe E (2008) Hydroxamate siderophores produced by Streptomyces acidiscabies E13 bind nickel and promote growth in cowpea (Vigna unguiculata L.) under nickel stress. Can J Microbiol 54:163–172

Dimkpa CO, Merten D, Svatoš A, Büchel G, Kothe E (2009a) Siderophores mediate reduced and increased uptake of cadmium by Streptomyces tendae F4 and sunflower (Helianthus annuus), respectively. J Appl Microbiol 107:1687–1696

Dimkpa CO, Merten D, Svatoš A, Büchel G, Kothe E (2009b) Metal-induced oxidative stress impacting plant growth in contaminated soil is alleviated by microbial siderophores. Soil Biol Biochem 41:154–162

Din IU, Muhammad S, Rehman IU (2023) Heavy metal(loid)s contaminations in soils of Pakistan: a review for the evaluation of human and ecological risks assessment and spatial distribution. Environ Geochem Health 45:1991–2012. https://doi.org/10.1007/s10653-022-01312-x

Dushenkov V, Kumar PN, Motto H, Raskin I (1995) Rhizofiltration: the use of plants to remove heavy metals from aqueous streams. Environ Sci Technol 29:1239–1245 Eapen S, D'Souza SF (2005)

Erakhrumen AA (2007) Phytoremediation: an environmentally sound technology for pollution prevention, control and remediation in developing countries. Educ Res Rev 2:151–156

Ernst WHO (1996) Bioavailability of heavy metals and decontamination of soil by plants. Appl Geochem 11:163–167

Flathman PE, Lanza GR (2010) Phytoremediation:currentviewsonan emerginggreentechnology. J Soil Contam 7:415–432

Gadd GM (2010) Metals: minerals and microbes: geomicrobiology and bioremediation. Microbiology 156:609–643
Gall JE, Boyd RS, Rajakaruna N (2015) Transfer of heavy metals through terrestrial food webs: a review. Environ Monit Assess 187(4):1–21
Garbisu C, Alkorta I (2001) Phytoextraction: A cost-effective plant-based technology for the removal of metals from the environment. Bioresour Technol 77(3):229–236
Garbisu C, Hernández-Allica J, Barrutia O, Alkorta I, Becerril JM (2002) Phytoremediation: a technology that uses green plants to remove contaminants from polluted areas. Rev Environ Health 17(3):173–188
Ginocchio R, Baker AJM, Cucuzza J (2004) Phytoremediation. J Mining Environ Manag November 7–10
Glick BR (2010) Using soil bacteria to facilitate phytoremediation. Biotechnol Adv 28:367–374
Glick BR, Bashan Y (1997) Genetic manipulation of plant growth-promoting bacteria to enhance biocontrol of phytopathogens. Biotechnol Adv 15:353–378
Glick BR (2012) Plant growth-promoting bacteria: mechanisms and applications. Scientifica Article ID: 963401
Grčman H, Velikonja Bolta S, Vodnik D, Kos B, Lestan D (2001) EDTA enhanced phytoextraction: metal accumulation, leaching and toxicity. Plant Soil 235:105–114
Guo GL, Zhou QX, Ma LQ (2006) Availability and assessment of fixing additives for the in-situ remediation of heavy metal contaminated soils: a review. Environ Monit Asses 116:513–528. https://doi.org/10.1007/s10661-006-7668-4
Gupta N, Gedam VV, Moghe C, Labhasetwar P (2019) Comparative assessment of batch and column leaching studies for heavy metals release from Coal Fly Ash Bricks and Clay Bricks Environ. Technol Innov 16:100461
Hall JL (2002) Cellular mechanisms for heavy metal detoxification and tolerance. J Exp Bot 53:1–11
Henry JR (2000) In an overview of phytoremediation of lead and mercury. NNEMS Report. Washington, D.C., pp 3–9
Hodson ME, Jones ÉV, Howells JDC (2000) Bonemeal additions as a remediation treatment for metal contaminated soil. Environ Sci Technol 34:3501–3507. https://doi.org/10.1021/es990972a
Huang C, Huang C, Morehart AL (1990) The removal of copper from dilute aqueous solutions by Saccharomyces cerevisiae. Water Res 24:433–439
Islam MN, Taki G, Xuan PN, Jo YT, Kim J, Park JH (2017) Heavy metal stabilization in contaminated soil by treatment with calcined cockle shell. Environ Sci Pollut Res 24:7177–7183. https://doi.org/10.1007/s11356-016-8330-5
Jankaite A, Vasarevicius S (2005) Remediation technologies for soils contaminated with heavy metals. J Environ Eng Landscape Manage 13:109–113. https://doi.org/10.1080/16486897.2005.9636854
Jiang W, Tao T, Liao ZM (2011) Removal of heavy metal from contaminated soil with chelating agents. Open J Soil Sci 1:70–76. https://doi.org/10.4236/ojss.2011.12010
Kärenlampi S, Schat H, Vangronsveld J, Verkleij JAC, van der Lelie D, Mergeay M, Tervahauta AI (2000) Genetic engineering in the improvement of plants for phytoremediation of metal polluted soils. Environ Pollut 107:225–231
Kavamura VN, Esposito E (2009) Biotechnological strategies applied to the decontamination of soils polluted with heavy metals. Biotechnol Adv 28(1):61–69
Kim GN, Jung YH, Lee JJ, Moon JK, Jung CH (2008) An analysis of a flushing effect on the electrokinetic-flushing removal of cobalt and cesium from a soil around decommissioning site. Sep Purif Technol 63:116–121. https://doi.org/10.1016/j.seppur.2008.04.006
Kim GN, Kim SS, Moon JK, Hyun JH (2015) Removal of uranium from soil using full–Sized washing electrokinetic separation equipment. Ann Nucl Energy 81:188–195
Krauter P, Martinelli R, Williams K, Martins S (1996) Removal of Cr(VI) from ground water by *Saccharomyces cerevisiae*. Biodegradation 7:277–286

Kumar CP, Mms AK, Chaturvedi AA, Shabnam G, Subrahmanyam R, Mondal KK, Yadav (2020) Lead toxicity: health hazards, influence on food chain, and sustainable remediation approaches. Int J Environ Res Public Health 17(7):2179. https://doi.org/10.3390/ijerph17072179

Lasat MM (2000) Phytoextraction of metals from contaminated soil: a review of plant /soil/metal interaction and assessment of pertinent agronomic issues. J Hazard Subs Res 2:1–25

Lestan D, Luo CL, Li XD (2008) The use of chelating agents in the remediation of metal-contaminated soils: a review. Environ Pollut 153:3–13. https://doi.org/10.1016/j.envpol.2007.11.015

Lim JE, Ahmad M, Lee SS, Shope CL, Hashimoto Y, Kim KR, Usman ARA, Yang JE, Ok YS (2013) Effects of lime-based waste materials on immobilization and Phyto availability of cadmium and lead in contaminated soil. Acta Hydrochim Et Hydrobiol 41:1235–1241. https://doi.org/10.1002/clen.201200169

Liu R, Zhao D (2007) Reducing Leachibility and Bioaccessibility of Lead in Soils Using a New Class of Stabilized Iron Phosphate Nanoparticle 12:41. https://doi.org/10.1016/j.watres.2007.03.026

Liu Z, Li Z, Zhong H, Zeng G, Liang Y, Chen M, Shao B (2017) Recent advances in the environmental applications of biosurfactant saponins: a review. J Environ Chem Eng 5(6):6030–6038

Ma Y, Prasad MNV, Rajkumar M, Freitas H (2011) Plant growth promoting rhizobacteria and endophytes accelerate phytoremediation of metalliferous soils. Biotechnol Adv 29:248–258

Macaskie LE, Dean ACR (1989) Microbial metabolism, desolubilisation and deposition of heavy metals: metal uptake by immobilized cells and application to the detoxification of liquid wastes. Adv Biotechnol Proces 12:159–172

Mahar A, Wang P, Ali A, Awasthi MK, Lahori AH, Wang Q, Li R, Zhang Z (2016) Challenges and opportunities in the phytoremediation of heavy metals contaminated soils: a review. Ecotoxicol Environ Saf 126:111–121

Maity JP, Huang YM, Fan CW, Chen CC, Li CY, Hsu CM, Chang YF, Wu CI, Chen CY, Jean JS (2013) Evaluation of remediation process with soapberry derived saponin for removal of heavy metals from contaminated soils in Hai-Pu. Taiwan J Environ Sci 25:1180–1185. https://doi.org/10.1016/S1001-0742(12)60162-4

Malik A (2003) Metal bioremediation through growing cells. Environ Int 30(2):261–278

Mao X, Jiang R, Xiao W, Yu J (2015) Use of surfactants for the remediation of contaminated soils: a review. J Hazard Mater 21(285):419–435

Mesjasz-Przyby OJ, Nakonieczny M, Migula P, Augustyniak M, Tarnawska MM, Reimold WU, Koeberl C, Przyby OW, Owacka EG (2004) Uptake of cadmium, lead, nickel and zinc from soil and water solutions by the nickel hyperaccumulator Berkheya coddii. Acta Biol Cracov Ser Bot 46:75–85

Mosa KA, Saadoun I, Kumar K, Helmy M, Dhankher OP (2016) Potential biotechnological strategies for the cleanup of heavy metals and metalloids. Front Plant Sci 7:1–14. https://doi.org/10.3389/fpls.2016.00303

Mulligan CN, Sharma SK, Mudhoo A, Makhijani K (2014) Green chemistry and biosurfactant research. In: Mulligan CN, Sharma SK, Mudhoo A (eds) Biosurfactants: research trends and applications. (Boca Raton, FL: CRC Press), pp 1–30

Navarro A (2012) Effect of sludge amendment on remediation of metal contaminated soils. Minerals 2:473–492. https://doi.org/10.3390/min2040473

Naz T, Khan MD, Ahmed I, Rehman SU, Rha ES, Malook I, Jamil M (2016) Biosorption of heavy metals *by Pseudomonas* species isolated from sugar industry. Toxicol Ind Health 32(9):1619–1627

Neubauer U, Furrer G, Kayser A, Schulin R (2000) Siderophores, NTA, and citrate: potential soil amendments to enhance heavy metal mobility in phytoremediation. Int J Phytoremed 2:353–368

Ng YS, Gupta BS, Hashim MA (2014) Performance evaluation of two-stage electrokinetic washing as soil remediation method for lead removal using different wash solutions. Electrochim Acta 147:9–18. https://doi.org/10.1016/j.electacta.2014.08.124

Otero M, Cutillasbarreiro L, Nóvoamuñoz JC, Arias-Estévez M, Fernández-Sanjurjo MJ, Álvarez-Rodríguez E, Núñez-Delgado A (2015) Cr (VI) sorption/desorption on untreated and mussel-shell-treated soil materials: fractionation and effects of pH and chromium concentration. Solid Earth 6:3419–3444. https://doi.org/10.5194/se-6-373-2015
Padmavathiamma PK, Li LY (2007) Phytoremediation technology: hyperaccumulation metals in plants. Water Air Soil Pollut 184:105–126
Paquin DG, Campbell S, Li QX (2004) Phytoremediation in subtropical Hawaii—a review of over 100 plant species. Remed J 14:127–139
Pivertz BE (2001) Ground water issue: phytoremediation of contaminated soil and ground water at hazardous waste sites, pp 1–36
Pollard AJ, Powell KD, Harper FA, Smith JAC (2002) The genetic basis of metal hyperaccumulation in plants. Crit Rev Plant Sci 21:539–566
Pollard AJ, Reeves RD, Baker AJ (2014) Facultative hyperaccumulation of heavy metals and metalloids. Plant Sci 217–218:8–17
Prasad MNV, Freitas HMO (2003) Metal hyperaccumulation in plants biodiversity prospecting for phytoremediation technology. Electron J Biotechnol 6(3):285–321
Prasad MNV, Freitas H, Fraenzle S, Wuenschmann S, Markert B (2010) Knowledge explosion in phytotechnologies for environmental solutions. Environ Pollut 158:18–23
Pulford ID, Watson C (2003) Phytoremediation of heavy metal contaminated land by trees-a review. Environ Int 29:529–540
Punshon T, Dickinson NM, Lepp NW (1996) The potential of Salix clones for bioremediating metal polluted soil. In: Glimmerveen I (ed) Heavy metals and trees. Proceedings of a discussion meeting, Glasgow. Institute of Chartered Foresters, Edinburgh, pp 93–104
Qiao JB, Sun HM, Luo XH, Zhang W, Mathews S, Yin XQ (2017) EDTA–Assisted leaching of Pb and Cd from contaminated soil. Chemosphere 167:422–428. https://doi.org/10.1016/j.chemosphere.2016.10.034
Radulescu C, Stihi C, Popescu IV, Dulama ID, Chelarescu ED, Chilian A (2013) Heavy metal accumulation and translocation in different parts of Brassica oleracea L. Rom J Physiol 58:1337–1354
Rai LC, Gaur JP, Kumar HD (1981) Phycology and heavy metal pollution. Biol Rev 56:99–151
Rajkumar M, Ae N, Prasad MNV, Freitas H (2010) Potential of siderophore- producing bacteria for improving heavy metal phytoextraction. Trends Biotechnol 28:142–149
Rajkumar M, Sandhya S, Prasad MNV, Freitas H (2012) Perspectives of plant-associated microbes in heavy metal phytoremediation. Biotechnol Adv 30:1562–1574
Rascio N, Navari-Izzo F (2011) Heavy metal accumulating plants: how and why do they do it? and what makes them so interesting? Plant Sci 180:169–181
Raskin I, Ensley BD (2000) Phytoremediation of toxic metals: using plants to clean up the environment. Wiley, New York, p 304
Robinson B, Fernandez JE, Madejon P, Maranon T, Murillo JM, Green S, Clothier B (2003) Phytoextraction: an assessment of biogeochemical and economic viability. Plant Soil 249:117–125
Rozas MA, Alkorta I, Garbisu C (2006) Phytoextraction and phytofilteration of arsenic. Rev Environ Health 21(1):43–56
Saadoun IMK, Al-Ghzawi ZD (2005) Bioremediation of petroleum contamination. Biorem Aquat Terr Ecosyst 173–212
Salt DE, Smith RD, Raskin I (1998) Phytoremediation. Ann Rev Plant Physiol PlantMol Biol 49:643–668
Salt DE, Baker AJM (2000) Phytoremediation of metals. In: Rehm HJ, Reeds G (eds) Biotechnology: environmental processes II, vol 11b, 2nd edn. Wiley-VCH Verlag Gmbh, Weinheim, pp 385–397
Schalk IJ, Hannauer M, Braud A (2011) New roles for bacterial siderophores in metal transport and tolerance. Environ Microbiol 13:2844–2854
Schmidt U (2003) Enhancing phytoextraction: the effect of chemical soil manipulation on mobility, plant accumulation, and leaching of heavy metals. J Environ Qual 32(6):1939–1954

Selvi A, Rajasekar J, Theerthagiri A, Ananthaselvam K, Sathishkumar J, Madhavan PK, Rahman SM (2019) Integrated remediation processes toward heavy metal removal/recovery from various environments-a review front. Environ Sci 7:66
Shah K, Nongkynrih JM (2007) Metal hyperaccumulation and bioremediation. Biol Plant 51:618–34
Shao H, Cu G, Bhat SA (2022) Phytoremediation: a green technology for treating heavy metal contaminated soil. In: Malik A, Kidwai MK, Garg VK (eds) Bioremediation of toxic metal(loid)s, 1st edn. CRC Press, pp 81–94
Shilev SI, Ruso J, Puig A, Benlloch M, Jorrin J, Sancho E (2001) Rhizospheric bacteria promote sunflower (Helianthus annus L.) plant growth and tolerance to heavy metals. Minerva Biotechnol 13:37–39
Sidhu GPS, Singh HP, Batish DR, Kohli RK (2017a) Appraising the role of environment friendly chelants in alleviating lead by Coronopusdidymus from Pb-contaminated soils. Chemosphere 182:129–136
Sidhu GPS, Singh HP, Batish DR, Kohli RK (2017b) Tolerance and hyperaccumulation of cadmium by a wild, unpalatable herb Coronopusdidymus (L.) Sm. (Brassicaceae). Ecotoxicol Environ Saf 135:209–215
Singh OV, Labana S, Pandey G, Budhiraja R, Jain RK (2003) Phytoremediation: an overview of metallicion decontamination from soil. Appl Microbiol Biotechnol 61:405–412
Smith LA, Alleman BC, Copley-Graves L (1994) Biological treatment options. In: Means JL, Hinchee RE (eds) Emerging technology for bioremediation of metals. Lewis Publishers, Boca Raton, pp 1–12
Soares MA, Quina MJ, Quinta-Ferreira RM (2015) Immobilization of lead and zinc in contaminated soil using compost derived from industrial eggshell. J Environ Manage 164:137–145. https://doi.org/10.1016/j.jenvman.2015.08.042
Song B, Zeng GM, Gong JL, Liang J, Xu P, Liu ZF, Zhang Y, Zhang C, Cheng M, Liu Y, Ye SJ, Yi H, Ren XY (2017) Evaluation methods for assessing effectiveness of in situ remediation of soil and sediment contaminated with organic pollutants and heavy metals. Environ Int 105:43–55. https://doi.org/10.1016/j.envint.2017.05.001
Sun HC, Wang HL, Qi JX, Shen L, Lian XY (2011) Study on surfactants remediation in heavy metals contaminated soils. Int Symp Water Resour Environ Prot IEEE 1862–1865. https://doi.org/10.1109/ISWREP.2011.5893616
Veglio F, Beolchini F, Gasbarro A (1997) Biosorption of toxic heavy metals: an equilibrium study using free cells of Arthrobacter spp. Process Biochem 32:99–105
Vidali M (2001) Bioremediation an overview. Pure Appl Chem 73:1163–1172
Wani L, Anjum ARA, Usmani JA (2015) Lead toxicity: a review. Interdiscip Toxicol 8(2):55. https://doi.org/10.1515/intox-2015-0009
Whiting SN, Reeves RD, Baker AJM (2002) Mining, metallophytes and land reclamation. Mining Environ Manag 10:11–16
Whiting SN, Reeves RD, Richards D et al (2004) Research priorities for conservation of metallophyte biodiversity and their potential for restoration and site remediation. Restor Ecol 12:106–116
Wilson B, Pyatt FB (2007) Heavy metal bioaccumulation by the important food plant, Olea europaea L. in an ancient metalliferous polluted area of Cyprus. Bull Environ Contam Toxicol 78:390–394
Xu J, Tian YS, Peng RH, Xiong AS, Zhu B, Hou XL, Yao QH (2010) Cyanobacteria MT gene SmtA enhances zinc tolerance in Arabidopsis. Mol Biol Rep 37:1105–1110
Yao ZT, Li JH, Xie HH, Yu CH (2012) Review on remediation technologies of soil contaminated by heavy metals. Proc Environ Sci 16:722–729. https://doi.org/10.1016/j.proenv.2012.10.099
Ye ZH, Baker AJM, Wong MH (2002) Problems of toxicities. In: Wong MH, Bradshaw AD (eds) The restoration and management of derelict land: modern approaches. World Scientific, Hackensack, pp 66–79
Yong SO, Oh SE, Ahmad M, Hyun S, Kim KR, Moon DH, Lee SS (2010) Effects of natural and calcined oyster shells on Cd and Pb immobilization in contaminated soils. Environ Earth Sci 61:1301–1308. https://doi.org/10.1007/s12665-010-0674-4

Zacchini M, Pietrini F, Mugnozza GS, Iori V, Pietrosanti L, Massacci A (2009) Metal tolerance, accumulation and translocation in poplar and willow clones treated with cadmium in hydroponics. Water Air Soil Pollut 197:23–34
Zayed A, Gowthaman S, Terry N (1998) Phytoaccumulation of trace elements by wetland plants: I. Duckweed J Environ Qual 27:715–721
Zenk MH (1996) Heavy metal detoxification in higher plants — a review. Gene 179:21–30
Zhuang P, Yang Q, Wang H, Shu W (2007) Phytoextraction of heavy metals by eight plant species in the field. Water Air Soil Pollut 184:235–242

Chapter 14
Oxidative Stress in Lead Toxicity in Plants and Its Amelioration

Neetu Jagota, Swapnil Singh, Harleen Kaur, Ravneet Kaur, and Ashish Sharma

Abstract Heavy metals, also known as trace elements, are hazardous even at low concentrations and are becoming a growing issue in many countries, including India. Human activities ranging from agriculture to mining to power generation in power plants are the major sources of heavy metals in environment. Lead (Pb) is the second most harmful environmental toxin after arsenic, and its deposition in the environment is steadily increasing due to anthropogenic activity. Pb significantly harms plants, affecting their morphophysiological and biochemical traits, such as irregular cell division during mitosis, subordinate growth of seedlings, and chlorosis. The latter changes the biochemistry of fruits and flowers, which negatively affects the rate of photosynthetic activity. Pb also damages nutrient interactions, photosynthesis, respiration, oxidative damage, and antioxidant defence mechanisms in various plant species. Soil remediation methods, such as biochar supplements and phytoremediation technology, can help address Pb-contaminated soils. Hyperaccumulating plants have developed molecular processes that enable their use in environmental bioremediation. However, efficacy of these methods still needs to be evaluated by rigorous research activities, and new ecologically acceptable remediation techniques need to be developed to reduce lead toxicity.

Keywords Toxicity · Heavy metals · Environment · Plants · Anthropogenic activities · ROS · Lead

14.1 Introduction

For living, a healthy form of life on the planet, air, water, soil, and the whole environment is an essential component since pollutants can hazard the health of citizens. Simply due to the sheer number of heavy metals released into the environment as a result of anthropogenic activities, it has led to the environmental contamination

N. Jagota · S. Singh · H. Kaur · R. Kaur · A. Sharma (✉)
Department of Botany and Environment Science, DAV University, Jalandhar, Punjab, India
e-mail: ashish10210@davuniversity.org

N. Kumar and A. K. Jha (eds.), *Lead Toxicity Mitigation: Sustainable Nexus Approaches*, Environmental Contamination Remediation and Management,
https://doi.org/10.1007/978-3-031-46146-0_14

by either directly or indirectly interacting with these hazardous chemicals (Sajid et al. 2018). Heavy metals are particularly dangerous environmental toxins for the ecosystem due to their chronic toxicity, non-biodegradability, and environmental bioaccumulation (Valdés et al. 2014). The health of people is significantly threatened by the biomagnification and transfer of heavy metals through food chains (Liu et al. 2018). Heavy metals contaminate the soil at low concentrations; they are the most hazardous elements that is affecting the environment of plants, animals, and humans because they are not biodegradable elements (Wang et al. 2020). In some circumstances, high concentrations of heavy metals affect the ecosystem and are directly linked to environmental contamination.

14.1.1 Heavy Metals

HMs are natural metals and metalloid elements with a high atomic weight that are abundantly available on earth. It is the high relative density and toxicity of heavy metals that makes them significant pollutants (Wang et al. 2020). With atomic densities of 4 g/cm^3 or higher, or five times or more than H_2O, even modest concentrations of heavy metals are toxic (Ferrey et al. 2018). Heavy metal toxicity has increased in recent years for nutritional, ecological, and environmental reasons because of its chronic toxicity and non-biodegradability (Valdés et al. 2014). Heavy metals are also observed as trace elements since they exist in a variety of environmental matrices in trace concentrations (ppb range to less than 10 ppm) (Kabata-Pendia 2001). The general collective term of heavy metals mentions any metallic compound that has a toxic nature even in the low concentrations with greater atomic density (Lenntech 2004). In contrast to their density, the heavy metals' chemical characteristics have the most of an impact. Lead (Pb), cadmium (Cd), nickel (Ni), cobalt (Co), iron (Fe), zinc (Zn), chromium (Cr), arsenic (As), silver (Ag), and the elements of the platinum group are examples of heavy metals. Most heavy metals are dispersed throughout rock formations. One of the major issues is that industrialization and urbanization have enlarged the number of heavy metals that people create and release into the environment. Heavy anthropogenic activities, including industrial production and use, domestic and agricultural use of metals and metal-containing compounds, pharmaceuticals, mining, foundries and smelters, and other metal-based industrial operations, are some of the sources of heavy metal pollution on the planet (Srivastava et al. 2018). Heavy metals are more readily available in soil and aquatic habitats than in the atmosphere as particles, where they are scarcer. The toxicity of heavy metals in plants varies with plant species, the particular metal, concentration, chemical form, soil composition, and pH because many heavy metals are regarded to be essential for plant growth. Some of the heavy metals, like Cu and Zn, either possess a catalytic property, like the prosthetic group in metalloproteins, or act as cofactors and activators of enzyme processes, like in the case of the complex between the enzymes and the substrate metal (Lin and Mark 2012). These crucial trace metal elements participate in electron transfer, redox reactions, and structural processes in the metabolism of nucleic acids. Metal-sensitive enzymes can become very toxic when exposed to certain heavy metals, such as Cd, Hg, and As, which can lead to the death of organisms. Highly toxic elements like Hg, Ag, Pb, and Ni are categorized as class B metals under the category of non-essential trace elements under a distinct categorization of metals based on their coordination chemistry (Nieboer and Richardson 1980). Because some of these heavy metals are bio-accumulative, they cannot be easily metabolized and do not degrade in the environment. These metals are ingested at the consumer level after being absorbed at the primary producer level, where they accumulate in the ecological food chain. Due to the sedentary nature of plants, the roots are the primary site of interaction for heavy metal ions. In aquatic environments, the entire plant body is exposed to these ions. Heavy metals are also straight absorbed into the leaves as a result of particles that are deposited on the foliar surfaces.

The physiology and biochemistry of both plants and animals are impacted by the important heavy metals. They play important roles in a variety of oxidation–reduction reactions and are necessary building blocks for a number of vital enzymes (WHO/FAO/IAEA 1996). Despite the fact that evidence suggests that heavy metals had an impact on plants' biological systems at higher concentrations, these effects had an effect on cell membranes, mitochondria, lysosomes, endoplasmic reticulum, nuclei, and several enzymes involved in metabolism, detoxification, and damage repair (Wang and Shi 2001). Non-essential metals are those that exhibit high levels of bioaccumulation and have no known biological functions. These substances include gold (Au), indium (In), lead (Pb), lithium (Li), mercury (Hg), nickel (Ni), antimony (Sb), arsenic (As), barium (Ba), beryllium (Be), bismuth (Bi), cadmium (Cd), gallium (Ge), and germanium (Ge) (Chang et al. 1996). The heavy metals with the biggest effects on the environment and the survival of the majority of species are mercury (Hg), arsenic (As), nickel (Ni), cobalt (Co), copper (Cu), cadmium (Cd), chromium (Cr), zinc (Zn), mercury (Hg), silver (Ag), iron (Fe), and platinum (Pt) (Rahman and Singh 2019). There are three classes of metals and metalloid ions. The first category contains metals that can be harmful even in small levels, such as lead, cadmium, and mercury. Whilst zinc, cobalt, copper, iron, and selenium are amongst the necessary metals that are only poisonous over a specific quantity in the body and are employed in several chemical or biochemical processes, bismuth, indium, arsenic, thallium, and antimony are amongst the less dangerous metals (Odobasic et al. 2019).

14.1.1.1 Essential Heavy Metals

According to Reeves and Baker (2000), both plants and animals harbour a requirement for a variety of heavy metals, in which significant micronutrients, including Fe, Zn, Mn, Cu, Co, Ni and Mo as well as other heavy metals whose uptake exceeds the needs of the plant and has detrimental effects, are frequently found in rhizosphere and are also taken up by the plants (Monni et al. 2000). Many of these heavy metals like Zn and Fe are physiologically important in both plants and animals. Essential heavy metals provide the following two main purposes: involvement in redox reactions and direct involvement as a component of many enzymes, respectively.

14.1.1.2 Source of Heavy Metals

Inorganic elements form the part of basic requirement for all groups of organisms including animals, plants, and humans, which can be divided into major and minor elements based on their concentration taken up and required by the organism. However, minor elements (also called trace elements) comprises both essential and optional substances (Blaser et al. 2000). In the recent times, one of the major causes of heavy metal toxicity are the anthropogenic activities. Some examples of HMs' natural origins include wind-borne soil debris, forest fires, volcano eruptions, biogenic processes, and marine salt (Muhammad et al. 2011). Along with these, agricultural activities are the major anthropogenic contributor to the concentration of heavy metals in environment. Field procedures, such as use of sewage and industrial wastewater for irrigation and application of fertilizers and pesticides, are the major contributor of heavy metals in environment (Srivastava et al. 2018). Fertilizers with trace amounts of heavy metals are significant contributors to these pollutants in our diet. Poor management of industrial waste, traffic pollution, lead (Pb) use as fuel antiknock, aerosol cans, metallurgy and smelting, sewage discharge, and the use of construction materials are examples of anthropogenic practices that contaminate HMs (Srivastava et al. 2017). Mercury (Hg) is released into the atmosphere by a number of sectors, including the production of pharmaceuticals, the preservation of paper and pulp, agriculture, and the production of chlorine and caustic soda (Ibrahim et al. 2019). Cadmium is present in soils, rocks, coal, and mineral fertilizers to varying degrees. Many products, including batteries, pigments, textiles, and metal coatings, use cadmium (Cd) in electroplating (Saini and Dhania 2020). The increased environmental pollution of HMs is caused by all of these actions. The presence of heavy metals in the environment is caused by a number of different sources, including atmospheric sources, agricultural sources, industrial sources, residential effluent, and other sources. Both man-made and natural processes can lead to heavy metal pollution. Agriculture and various enterprises like mining and smelting have contaminated large parts of the earth (Fig. 14.1).

14.1.1.3 Heavy Metals Toxicity

Due to the massive arrangement of cells that heavy metal poisoning can affect, negative effects naturally appear in the cells that are first exposed to the metal or those that absorb it fastest. Heavy metals disrupt ionic equilibrium and enzyme function in physiological processes requiring specific organs such as nutrition intake by the roots. Then, these impacts can be observed in a wider range of processes that include germination and growth of the plants, photosynthesis, primary metabolism, and reproduction that are all vital processes in plants. Toxicity of heavy metals in

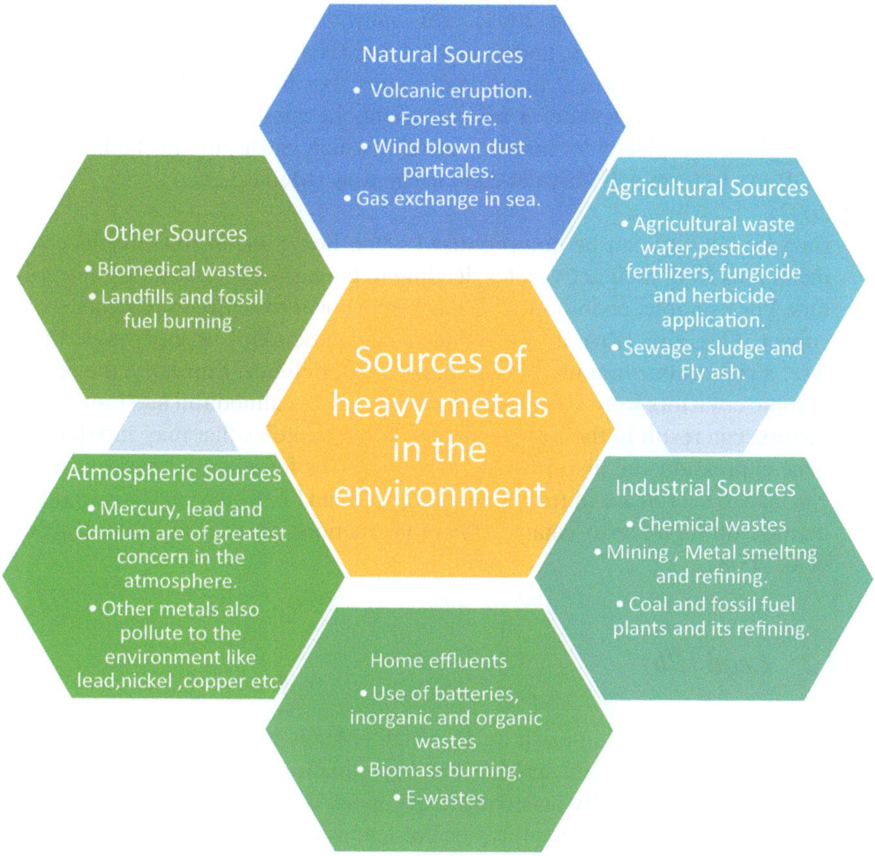

Fig. 14.1 Sources of heavy metals

plants is usually manifested by chlorosis and rolling of leaves followed by necrosis; low biomass production leading to stunted growth followed by senescence and death.

For HMs harmfulness, there are four main mechanisms that explain why plants become hazardous to heavy metals:

(a) Creating oxidative stress and altering the permeability and integrity of cell membranes: Numerous heavy metals, either directly or as a result of their toxicity to cells, result in the generation of ROS which harms the plant cells, including H_2O_2, O_2, and OH, e.g. via increasing lipid peroxidation and inhibiting the action of water channel and transporter proteins. The latter limits membrane-dependent processes, such as electron flow in mitochondria and chloroplasts, by altering the structure, fluidity, and stability of the membrane. To counteract ROS, antioxidant enzymes like SOD, APX, GPX, CAT, and GSR are triggered.

(b) Sulfhydryl group reaction (–SH): Due to their high affinity for –SH groups, heavy metals bind to structural proteins and enzymes that contain them

(Schützendübel and Polle 2002). This may disrupt enzyme-mediated redox control, catalytic activity, and proper folding (Hall 2002).

(c) Similarity to biochemical functional groups: For instance, As (V) in arsenate competes with phosphate, a micronutrient, in numerous cellular processes. Phosphate is replaced by arsenate in ATP, which causes the unstable complex ADP–As to form and obstruct cell energy flow (Meharg and Hartley-Whitaker 2002).

(d) Enzyme activity can be compromised when heavy metal ions displace other metal ions from their active sites, leading to the loss of essential cationic cofactors. For example, cadmium (Cd) has been shown to replace cofactors such as copper (Cu), zinc (Zn), iron (Fe), and manganese (Mn) in superoxide dismutase, a critical enzyme involved in antioxidant defence. Similarly, the removal of ionic cofactors from signalling proteins, such as calmodulin and transcription factors, can result in the formation of aberrant proteins that may interfere with gene expression and disrupt homeostatic systems for essential metal ions. This can lead to the release of free ions, such as Cu and Fe, which can cause oxidative damage through Fe/Cu-catalyzed Fenton reactions (Roth et al. 2006; DalCorso et al. 2008).

14.1.2 Lead (Pb)

Pb is a typical HM found in earth crust, which belongs to the periodic table's class IVA, has a high atomic weight, and has a stable oxidation state ($Pb^{2+)}$ that produces a divalent ion. Pb is a naturally occurring heavy metal with an atomic weight of 207.2 (Arias et al. 2010). When a metallic piece of lead is exposed to air or water, minute layers of lead compounds develop, such as oxides and carbonates, shielding it from further attacks: This metal is resistant to corrosion. Because of its ease in shaping and moulding as well as its corrosion resistance, it has been extensively utilized for centuries. Lead comes in three varieties: metallic, inorganic, and organic. Instead of its elemental form, lead is present in the form of ores in its oxidized state Pb^{2+} in the environment. But due to its persistent nature, lead finds its way into air, water, soil, and dust (Patrick 2006). In addition, trace amounts of lead can be found in coal, oil, and wood. However, the majority of lead released into the environment comes from anthropogenic sources, as opposed to natural sources such as volcanoes, wind-blown dust, and erosion. Lead exists in the atmosphere in the form of particles and can be removed through precipitation or gravity settling. The solubility of lead compounds in water is affected by various factors, including pH, hardness, salinity, and the presence of humic substances. Water that is soft and acidic has a higher solubility of lead. The soil and sediment serve as lead sinks, and due to its high adsorption capacity, it tends to accumulate in the upper soil layers and does not penetrate deeply into the subsoil or groundwater (Rabinowitz 2017).

Lead (Pb) is a harmful element that is widely distributed in soil, and it is one of the most common heavy metals produced by human activities such as mining, smelting, fuel and explosive use, and the discharge of municipal sewage sludge that has been enriched in Pb. Pb and Cd are considered to be amongst the most hazardous substances to human health as they are readily taken up by plants and can easily enter the food chain. Symptoms of lead poisoning in plants are similar to those caused by other heavy metal toxicities, including chlorosis, growth inhibition, and in severe cases, death. Pb uptake by roots leads to slower growth and altered branching patterns. Lead toxicity negatively affects plant morphology, growth, and photosynthetic processes, resulting in stunted growth, reduced biomass of roots and shoots, disrupted mineral uptake, inhibited cell division, and impaired photosynthesis (Sharma and Dubey 2005; Maestri et al. 2010).

14.1.2.1 Source of Lead

Lead can be introduced into the environment by a variety of mechanisms that include mining and smelting activities for lead ore, manufacturing of lead containing products, burning of coal and oil, and incineration of waste (Zheng et al. 2011). Use of many products that contained lead has been banned so as to prevent environmental pollution by lead, and some of these products were leaded petrol, lead-based paints, and lead-based herbicides. These previous applications create a legacy of increased lead concentrations in the environment because lead does not decay. Figure 14.2 shows a list of a variety of anthropogenic and natural processes that contributed to the environmental pollution of Pb.

14.1.2.2 Lead's Effects on Plants

Physiological and Biochemical Processes

According to Munzuroglu and Geckil (2002), Pb is known to negatively impact a number of aspects of plants (Fig. 14.3), including the rate at which seeds germinate, the growth of seedlings, the mass of dry roots and shoots, photosynthesis, respiration, the relationship between plants and water, mineral nutrition, and a number of enzymatic activities (Table 14.1). At larger dosages and over a longer period of time, Pb effects are inherently more distinct. However, even a modest amount of metal can trigger a number of biological reactions (Gomes 2011).

The damaging effects of lead on plants can be manifested in the form of various symptoms that include chlorosis and necrosis on the leaf surface and senescence of the leaf followed by restricted growth. At increasing concentrations, seeds become significantly more sensitive to germination. During the plantlet stage, when roots are more vulnerable to this action, root and shoot growth is also suppressed. Lead can also impact the plant by inhibiting the ability of the roots to take up nutrients and also be blocking the movement of nutrients such as Fe, Zn, Ca, and P inside plants (Xiong 1997). Hence, there exists sufficient evidence to confirm that lead can significantly impact physiological and metabolic activities in plants (Gomes 2011).

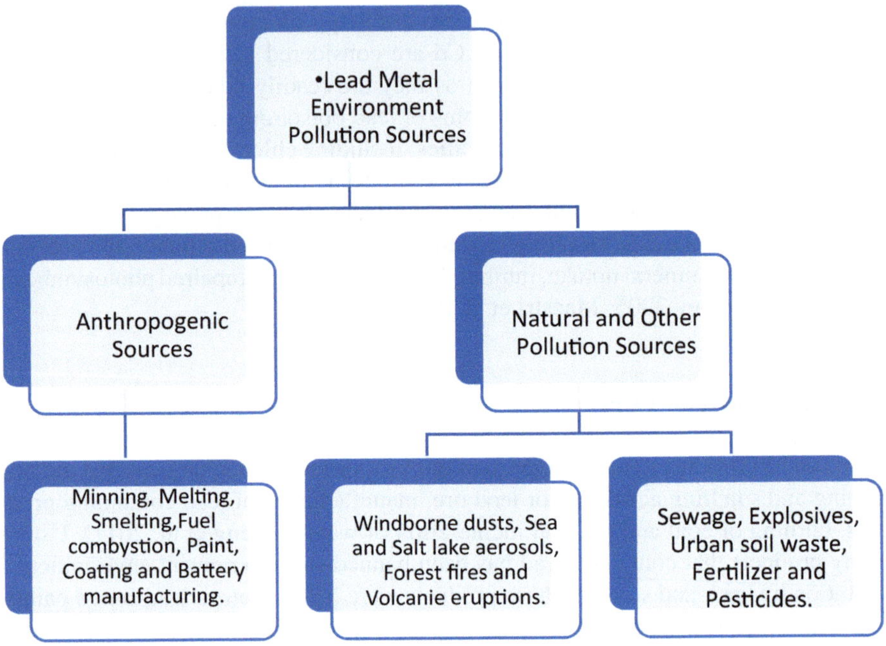

Fig. 14.2 Causes of lead pollution in the environment. Both natural and anthropogenic activities have contributed to the environmental lead pollution

Plant Growth Germination

Chlorosis, a quick slowing of root growth, stunted plant growth, and a darkening of the root structure are all apparent signs of lead toxicity. Too much lead can eventually result in toxic chemicals that stop seeds from germinating, slowing down seedling growth, and lowering the length, tolerance index, and dry mass of roots and shoots. Even at micromolar concentrations, lead exposure can harm plant germination and growth (Kopittke et al. 2007). Even uncertain amounts of Pb^{2+} substantially delays germination (Islam et al. 2007). Moreover, lead exposure in plants severely restricts seedling growth and sprouting (Gichner et al. 2008).

High levels of lead in plant roots can lead to increased inhibition, as noted by Liu et al. (2008). Other signs of lead toxicity include short, twisted, and stubby roots, as well as increased number of secondary roots per unit length, as reported by Kopittke et al. (2007). Furthermore, exposure to lead can reduce plant biomass, leading to significant growth retardation, fewer and smaller leaves, and dark purplish abaxial surfaces, as observed by Gupta et al. (2009).

According to Gopal and Rizvi (2008) and Islam et al. (2008), the detrimental effects of lead on plant growth may be attributed to altered nutrient metabolism and interrupted photosynthesis. Gupta et al. (2009, 2010) also found that the degree of lead toxicity on plant growth is often dependent on the dose and time of exposure.

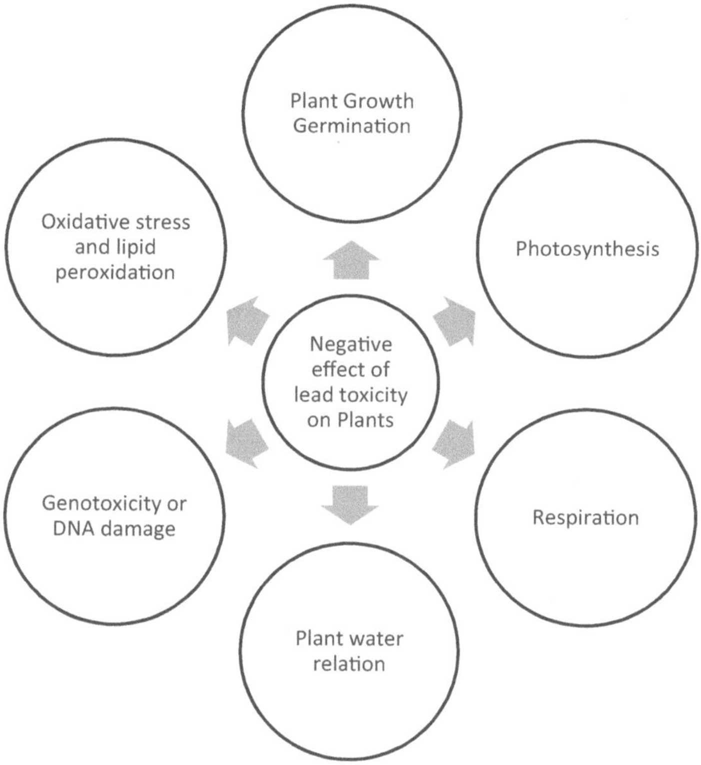

Fig. 14.3 Lead toxicity effects on the plant

Nonetheless, low levels of lead exposure may still have an impact, and growth inhibition does not necessarily equate to a reduction in biomass (Yan et al. 2010). Moreover, different plant species may have varying responses to lead toxicity; for example, hyperaccumulators are naturally more tolerant of lead poisoning than sensitive plants (Arshad et al. 2008). It was also shown that in lead-spiked soil, the performance of sorghum was severely affected in terms of germination index, plumule and radical length, vigour index, and tolerance index (Osman and Fadlallah 2023).

Photosynthesis

Lead toxicity has a negative impact on the rate of photosynthesis due to decreased crop productivity (Singh et al. 2010). Plants exposed to lead ions show decrease in photosynthetic rate as a result of reduced chloroplast ultrastructure, constrained chlorophyll, plastoquinone, and carotene synthesis, hindered electron transport, impeded Calvin cycle enzyme activity, and a CO_2 deficit as a result of stomatal closure (Qufei and Fashui 2009). Thylakoid membrane lipid content is also altered by lead treatment

Table 14.1 Growth stages in different plant species that are most affected by lead excess

Plant name	Pb concentration	Lead effect	References
Spartina alterniflora	10 ppm	Inhibit seed germination	Morzck and Funicclli (1982)
Pinus halepensis	1500 ppm	Inhibit seed germination	Nakos (1979)
Oryza sativa	10^{-1} M	Inhibit protease and amylase	Mukherji and Maitra (1976)
Glycine max	300 μM	Early seedling growth inhibition	Huang et al. (1974)
Zea mays	250 μg Pb/g of soil	Early seedling growth inhibition	Miller et al. (1975)
Hordeum vulgare	100–1000 μm	Early seedling growth inhibition	Stiborova et al. (1987)
Solanum lycopersicum, Solanum melongena	600 ppm	Early seedling growth inhibition	Khan and Khan (1983)
Fabaceae	2.0 ppm	Early seedling growth inhibition	Sudhakar et al. (1992)
Allium porrum	10^{-5} M	Prevented leaf growth and root and stem elongation	Gruenhage and Jager (1985)
Hordeum vulgare, Raphanus sativas	0.01 mM	Prevented leaf growth and root and stem elongation	Juwarkar and Shende (1986)
Sesamum indicum	0.04–1.9 mM	Prevention of root expansion	Kumar et al. (1992)
Pisum sativum	1.0 mM	Irregular radial thickening in roots, endodermis cell walls, and cortical parenchyma lignification	Paivoke (1983)
Beta vulgaris	0.5 M	Decrease in growth and chlorosis	Hewilt (1953)
Lactuca sativa; Daucus carota	5 mg/l	Reduction in growth	Baker (1972)
Picea abies	0.5 μm	Secondary root development and emergence are extremely delicate processes	Godbold and Kettner (1991)

(continued)

Table 14.1 (continued)

Plant name	Pb concentration	Lead effect	References
Zea mays	200 μm	Changes the root meristem's microtubule network structured, causing the branching zone to be shorter and the lateral roots to emerge more compactly and close to the root tips	Eun et al. (2000)
Cucumis sativus, Glycine max	10^{-5} to 2×10^{-4} M	Nitrate reductase and glutamine synthetase activity inhibition	Sharma and Dubey (2005)
Brassica juncea	1000 μm	Prevent the growth	Zaier et al. (2010)
Zea mays, Pisum sativum	400 mg kg^{-1}	Root, shoot, and leaf growth; drastically reduced fresh and dry biomass	Cimrin et al. (2007)
Glycine max	40 mg dm^{-3}	The vascular bundles' xylem and phloem were diminished, the leaf blades became thin, and the diameter of the xylem vessels was also decreased	Elzbieta and Minoslawa (2005)
Solanum lycopersicum, Pisum sativum, Zea mays, Paspalum distichum, Cynodon dactylon, Lycopersicon esculentum, Ipomoea aquatica, Phaseolus vulgaris and *Lens culinaris*	400 mg kg^{-1}, 1×10^{-7} mol dm^{-3}, 40 mg l^{-1}, 1%, 600 μm, 250 ppm	Deleterious impact on the growth development, fresh–dry biomass, and growth tolerance index of roots, shoots, and leaves	Cimrin et al. (2007), Kevresan et al. (2001), Shua et al. (2002), Jaja and Odoemena (2004), Gothberg et al. (2004), Haider et al. (2006)
Hordeum vulgare, Elsholtzia argyi, Spartina alterniflora, Pinus halepensis, Oryza sativa, and *Zea mays*	400 μm, 20,000 μg/g, 2.5 mg dm^{-1}, 0.05%, 250 μg/g	Preventing the germination of seeds	Islam et al. (2007), Baker (1981), Wierzbicka (1987), Tomulescu et al. (2004), Sengar et al. (2009)
Oryza sativa	0.5 μm	Germination of rice seeds reduction	Gidlow (2015)
Raphanus raphanistrum	0.05%	The negative effects of Pb on plant development in the early stages	Tomulescu et al. (2004)

(continued)

Table 14.1 (continued)

Plant name	Pb concentration	Lead effect	References
Prosopis juliflora	100 mg L^{-1}	The root's length is severely suppressed	Arias et al. (2010)
Allium sativum	10^{-5} M	Injured plasma membrane, deep-coloured nuclei, loss of cristae, endoplasmic reticulum and dictyosome vacuolization, and swelling of the mitochondria	Jiang et al. (2010)
Honey mesquite	100 mg L^{-1}	Inhibited root elongation	Arias et al. (2010)
Helianthus annus L.	10^{-6} M	Deeply coloured nuclei, damaged plasma membranes, lack of cristae, endoplasmic reticulum vacuolization, and dictyosomes that are swollen within the mitochondria	Kastori et al. (2008)
Allium cepa	1–1000 mM	A reduction in the rate of mitosis in root cells	Patra et al. (2004)
Vicia faba	1–1000 mM	Increased interphase length and decreased mitotic length, prolonging the cell cycle	Patra et al. (2004)

by damaging the working enzymes that are involved in the carbon dioxide fixation during photosynthesis (Mishra et al. 2006; Qufei and Fashui 2009). Lead prevents the production of chlorophyll by preventing plants from absorbing vital nutrients like magnesium and iron. It damages the photosynthetic machinery because it is attracted to protein N- and S-ligands (Islam et al. 2007).

Lead exposure has harmful effects on plants because it increases oxidative stress and chlorophyllase activity, which hastens the destruction of chlorophyll (Liu et al. 2008). According to reports, lead treatment has a greater impact on chlorophyll b than chlorophyll a (Harpaz-Saad et al. 2007). Lead impacts (PS I) have been observed at the cytochrome b/f complex, photosynthesis 1, and photosynthesis 2 donor and acceptor sites. It is generally accepted that PS II electron transport during photosynthesis 2 is more susceptible to lead suppression than PS I electron transport.

Respiration

Photosynthetic plants exposed to lead (Pb) experience negative effects on their composition of adenosine triphosphate (ATP) and respiration. Whilst the impact of Pb on respiration has been mainly studied in relation to leaves, its effects on roots are not yet known. According to Assche and Clijsters (1990), Pb has no effect on the oxygenase activity of C3 plants, but it does affect the assimilation of CO_2 by ribulose bisphosphate carboxylase. Divalent cations such as Pb, Zn, Cd, and Ni bind to mitochondrial membranes and interfere with electron transport, which can lead to phosphorylation decoupling, as noted by Romanowska et al. (2006). Pb has been found to inhibit the Hill reaction and photophosphorylation in spinach chloroplasts, with a more significant impact on cyclic photophosphorylation than non-cyclic photophosphorylation (Romanowska et al. 2008). The adhesion of Pb to membranes has various physiological effects. Mitochondria from Pb-treated pea leaves oxidize substrates such as glycine, succinate, and malate more rapidly than mitochondria from untreated pea leaves, according to Romanowska et al. (2002).

Plant–Water Relation

Turgor pressure is another major plant component that is affected by Pb stress (Rucińska-Sobkowiak et al. 2013). Pb decreases the plant's closed cell wall's flexibility, which lowers the guard cells' turgor pressure (Pinho and Ladeiro 2012). Sugar, amino acids, and other chemicals necessary for sustaining the cell's turgor pressure are reduced in plants exposed to Pb (Barceló and Poschenrieder 1990). Moreover, Pb toxicity has been linked to the increased production of ABA in plants, leading to stomatal closure and reduction in transpiration rate (Atici et al. 2005). Water relations in a number of crop plants have been reported to be affected by Pb excess (Sharma and Dubey 2005). Plants with higher stomatal densities can withstand such effects (Elzbieta and Minoslawa 2005). Respiration via the leaves is also impeded in plants under Pb stress due to the accumulation of a waxy coating (Elzbieta and Minoslawa 2005).

Effects on Nucleic Acid

Genotoxic agents, also referred to as mutagens, are substances that harm cell's DNA or other genetic material whether it occurs inside the nucleus or outside the nucleus. One such mutagen is Pb, which makes people develop cancer (Shahid et al. 2011). It disrupts the spindle fibres during mitosis by functioning as a potent toxin (Patra et al. 2004). The disruption of the cytoskeleton and nucleus, DNA strand breakage, the formation of micronuclei, chromosomal abnormalities, variability in simple sequence repeats, and depolymerization of microtubules are only a few of the negative consequences of Pb in plant cells (Kumar et al. 2017). Low concentrations of Pb do not affect mitosis, although they do cause some abnormalities, such as chromosome breakage, damage to central DNA segments during meiotic division, and the emergence of micronuclei (Shahid et al. 2011). Pb can attach to either DNA or proteins when it enters the nucleus (Małecka et al. 2008). Pb interferes with DNA replication and repairs when it binds to DNA. Unless it binds to bare DNA, it has no direct genotoxic effects. Enzymes responsible for polymerizing nucleotides during DNA synthesis also have their conformation disturbed, which has an impact on how they function (Pourrut et al. 2011). Cenkci et al. (2010) has demonstrated the influence of Pb on the consistency of the template DNA strand in *Brassica rapa* using the RAPD assay.

Lipid Peroxidation and Oxidative Stress

Chloroplasts produce reactive oxygen species (ROS) as a by-product of normal cellular metabolism, which includes hydrogen peroxide (H_2O_2), hydroxyl radicals (•OH), and superoxide radicals ($O_2^{\bullet -}$). These ROS are also produced as a result of exposure to environmental pollutants. Hydrogen peroxide is capable of crossing cell membranes and has a direct impact on cell signalling (Pitzschke et al. 2006). The generation of ROS causing oxidative stress in plant cells is a characteristic of harmful heavy metals, such as Pb (Grover et al. 2010). Once cellular antioxidant reserves are depleted, ROS can quickly attack and damage various types of biomolecules, including nucleic acids, proteins, and lipids (Yadav 2010). These attacks can cause metabolic dysfunction and cell death. Pb is known to significantly alter the lipid composition of many cell membranes, including those containing polyunsaturated fatty acids and esters that are highly susceptible to ROS (Gupta et al. 2009; Yan et al. 2010). ROS can destroy the lipid bilayer by removing the hydrogen from unsaturated fatty acids, resulting in the production of sensitive lipid radicals and aldehydes (Mishra et al. 2006). Redox enzyme activity has been observed to decrease in the presence of sufficient Pb (Lamhamdi et al. 2013). Pb significantly modifies the lipids of the plasma membrane, resulting in an abnormal cell structure (Gupta et al. 2009; Grover et al. 2010; Yan et al. 2010). Pb-induced modifications to lipid content and K^+ ion seepage have been documented in *Z. mays* (Małkowski et al. 2002). Pb ions increase the levels of unsaturated fatty acids whilst reducing the amount of saturated fatty acids in several plant species (Singh et al. 2010).

14.1.2.3 Lead in Oxidative Stress: Activation of Reduced Oxygen Forms and Resulting Metabolic Harm

Molecular oxygen creates oxygen free radicals when it receives electrons from other molecules. Various intracellular activities that consume oxygen produce superoxide (O_2^-) or hydrogen peroxide (H_2O_2). Hydroxyl radicals (–OH), which can be produced even by relatively unreactive substances, are believed to cause the majority of oxidative damage in biological systems (Halliwell and Cutteridge 1990). Mishra et al. (2007) hypothesized that Pb produces reactive oxygen species (ROS) and increases the activity of antioxidant enzymes in plants. ROS generated due to oxidative stress in plant cells can inhibit photosynthetic activity, decrease ATP synthesis, cause lipid peroxidation, and damage DNA (Ruley 2004). The production of ROS is one of the primary impacts of heavy metal stress, and it damages cell membranes, nucleic acids, and chloroplast pigments (Tewari et al. 2002). Heavy metal stress in plants produces several ROS, including the superoxide anion (O_2^-), singlet oxygen, hydroxyl radical (OH), and hydrogen peroxide (H_2O_2). However, when these ROS are produced in higher concentrations, they can be extremely harmful. An imbalance between the production of excess ROS and the antioxidative enzyme activity, which includes enzymatic antioxidants such as superoxide dismutase (SOD), catalase (CAT), ascorbate peroxidase (APX), and peroxidase (POX), may result in excessive ROS production in heavy metal-stressed plants (Jiang et al. 2010). In stressed plants, superoxide dismutase scavenges the superoxide radical (O_2^-) and converts it to hydrogen peroxide (H_2O_2) (Reddy et al. 2005). Catalase immediately scavenges H_2O_2, producing H_2O and O_2. Ascorbate peroxidase and other peroxidases can scavenge H_2O_2 indirectly by combining it with other antioxidant molecules (Yingli et al. 2011). Plants have several mechanisms for adapting to changes in the concentration of heavy metals in polluted environments by altering their levels. Heavy metal accumulation in plants exceeds the capacity of plant tissues to detoxify, making them toxic (Zhang et al. 2007). **In another study,** increase in CAT and SOD activity has been correlated with the decrease in electrolyte leakage, MDA, and H_2O_2 content in wheat plants when treated with biochar to control lead toxicity (Mahmood et al. 2023). In another study, supplementation of maize crop with sugarcane bagasse biochar resulted in significantly improved growth of plants even under 800 mg Pb kg^{-1} soil (Rassaei 2023).

14.1.2.4 Plant Defence Mechanisms Against Lead Toxicity

Different mechanisms have been observed in plants that are used for prevention from lead toxicity. Some of the more effective ones include selective absorption of metals, sequestration at root level, sequestration at cell wall, or production of antioxidants. Non-enzymatic antioxidants like glutathione, ascorbic acid, and proline and enzymatic antioxidants like glutathione reductase, superoxide dismutase, catalase, etc. have commonly been reported to be produced by the plants. Also, species, dosage exposed to, and circumstances of exposure are important determinants of plants response to lead excess.

Passive Mechanisms

Lead interacts with cellular elements and thickens cell walls even when tiny levels of lead enter root cell membranes (Krzesłowska et al. 2010). Pectin, a cell wall component, can form complexes with lead and prevent its entry into the cell, thereby protecting plant cells against damage by lead excess (Jiang and Liu 2010). Krzesłowska et al. (2009) reported that binding of lead to JIM5-P creates a physical barrier to prevent lead access to the plasma membrane in *F. hygrometrica* protonemata. However, it was also reported that lead bound to JIM5-P can be picked up and taken into cell by endocytosis (Krzesłowska et al. 2010).

Inducible Mechanism

Detoxification by excretion of metal ions by specific transporters is also an important mechanism (Maestri et al. 2010). Lead resistance was simultaneously mediated by a number of ATP-binding cassette carriers, including AtATM3 and AtADPR12 at ATP-binding sites in *Arabidopsis* (Cao et al. 2008). Although it has not been proven, this detoxification procedure might help remove lead. According to transcriptome research, lead-mediated expression of these carriers' genes was observed (Liu et al. 2009). Heavy metal detoxification and plant metal homeostasis are both considered to depend heavily on cellular sequestration (Maestri et al. 2010). Vacuoles (Meyers et al. 2008), dictyosome (Malone et al. 1974), endoplasmic reticulum (Wierzbicka et al. 2007), or plasmatubules (Vadas and Ahner 2009) are only a few of the compartments in plant cells where the lead is concealed.

Plants contain non-enzymatic antioxidants such as cysteine and glutathione. When exposed to lead toxicity, *Arabidopsis thaliana* increases its cysteine content, whilst glutathione protects the plant by neutralizing lead-induced ROS. Glutathione is also involved in the formation of phytochelatin, a protein essential for heavy metal detoxification and homeostasis. These phytochelatins are mainly involved in detoxification of cadmium, arsenic, lead, mercury, copper, and zinc (Seregin and Kozhevnikova 2023). The activation of glutathione genes, such as glutathione synthetase, glutathione peroxidase, glutathione reductase, and glutamylcysteine synthetase, is triggered by lead exposure. Glutathione can also boost proline accumulation, which can decrease protein and membrane damage caused by stress.

GSH was found to be essential for lead detoxification even without the presence of PCs, according to Gupta et al. (2010). PCs and metallothioneins are metal-binding ligands in plant cells that are cysteine-rich proteins that bind heavy metals. PCs are low molecular weight, metal-binding proteins that are vital for plants to detoxify metals as they can form mercaptide interactions with various metals. These thiols protect plant cells from oxidative damage and are produced by phytochelatin synthase (PCS) from GSH, with the typical structure of (-glutamyl-cys)nGly, where $n = 2–11$. Lead increases PC production and activates PCS, which has been hypothesized to play a role in the cellular detoxification and accumulation of several metals, including lead. The mechanism that directs the movement of the lead–PC complex through the tonoplast is still unknown. Gisbert et al. (2003) demonstrated that lead and Cd tolerance and absorption were significantly increased when a wheat gene (TaPCS1) coding for phytochelatin synthase was induced and overexpressed in *Nicotiana glauca*.

14.1.2.5 Antioxidant Enzymes

ROS-scavenging system is inherent to the plants that deals with the increased ROS content and protects the cells against oxidative damage (Gupta et al. 2010). However, lead excess may block or inhibit the activity of enzymes involved in the process (Table 14.2) (Singh et al. 2010). Lead typically has inactivation constants between 10 and 5, which means that in this range, inhibition of enzymatic activities can be up to 50%. In general, lead blocks enzymatic activity (Seregin and Ivanov 2001). The lead inhibits the enzyme because of its affinity for the –SH groups of the enzyme (Gupta et al. 2009). More than 100 enzymes can be inhibited in this manner including many important enzymes such as ribulose-1,5-bisphosphate carboxylase oxygenase and nitrate reductase. Lead can also modify the tertiary structure of the enzymes by interacting with the active site of the enzyme. Lead can also be bound to protein-COOH groups to produce the same outcome (Gupta et al. 2010). Lead and metalloid enzymes also interact. Plant's ability to take up minerals can also be affected by lead, particularly for zinc, iron, and manganese. Lead and other divalent cations can perform the same function as ALAD in inactivating enzymes as these metals (Cenkci et al. 2010). Lead exposure also affects ROS, which has an effect on how proteins behave (Gupta et al. 2010). Additionally, lead exposure is known to increase the activity of numerous enzymes (Table 14.2), whilst the precise mechanisms are still understood. Several studies have reported that lead can increase the activity of certain enzymes by altering gene expression or inhibiting enzyme inhibitors (Seregin and Ivanov 2001). In response to metal toxicity, antioxidant enzymes are responsible for scavenging ROS produced in excess. One of these enzymes is superoxide dismutase, which is a metalloenzyme present in various cell compartments and is considered the first line of defence against oxidative stress (Mishra et al. 2006). It maintains steady-state levels of superoxide radicals by catalyzing the dismutation of two superoxide radicals into H_2O_2 and oxygen (Gupta et al. 2009). The swift removal of H_2O_2 is necessary to prevent oxidative damage. The APX enzyme of the ascorbate–glutathione cycle or the GPX and CAT enzymes of the cytoplasm and other cell compartments accomplish this (Mishra et al. 2006). GSH and glutathione reductase play an established role in the H_2O_2-scavenging process in plant cells through the Halliwell–Asada enzyme pathway (Piechalak et al. 2002). Additionally, an increase in the concentration of their substrates may activate antioxidant enzymes, rather than a direct reaction with lead (Islam et al. 2008).

14.1.2.6 Uptake, Translocation, and Accumulation of Lead

The amount of lead that dissolves in the soil and the ease with which plants may reach it can depend on a plant's root surface area, root exudates, transpiration pull, mycorrhizal connections in the rhizosphere, and the variety of root types it generates (Davies 1995). The roots, which receive lead exposure initially, have the potential to store lead or act as a conduit for the export of lead ions from the soil to the aboveground plant parts (Fangmin et al. 2006). Resting, the mechanisms for the

Table 14.2 Effect of lead on the enzymatic activity in different plants species

Species of plants	Enhance of enzyme activity	Reduction of enzyme activity	References
Taxithelium nepalense	APX, GPOX, CAT		Choudhury and Panda (2004)
Cicer arietinum	SOD, CAT, POD, GR, GST		Reddy et al. (2005)
Macrotyloma uniflorum	SOD, CAT, POD, GR, GST		Reddy et al. (2005)
Ceratophyllum demersum	SOD, GPX, APX, CAT, GR	SOD, GPX, APX, CAT, GR	Mishra et al. (2006)
Helianthus annuus	GR	CAT	Garcia et al. (2006)
Kandelia candel	SOD, POD, CAT		Zhang et al. (2007)
Bruguiera gymnorrhiza	SOD, POD, CAT		Zhang et al. (2007)
Potamogeton crispus	SOD, POD, CAT	SOD, CAT	Hu et al. (2007)
Cassia angustifolia	SOD, APX, GR, CAT		Qureshi et al. (2007)
Raphanus sativus	POD, ribonuclease	CAT	Gopal and Rizvi (2008)
Elsholtzia argyi	CAT	SOD, GPX	Islam et al. (2008)
Wolffia arrhizal	CAT, APX		Piotrowska et al. (2009)
Lathyrus sativus	APX, GR, GST		Brunet et al. (2009)
Zea mays	SOD, CAT, POD		Gupta et al. (2009)
Jatropha curcas	SOD, POD, PAL		Gao et al. (2009)
Najas indica	SOD, GPX, APX, CAT, GR		Singh et al. (2010)
Sedum alfredii	SOD, APX, POD		Gupta et al. (2010)
Spinacia oleracea L.	SOD, CAT, POD		Lamhamdi et al. (2010)
Gossypium hirsutum	SOD, GPX APX, CAT		Bharwana et al. (2013)
Oryza sativa	SOD, GPX, APX, CAT, GR	GR, CAT, MDA	Rao et al. (2018)
Triticum aestivum	SOD, LOX, GPX and APX	CAT	Navabpour et al. (2020)
Brassica juncea L.	SOD, POD, GST, GR CAT, DHAR		Singh et al. (2020)
Brassica chinensis L.	CAT, SOD, APX		Ji et al. (2022)

Lead can either activate or inhibit the activity of several enzymes
POD Superoxide dismutase; *APX* ascorbate peroxidase; *GPX* guaiacol peroxidase; *CAT* catalase; *GR* glutathione reductase; *AsA* ascorbic acid; *GST* GSH S-transferase; *GSH* glutathione; *POD* peroxidase

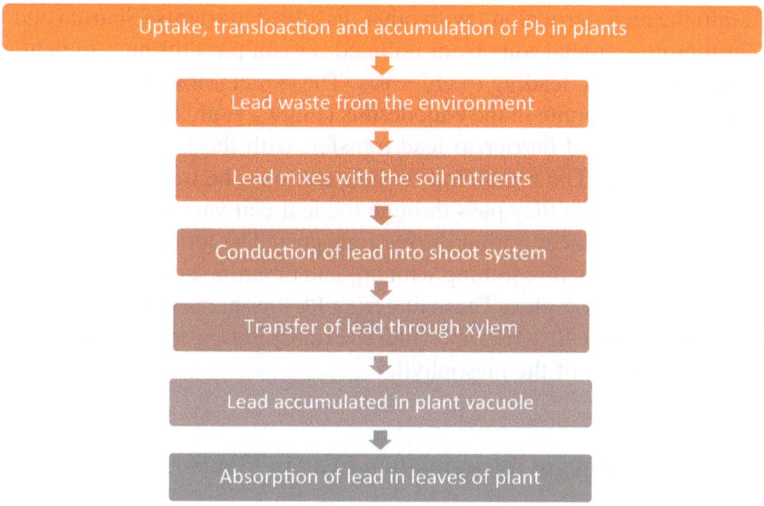

Fig. 14.4 Schematic flow chart to the build-up of lead concentration in a plant species

absorption, transport, and storage of poisonous metals in plants whose chemical characteristics mimic those of vital nutrients are identical to those for the uptake of micronutrients from the soil. The same mechanism is used for Pb uptake (Hseu et al. 2010). However, as depicted in Fig. 14.4, there are a variety of variables that influence the lead uptake from the soil. These variables include the pH of the soil, the amount of organic matter in the soil, redox processes, and plant-derived chelating agents that solubilize the micronutrients in the rhizosphere.

Bioaccumulative Potential

Through absorption, bioavailability, bioconcentration, and biomagnifications, the bioaccumulation process occurs. Importer complexes construct a translocation channel via the lipid bilayer during bioaccumulation, a metabolically active process by which living things take up HM in their intracellular space. Proteins and peptide ligands may sequester the HM once it has entered the intracellular environment (i.e. storage system as plants).

Uptake and Translocation of Lead

Plants can take in free Pb ions from the atmosphere through cellular respiration or capillary action (Sharma and Dubey 2005). Once lead is absorbed from the polluted soil by the highly developed root systems of plants, it is transported through the xylem channels (Engwa et al. 2019). This movement occurs in upward flow before being released into the endoderm along with other dissolved nutrients (Sharma and Dubey 2005). Furthermore, metal ions from polluted air can pass through the cuticle and stomata of plant leaves, leading to chlorosis. These metal ions then bind to the cell wall and plasma membrane in the endodermis (Engwa et al. 2019). The endodermal cells serve as a physical barrier to lead transfer, with the Casparian strip limiting the apoplast pathway. The remaining Pb ions can only be translocated through the symplast channel, and as they pass through the leaf cell vacuole, the barrier's job is to decrease the lead concentration during transport. The leftover Pb ions passively move and accumulate before joining with organic or amino acids in the transpiration stream to form a lead complex. These lingering Pb ions participate in photosynthesis as ions (Rahimzadeh et al. 2017) and enter the transpiration stream, passing through the intercellular spaces of the mesophyll.

14.1.2.7 Amelioration of Accumulation Lead in Plants

Phytoremediation

Heavy metal-contaminated soils can be cleaned up using a variety of phytoremediation approaches (Fig. 14.5). Here, we focus on the different types often used phytoremediation methods for removing heavy metals from soil: phytostabilization, phytoextraction, and phytovolatilization, etc (Table 14.3).

(i) Phytostabilization: It employs vegetation to lower the soil's bioavailability of heavy metals.
(ii) Phytoextraction: It comprises using plants to remove heavy metals from the soil.
(iii) Phytovolatilization: Using plants to absorb heavy metals from soil and release them as volatile chemicals into the environment.
(iv) Phytofiltration: It employs with two phytoremediation methods, namely rhizodegradation and phytodegradation, which destroy organic pollutants (Marques et al. 2009). Here, we concentrate on the four most popular phytoremediation techniques, i.e. phytostabilization, phytoextraction, phytovolatilization and phytofilteration—for cleaning up soil exhibiting heavy metal contamination.

Phytostabilization

Phytostabilization is a technique that utilizes metal-tolerant plant species to immobilize heavy metals and reduce their bioavailability, thereby lowering the risk of these metals entering the food chain and settling into the ecosystem (Marques et al. 2009). This process can be achieved through heavy metal precipitation, a reduction in metal valency in the rhizosphere, absorption and sequestration within root tissues, or adsorption onto root cell walls (Gerhardt et al. 2017). Plant growth also plays a

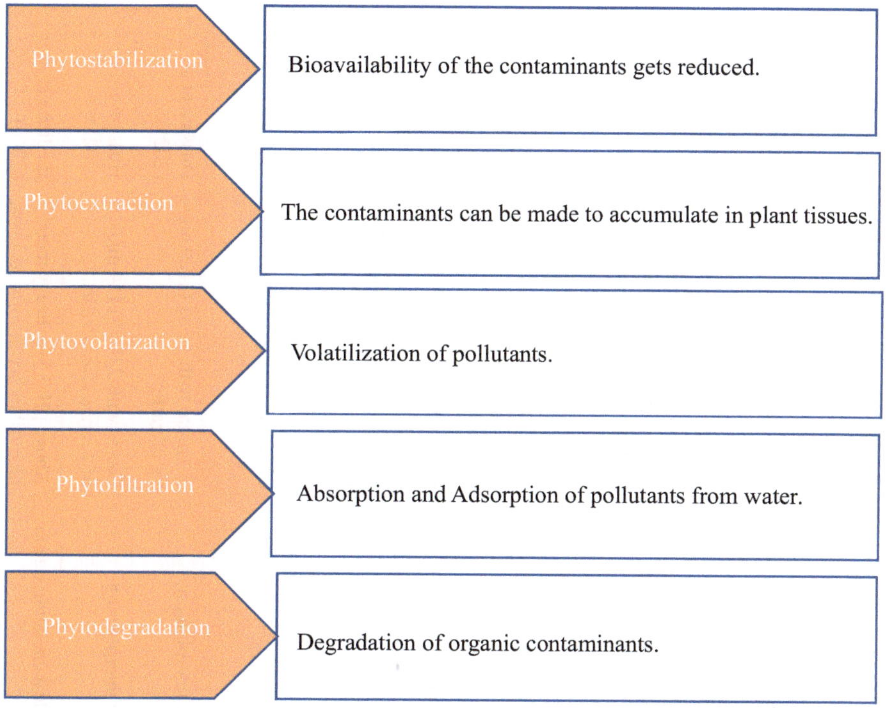

Fig. 14.5 Conceptual schematic representation of different methods used for amelioration of lead

crucial role in preserving soil health in areas with heavy metal contamination, and the resulting plant cover can help prevent the spread of heavy metal-bearing soil fragments by wind and stabilize heavy metals underground, limiting their leakage into groundwater (Mench et al. 2010; Kumar and Singh 2023).

One advantage of phytostabilization over phytoextraction is that it does not require the disposal of hazardous biomass (Wuana and Okieimen 2011). However, the selection of the appropriate plant species is critical for effective phytostabilization. Plants must be tolerant to the conditions caused by heavy metals to be effective, and large root systems are necessary for immobilizing heavy metals, stabilizing soil structure, and reducing soil erosion. Additionally, plants should be capable of producing a significant amount of biomass and growing rapidly to establish a vegetative cover on the site quickly. Furthermore, the plant cover must be easy to maintain in outdoor situations (Marques et al. 2009).

Phytoextraction

Utilizing plants to move and collect contaminants in their aboveground biomass by absorbing them from soil or water is known as phytoextraction (Jacob et al. 2018).

Table 14.3 A brief summary of the different plant species used for bioremediation and the amount of lead accumulated in their tissues

Species name	Plant part in which lead accumulated	Concentration of Pb (mg/kg)	Method	Removed Pb quantity	References
T. rotundifolium	Whole plant	8200	Phytoremediation method	4000 (kg/ha)	Cunningham and Ow (1996)
Euphorbia cheradenia	Shoot	13,500	Phytoremediation method	13,249.25	Chehregani and Malayeri (2007)
Tetraena qataranse	Root	2784	Phytochelatin method	Metals tend to precipitate in soil at pH levels above 8, and 98.13% of them may be eliminated	Usman et al. (2020)
	Shoot	1141.6			
Helianthus annus	Roots	7.76 ± 0.008	Phytostabilization method	At higher concentrations, removal efficiency ranges from 60 to 80%	Kumar et al. (2020)
Ocimum sanctum	Leaves	4.59 ± 0.0017	Phytoaccumulation method	40 kg/ha/yr of removal capacity	Inam et al. (2013)
Athyrium esculentum	Leaves	0.26 ± 0.03	Phytovolatilization method	Up to 80% removal effectiveness at higher metal levels	Sulaiman and Hamzah (2018)
	Stem	0.33 ± 0.04			
	Root	3.79 ± 0.19			
Chromolaena odorata	Leaves	0.33 ± 0.06	By *Eichhornia crassipes*	For heavier metals, removal effectiveness can reach 80%	Sulaiman and Hamzah (2018)
	Stem	0.69 ± 0.07			
	Root	2.64 ± 0.43			
P. vittata	Roots	2.545 ± 0.067	Garlic interplanting	*P. vittata* with 1.382 0.120 mg/kg of garlic	Hussain et al. (2021)
	Shoots	11.964 ± 1.25			
Conyza canadensis	Roots	1.308 ± 0.043	Garlic interplanting	Garlic with 2.081 0.154 mg/kg of *Conyza Canadensis*	Hussain et al. (2021)
	Shoots	8.517 ± 0.136			
Oryza sativa	Shoots	0.34	Include a chelating ligand	Above pH 5.5–6, heavy metal bioavailability in soils declines, necessitating the use of a chelating agent	Ihedioha et al. (2016)

Conflicting to phytostabilization, which only temporarily retains heavy metals that still remain underground, phytoextraction is a long-term approach for eliminating heavy metals from polluted soil. Therefore, it is more suited for commercial application. Heavy metal phytoextraction involves the following steps: (a) heavy metal mobilization in the rhizosphere, (b) heavy metal absorption by plant roots, (c) the movement of heavy metal ions from plant roots to aerial portions, and (d) heavy metal ion compartmentalization and sequestration in plant tissues (Ali et al. 2013).

Careful plant species selection is necessary for effective phytoextraction. The following characteristics should be present in the plant species utilized for phytoextraction: (a) strong tolerance for heavy metals' harmful effects; (b) high extraction capacity and high quantities of heavy metals accumulating in aboveground components; (c) rapid development and substantial biomass production; (d) a profusion of shoots and a deep root system; (e) good environmental sensitivity; (f) great capacity to flourish in arid conditions; simple cultivation and harvest; and (g) extremely resilient to illnesses and pests (Seth 2012; Ali et al. 2013).

The following characteristics of the plant species utilized for phytoextraction should be present: (a) strong tolerance for heavy metals' harmful effects; (b) high extraction capacity and high quantities of heavy metals accumulating in aboveground components; (c) rapid development and substantial biomass production; (d) a profusion of shoots and a deep root system; (e) good environmental sensitivity; (f) a powerful capacity to flourish in subpar soils; simple cultivation and harvest; and (g) extremely resilient to illnesses and pests (Seth 2012; Ali et al. 2013).

Plants that accumulate extraordinarily high amounts of heavy metals in their aboveground parts and do not show the symptoms of toxicity are called "hyperaccumulators" (Van der Ent et al. 2013). Up to 100 times, more heavy metals have been reported in natural hyperaccumulators in comparison to non-accumulating species under similar conditions (Rascio and Navari-Izzo 2011). More than 45 angiosperm families like Brassicaceae, Fabaceae, Euphorbiaceae, Asteraceae, Lamiaceae, and Scrophulariaceae have been reported to include around 450 hyperaccumulating plant species (Suman et al. 2018). These species include permanent trees and shrubs as well as annual herbs (Dushenkov 2003). Some species, like *Sedum alfredii*, can hyperaccumulate more than two elements, including Zn, Pb, and Cd (Yang et al. 2002).

Phytovolatilization

Phytovolatilization is a process where plants uptake pollutants from the soil, convert them into less toxic volatiles, and release them into the atmosphere through transpiration. This technique can be effective in detoxifying heavy metals such as Se, Hg, and As, as well as organic pollutants (Mahar et al. 2016). Brassicaceae family members, such as *Brassica juncea*, are particularly efficient in this process (Banuelos et al. 1993). Unlike traditional phytoremediation methods, phytovolatilization does not require plant harvesting and disposal since heavy metal contaminants are removed from the site and disseminated as gaseous components. However, it is essential to conduct a risk assessment before deploying this technique in the field. Although phytovolatilization reduces the amount of contaminants in the soil, some pollutants may remain in the environment, as they are still present in the air. Additionally, rainfall may redeposit them in the soil (Vangronsveld et al. 2009). Therefore, it is crucial to evaluate the potential risks of this technique before implementing it.

Phytofiltration

The use of plants to decontaminate polluted surface waters or waste waters is called as phytofiltration (Mesjasz-Przybyłowicz et al. 2004). Uptake and/or adsorption of heavy metals on root surface in a response to alteration of rhizosphere pH by the action of root exudates is categorized under rhizofiltration (Javed et al. 2019), lowering the flow of heavy metals into groundwater even more. In order to prepare the plants for rhizofiltration, hydroponically grown plants must first be placed in clean water for development of root system and then relocated to polluted water. After this acclimation step, the plants should be transferred to the contaminated location for removal of heavy metals. After the process is complete, the roots of these plants should be gathered and discarded (Wuana and Okieimen 2011). Dense root system, high biomass production, and resistance to heavy metals are some prerequisites for any plant to be used for rhizofilteration. Rhizofilteration can be done with either aquatic or terrestrial plants. Wetland water remediation typically uses aquatic plants, such as hyacinth, Azolla, duckweed, cattail, and poplar, due to their high accumulation of heavy metals, high tolerance, or rapid growth and high biomass output (Hooda 2007). Compared to aquatic plants, terrestrial plants like Indian mustard (*B. juncea*) and sunflower (*H. annuus*) have longer and hairier root systems. They also clearly exhibit the capacity to acquire heavy metals during rhizofiltration (Rezania et al. 2016; Dhanwal et al. 2017).

Genetic Engineering

Genetic engineering is a promising technique for improving a plant's ability to phytoremediate heavy metal contamination by transferring foreign genes from other organisms, such as bacteria, plants, or animals, into the target plant's genome. Compared to traditional breeding, genetic engineering is faster and allows the transfer of desired genes from hyperaccumulators to incompatible plant species (Marques et al. 2009). However, modifying multiple genes to improve heavy metal detoxification and accumulation is complex and time-consuming, and there are safety concerns regarding the field testing of genetically modified plants.

To guide gene selection for genetic engineering, it is crucial to understand how plants acquire and tolerate heavy metals. Heavy metal tolerance is often measured by the efficiency of oxidative stress defences since heavy metals can cause excessive ROS formation and oxidative stress. Therefore, overexpressing genes involved in the antioxidant system is a popular strategy for increasing heavy metal tolerance (Koźmińska et al. 2018).

To improve heavy metal absorption, translocation, and sequestration, genes involved in these processes can be transferred and overexpressed in target plants (Mani and Kumar 2014; Das et al. 2016). This includes genes encoding members of the ZIP, MTP, MATE, and HMA families of metal ion transporters. Additionally, encouraging the production of metal chelators through genetic engineering can increase heavy metal bioavailability, uptake, and translocation. This can be achieved by overexpressing genes for natural chelators (Wu et al. 2010; Ghuge et al. 2023).

However, if genetic engineering is not practical, other techniques must be employed to improve plant performance in phytoremediation. For instance, hyperaccumulators with high heavy metal tolerance and accumulation capacities can be bred with fast-growing, high biomass plants to produce high quantities of biomass. Furthermore, understanding the mechanisms of heavy metal detoxification and accumulation can help identify and select plants that are more efficient at phytoremediation.

Use of Plant-Associated Microbes

Another important strategy for phytoremediation is the use of plant-associated microbes for the improvement of plant performance. The microbial community present in the rhizosphere can directly improve plant's fitness by improving their capacity to tolerate heavy metals (Fasani et al. 2018) (Table 14.3). It has been established that rhizobacteria that encourage plant growth have a substantial potential to improve the efficacy of phytoremediation. The absorption and transport of heavy metals by plants can both be enhanced by PGPR. Additionally, it can increase a plant's resistance to heavy metals and shield plants from infections (Ma et al. 2011). This is made possible by the production of many chemicals, including organic acids, siderophores, antibiotics, enzymes, and phytohormones (Ma et al. 2011). One of the beneficial products of plant growth-promoting rhizobacteria (PGPR) is the production of the enzyme ACC deaminase which can degrade the ethylene precursor 1-aminocyclopropane-1-carboxylate (ACC) (Glick 2014). The presence of ACC deaminase can reduce ethylene synthesis and promote plant growth, leading to increased biomass production and enhancing the efficiency of phytoremediation and heavy metal uptake (Huang et al. 2004; Arshad et al. 2007). Moreover, PGPR can also produce bacterial auxin (IAA), which stimulates the formation of lateral roots and root hairs, further promoting plant growth and phytoremediation (Glick 2010; DalCorso et al. 2019).

Another useful microbial community in phytoremediation is arbuscular mycorrhizal fungus (AMF). AMF can expand the absorptive surface area of plant roots by forming an extensive hyphal network in the rhizosphere, thus enhancing water and nutrient uptake as well as heavy metal bioavailability (Gohre and Paszkowski 2006). In addition, AMF can produce phytohormones that aid in phytoremediation and promote plant growth (Vamerali et al. 2010).

14.2 Conclusion

Heavy metals are necessary for the biological and physiological processes of plants, including the synthesis of chlorophyll, secondary metabolites, stress tolerance, and the biosynthesis of nucleic acids and proteins. However, in specific concentrations that vary from species to species, these metal elements can become poisonous to plants, obstructing electron transport and producing ROS and free radicals, which cause oxidative damage to cellular components. Nonetheless, plants have enzymatic and non-enzymatic mechanisms, including CAT, POD, APX, SOD, and others, to activate the cell and support it in adjusting its metabolism to metal stress. These metals primarily come from emerging industrialized regions and human activities that harm ecosystems by dumping sludge that has not been properly processed. Phytoremediation, phytostabilization, phytoextraction, phytovolatilization, and phytofiltration are some ameliorative techniques that can be employed to remediate metals and prevent pollution in the environment, lessening the effects of heavy metals. Lead, the main inorganic contaminant emerging amongst heavy metals, damages the soil and atmosphere. The build-up of lead from the soil in the various parts of plants and living things has a negative impact on physiological and metabolic processes since it contains non-essential elements for plants. Lead build-up in the intercellular gaps slows down plant activity. Once it enters the plants through the soil, roots, and leaves, it affects their metabolic processes and stunts growth and crop yield. The amelioration methods produce improved crop types with lower levels of toxicity and oxidative stress in the plants.

References

Ali H, Khan E, Sajad MA (2013) Phytoremediation of heavy metals-concepts and applications. Chemosphere 91:869–881

Arias JA, Peralta-Videa JR, Ellzey JT et al (2010) Effects of Glomus deserticola inoculation on Prosopis: enhancing chromium and lead uptake and translocation as confirmed by X-ray mapping, ICP–OES and TEM techniques. Environ Exp Bot 68(2):139–148

Arshad M, Saleem M, Hussain S (2007) Perspectives of bacterial ACC deaminase in phytoremediation. Trends Biotechnol 25:356–362

Arshad M, Silvestre J, Pinelli E, Kallerhoff J, Kaemmerer M, Tarigo A, Shahid M, Guiresse M, Pradere P, Dumat C (2008) A field study of lead phytoextraction by various scented Pelargonium cultivars. Chemosphere 71(11):2187–2192

Assche F, Clijsters H (1990) Effects of metals on enzyme activity in plants. Plant Cell Environ 13(3):195–206

Atici Ö, Ağarand G, Battal P (2005) Changes in phytohormone contents in chickpea seeds germinating under lead or zinc stress. Biol Plant 49(2):215–222

Baker WG (1972) Toxicity levels of mercury lead, copper and zinc in tissue culture systems of cauliflowers, lettuce, potato and carrot. Can J Bot 50:973–976

Baker AJM (1981) Accumulators and excluders-strategies in the response of plant to heavy metals. J Plant Nutr 3:326–329

Banuelos G, Cardon G, Mackey B, Ben-Asher J, Wu L, Beuselinck P et al (1993) Boron and selenium removal in boron—laden soils by four sprinkler irrigated plant species. J Environ Qual 22:786–792

Barceló JUAN, Poschenrieder C (1990) Plant water relations as affected by heavy metal stress: a review. J Plant Nutr 13(1):1–37

Bharwana SA, Ali S, Farooq MA, Iqbal N, Abbas F, Ahmad MSA (2013) Alleviation of lead toxicity by silicon is related to elevated photosynthesis, antioxidant enzymes suppressed lead uptake and oxidative stress in cotton. J Bioremed Biodeg 4:4

Blaser P, Zimmermann S, Luster J, Shotyk W (2000) Critical examination of trace element enrichments and depletions in soils: As, Cr, Cu, Ni, Pb, and Zn in Swiss forest soils. Sci Total Environ 249(1–3):257–280

Brunet J, Varrault G, Zuily-Fodil Y, Repellin A (2009) Accumulation of lead in the roots of grass pea (*Lathyrus sativus* L.) plants triggers systemic variation in gene expression in the shoots. Chemosphere 77(8):1113–1120

Cao X, Ma LQ, Singh SP, Zhou Q (2008) Phosphate-induced lead immobilization from different lead minerals in soils under varying pH conditions. Environ Pollut 152(1):184–192

Cenkci S, Cigerci IH, Yildiz M, Özay C, Bozdag A, Terzi H (2010) Lead contamination reduces chlorophyll biosynthesis and genomic template stability in *Brassica rapa* L. Environ Exp Bot 67(3):467–473

Chang HR, Chen SS, Chen TJ, Ho CK, Chiang HC, Yu HS (1996) Lymphocyte β2-adrenergic receptors and plasma catecholamine levels in lead-exposed workers. Toxicol Appl Pharmacol 139(1):1–5

Chehregani A, Malayeri BE (2007) Removal of heavy metals by native accumulator plants. Int J Agric Biol 9:462–465

Choudhury S, Panda S (2004). Toxic effects, oxidative stress and ultrastructural changes in moss *Taxithelium nepalense* (Schwaegr.) Broth. under chromium and lead phytotoxicity. Water Air Soil Pollut 167(1):73–90

Cimrin KM, Turan M, Kapur B (2007) Effect of elemental sulphur on heavy metals solubility and remediation by plants in calcareous soils. Fresenius Environ Bull 16(9):1113–1120

Cunningham SD, Ow DW (1996) Promises and prospects of phytoremediation. Plant Physiol 110:715

DalCorso G, Farinati S, Maistri S, Furini A (2008) How plants cope with cadmium: staking all on metabolism and gene expression. J Integr Plant Biol 50:1268–1280

DalCorso G, Fasani E, Manara A, Visioli G, Furini A (2019) Heavy metal pollutions: state of the art and innovation in phytoremediation. Int J Mol Sci 20:3412

Das N, Bhattacharya S, Maiti MK (2016) Enhanced cadmium accumulation and tolerance in transgenic tobacco overexpressing rice metal tolerance protein gene OsMTP1 is promising for phytoremediation. Plant Physiol Biochem 105:297–309

Davies BE (1995) Lead. In: Alloway BJ (ed) Heavy metals in soils. Blackie Academic, London, pp 206–223

Dhanwal P, Kumar A, Dudeja S, Chhokar V, Beniwal V (2017) Recent advances in phytoremediation technology. In: Kumar R, Sharma AK, Ahluwalia SS (eds) Advances in environmental biotechnology. Springer, Singapore, pp 227–241

Dushenkov S (2003) Trends in phytoremediation of radionuclides. Plant Soil 249:167–175

Elzbieta W, Minoslawa C (2005) Lead-induced histological and ultra–structural changes in the leaves of soyben (*Glycine max* (L.) Meee.). Soil Sci Plant Nutr 51(2):203–212

Engwa GA, Ferdinand PU, Nwalo FN, Unachukwu MN (2019) Mechanism and health effects of heavy metal toxicity in humans. In: Poisoning in the modern world-new tricks for an old dog, vol 10

Eun SO, Yon HS, Lee Y (2000) Lead distrurbs microtubule organization in the root meristem of *Zea mays*. Physiol Plant 110:357–365

Fangmin C, Ningchun Z, Haiming X, Yi L, Wenfang Z, Zhiwei Z, Mingxue C (2006) Cadmium and lead contamination in japonica rice grains and its variation among the different locations in southeast China. Sci Total Environ 359:156–166

Fasani E, Manara A, Martini F, Furini A, DalCorso G (2018) The potential of genetic engineering of plants for the remediation of soils contaminated with heavy metals. Plant Cell Environ 41:1201–1232

Ferrey ML, Coreen Hamilton M, Backe WJ, Anderson KE (2018) Pharmaceuticals and other anthropogenic chemicals in atmospheric particulates and precipitation. Sci Total Environ 612:1488–1497

Gao S, Li Q, Ou-Yang C, Chen L, Wang S, Chen F (2009) Lead toxicity induced antioxidant enzyme and phenylalanine ammonia-lyase activities in *Jatropha curcas* L. radicles. Fresenius Environ Bull 18

Garcia JS, Gratão PL, Azevedo RA, Arruda MAZ (2006) Metal contamination effects on sunflower (*Helianthus annuus* L.) growth and protein expression in leaves during development. J Agric Food Chem 54(22):8623–8630

Gerhardt KE, Gerwing PD, Greenberg BM (2017) Opinion: taking phytoremediation from proven technology to accepted practice. Plant Sci 256:170–185

Ghuge SA, Nikalje GC, Kadam US, Suprasanna P, Hong JC (2023) Comprehensive mechanisms of heavy metal toxicity in plants, detoxification, and remediation. J Hazard Mater 450. https://doi.org/10.1016/j.jhazmat.2023.131039

Gichner T, Znidar I, Száková J (2008) Evaluation of DNA damage and mutagenicity induced by lead in tobacco plants. Mutat Res Genet Toxicol Environ Mutagen 652(2):186–190

Gidlow DA (2015) Lead toxicity. Occup Med 65(5):348–356

Gisbert C, Ros R, De Haro A, Walker DJ, Pilar Bernal M, Serrano R, Navarro-Aviñó J (2003) A plant genetically modified that accumulates Pb is especially promising for phytoremediation. Biochem Biophys Res Commun 303(2):440–445

Glick BR (2010) Using soil bacteria to facilitate phytoremediation. Biotechnol Adv 28:367–374

Glick BR (2014) Bacteria with ACC deaminase can promote plant growth and help to feed the world. Microbiol Res 169:30–39

Godbold DL, Kettner C (1991) Lead influences root growth and mineral nutrition of *Picea abies* seedlings. J Plant Physiol 139:95–99

Gohre V, Paszkowski U (2006) Contribution of the arbuscular mycorrhizal symbiosis to heavy metal phytoremediation. Planta 223:1115–1122

Gomes EJR (2011) Genotoxicity and cytotoxicity of Cr (VI) and Pb^{2+} in *Pisum sativum.* Doctoral dissertation, Universidade de Aveiro (Portugal)

Gopal R, Rizvi AH (2008) Excess lead alters growth, metabolism and translocation of certain nutrients in radish. Chemosphere 70(9):1539–1544

Gothberg A, Greger M, Holm K et al (2004) Influence of nutrient levels on uptake and effects of mercury, cadmium, and lead in water spinach. J Environ Qual 33:1247–1255

Grover P, Rekhadevi PV, Danadevi K, Vuyyuri SB, Mahboob M, Rahman MF (2010) Genotoxicity evaluation in workers occupationally exposed to lead. Int J Hyg Environ Health 213(2):99–106

Gruenhage L, Jager IIJ (1985) Effect of heavy metals on growth and heavy metals content of *Allium Porrum* and *Pisum sativum.* Angew Bot 59:11–28

Gupta D, Nicoloso F, Schetinger M, Rossato L, Pereira L, Castro G, Srivastava S, Tripathi R (2009) Antioxidant defense mechanism in hydroponically grown *Zea mays* seedlings under moderate lead stress. J Hazard Mater 172(1):479–484

Gupta D, Huang H, Yang X, Razafindrabe B, Inouhe M (2010) The detoxification of lead in *Sedum alfredii* H. is not related to phytochelatins but the glutathione. J Hazard Mater 177(1–3):437–444

Haider S, Kanwal S, Uddin F et al (2006) Phytotoxicity of Pb II. Changes in chlorophyll absorption spectrum due to toxic metal Pb stress on *Phaseolus mungo* and *Lens culinaris*. Pak J Biol Sci 9(11):2062–2068

Hall JL (2002) Cellular mechanisms for heavy metal detoxification and tolerance. J Exp Bot 53:1–11

Halliwell B, Cutteridge JMC (1990) Role of free radicals and catalytic metal ions in human disease: an overview. Methods Enzymol 186:1

Harpaz-Saad S, Azoulay T, Arazi T, Ben-Yaakov E, Mett A, Shiboleth YM, Eyal Y (2007) Chlorophyllase is a rate-limiting enzyme in chlorophyll catabolism and is post translationally regulated. Plant Cell 19(3):1007–1022

Hewilt EJ (1953) Metal inter-relationships in plant nutrition. J Exp Bot 4:59–64

Hooda V (2007) Phytoremediation of toxic metals from soil and waste water. J Environ Biol 28(2 Suppl.):367–376

Hseu ZY, Su SW, Lai HY, Guo HY, Chen TC, Chen ZS (2010) Remediation techniques and heavy metal uptake by different rice varieties in metal-contaminated soils of Taiwan: new aspects for food safety regulation and sustainable agriculture. Soil Sci Plant Nutr 56:31–52

Hu J, Shi G, Xu Q, Wang X, Yuan Q, Du K (2007) Effects of $Pb2^+$ on the active oxygen scavenging enzyme activities and ultrastructure in *Potamogeton crispus* leaves. Russ J Plant Physl 54(3):414–419

Huang CV, Bazzaz FA, Venderhoef LN (1974) The inhibition of soya bean metabolism by cadmium and lead. Plant Physiol 34:122–124

Huang XD, El-Alawi Y, Penrose DM, Glick BR, Greenberg BM (2004) Responses of three grass species to creosote during phytoremediation. Environ Pollut 130:453–463

Hussain J, Wei X, Xue-Gang L, Shah SRU, Aslam M, Ahmed I, Azam T (2021) Garlic (*Allium sativum*) based interplanting alters the heavy metals absorption and bacterial diversity in neighboring plants. Sci Rep 11(1):1–13

Ibrahim ATA, Banaee M, Sureda A (2019) Selenium protection against mercury toxicity on the male reproductive system of *Clarias gariepinus*. Comp Biochem Physiol C Toxicol Pharmacol 225:108583

Ihedioha JN, Ujam OT, Nwuche CO, Ekere NR, Chime CC (2016) Assessment of heavy metal contamination of rice grains (*Oryza sativa*) and soil from Ada field, Enugu, Nigeria: estimating the human health risk. Hum Ecol Risk Assess Int J 22(8):1665–1677

Inam F, Deo S, Narkhede N (2013) Analysis of minerals and heavy metals in some spices collected from local market. J Pharm Biol Sci 8(2):40–43

Islam E, Yang X, Li T, Liu D, Jin X, Meng F (2007) Effect of Pb toxicity on root morphology, physiology and ultrastructure in the two ecotypes of *Elsholtzia argyi*. J Hazard Mater 147(3):806–816

Islam E, Liu D, Li T, Yang X, Jin X, Mahmood Q, Tian S, Li J (2008) Effect of Pb toxicity on leaf growth, physiology and ultrastructure in the two ecotypes of *Elsholtzia argyi*. J Hazard Mater 154(1–3):914–926

Jacob JM, Karthik C, Saratale RG, Kumar SS, Prabakar D, Kadirvelu K et al (2018) Biological approaches to tackle heavy metal pollution: a survey of literature. J Environ Manage 217:56–70

Jaja ET, Odoemena CSI (2004) Effect of Pb, Cu and Fe compounds on the germination and early seedling growth of tomato varieties. J Appl Sci Environ Manag 8(2):51–53

Javed MT, Tanwir K, Akram MS, Shahid M, Niazi NK, Lindberg S (2019) Chapter 20—phytoremediation of cadmium-polluted water/sediment by aquatic macrophytes: role of plant-induced pH changes. In: Hasanuzzaman M, Prasad MNV, Fujita M (eds) Cadmium toxicity and tolerance in plants. Academic Press, London, pp 495–529

Ji Y, Ren Y, Han C, Zhu W, Gu J, He J (2022) Application of exogenous glycinebetaine alleviates lead toxicity in pakchoi (*Brassica chinensis* L.) by promoting antioxidant enzymes and suppressing Pb accumulation. Environ Sci Pollut Res 1–13

Jiang W, Liu D (2010) Pb-induced cellular defense system in the root meristematic cells of *Allium sativum* L. BMC Plant Biol 10:40

Jiang N, Luo X, Zeng J, Yang ZR, Zheng LN, Wang ST (2010) Lead toxicity induced growth and antioxidant responses in *Luffa cylindrica* seedlings. Inter J Agric Biol 12:205–210

Juwarkar AS, Shende GB (1986) Interaction of Cd-Pb effect on growth yield and content of Cd, Pb in barley. Ind J Environ Health 28:235–243

Kabata-Pendia A (2001) Trace elements in soils and plants, 3rd edn. CRC Press, Boca Raton, FL

Kastori R, Petrovic M, Petrovic N (2008) Effect of excess lead, cadmium, copper, and zinc on water relations in sunflower. J Plant Nutr 15:2427–2439

Kevresan S, Petrovic N, Popovic M et al (2001). Nitrogen and protein metabolism in young pea plants as affected by different concentrations of nickel, cadmium, lead, and molybdenum. J Plant Nutr 24(10):1633–644

Khan S, Khan NN (1983) Influence of lead and cadmium on growth and nutrient concentration of tomato (*Lycopersicum esculentum*) and egg plant (*Solanum melongena*). Plant Soil 74:387–394

Kopittke PM, Asher CJ, Kopittke RA, Menzies NW (2007) Toxic effects of Pb^{2+} on growth of cowpea (*Vigna unguiculata*). Environ Pollut 150(2):280–287

Koźmińska A, Wiszniewska A, Hanus-Fajerska E, Muszyńska E (2018) Recent strategies of increasing metal tolerance and phytoremediation potential using genetic transformation of plants. Plant Biotechnol Rep 12:1–14

Krzeslowska M, Lenartowska M, Mellerowicz EJ, Samardakiewicz S, Wozny A (2009) Pectinous cell wall thickenings formation—a response of moss protonemata cells to lead. Environ Exp Bot 65(1):119–131

Krzesłowska M, Lenartowska M, Samardakiewicz S, Bilski H, Woźny A (2010) Lead deposited in the cell wall of *Funaria hygrometrica* protonemata is not stable—a remobilization can occur. Environ Pollut 158(1):325–338

Kumar K, Singh D (2023) Toxicity and bioremediation of lead: a critical review. Int J Environ Health Res. https://doi.org/10.1080/09603123.2023.2165047

Kumar G, Singh RP, Sushila (1992) Nitrate assimilation and biomass production in *Seasamum indicum* (L.) seedlings in lead enriched environment. Water Air Soil Pollut 215:124–215

Kumar A, Pal L, Agrawal V (2017) Glutathione and citric acid modulates lead-and arsenic-induced phytotoxicity and genotoxicity responses in two cultivars of *Solanum lycopersicum* L. Acta Physiol Plant 39(7):151

Kumar A, Mms CP, Chaturvedi AK, Shabnam AA, Subrahmanyam G, Mondal R, Yadav KK (2020) Lead toxicity: health hazards, influence on food chain, and sustainable remediation approaches. Int J Environ Res Public Health 17(7):2179

Lamhamdi M, Bakrim A, Aarab A, Lafont R, Sayahe F (2010) A comparison of lead toxicity using physiological and enzymatic parameters on spinach (*Spinacia oleracea* L.) and wheat (*Triticum aestivum* L.) growth. Moroccan J Biol 12/N:6–7

Lamhamdi M, El Galiou O, Bakrim A, Nóvoa-Muñoz JC, Arias-Estevez M, Aarab A, Lafont R (2013) Effect of lead stress on mineral content and growth of wheat (Triticum aestivum) and spinach (Spinacia oleracea) seedlings. Saudi J Biolog Sci 20(1):29–36

Lenntech Water Treatment and Air Purification (2004) Water treatment. Lenntech, Rotterdamseweg, Netherlands

Lin YF, Mark GMA (2012) The molecular mechanism of zinc and cadmium stress response in plants. Cell Mol Life Sci 69:3187–3206

Liu D, Li TQ, Jin XF, Yang XE, Islam E, Mahmood Q (2008) Lead induced changes in the growth and antioxidant metabolism of the lead accumulating and non-accumulating ecotypes of *Sedum alfredii*. J Integr Plant Biol 50(2):129–140

Liu T, Liu S, Guan H, Ma L, Chen Z, Gu H (2009) Transcriptional profiling of Arabidopsis seedlings in response to heavy metal lead (Pb). Environ Exp Bot 67(2):377–386

Liu Y, Liu G, Yuan Z, Liu H, Lam PKS (2018) Heavy metals (As, Hg and V) and stable isotope ratios (d13C and d15N) in fish from Yellow River Estuary, China. Sci Total Environ 613–614:462–471

Ma Y, Prasad M, Rajkumar M, Freitas H (2011) Plant growth promoting rhizobacteria and endophytes accelerate phytoremediation of metalliferous soils. Biotechnol Adv 29:248–258

Maestri E, Marmiroli M, Visioli G, Marmiroli N (2010) Metal tolerance and hyperaccumulation: costs and trade-offs between traits and environment. Environ Exp Bot 68(1):1–13

Mahar A, Wang P, Ali A, Awasthi MK, Lahori AH, Wang Q et al (2016) Challenges and opportunities in the phytoremediation of heavy metals contaminated soils: a review. Ecotox Environ Safe 126:111–121

Mahmood MN, Zhu S, Noman A, Mahmood A, Ashraf S, Aqeel M, Ibrahim M, Ashraf S, Liew RK, Lam SS, Irshad MK (2023) An assessment of the efficacy of biochar and zero-valent iron nanoparticles in reducing lead toxicity in wheat (*Triticum aestivum* L.). Environ Pollut 319. https://doi.org/10.1016/j.envpol.2022.120979

Małecka A, Piechalak A, Morkunas I, Tomaszewska B (2008) Accumulation of lead in root cells of *Pisum sativum*. Acta Physiol Plant 30(5):629–637

Małkowski E, Kita A, Galas W, Karczand W, Kuperberg JM (2002) Lead distribution in corn seedlings (*Zea mays* L.) and its effect on growth and the concentrations of potassium and calcium. Plant Growth Regul 37(1):69–76

Malone C, Koeppe DE, Miller RJ (1974) Localization of lead accumulated by corn plants. Plant Physiol 53(3):388–394

Mani D, Kumar C (2014) Biotechnological advances in bioremediation of heavy metals contaminated ecosystems: an overview with special reference to phytoremediation. Int J Environ Sci Technol 11:843–872

Marques AP, Rangel AO, Castro PM (2009) Remediation of heavy metal contaminated soils: phytoremediation as a potentially promising clean-up technology. Crit Rev Env Sci Technol 39:622–654

Meharg AA, Hartley-Whitaker J (2002) Arsenic uptake and metabolism in arsenic resistant and nonresistant plant species. New Phytol 154:29–43

Mench M, Lepp N, Bert V, Schwitzguébel J-P, Gawronski SW, Schröder P et al (2010) Successes and limitations of phytotechnologies at field scale: outcomes, assessment and outlook from COST Action 859. J Soil Sediment 10:1039–1070

Mesjasz-Przybyłowicz J, Nakonieczny M, Migula P, Augustyniak M, Tarnawska M, Reimold U et al (2004) Uptake of cadmium, lead nickel and zinc from soil and water solutions by the nickel hyperaccumulator *Berkheya coddii*. Acta Biol Cracoviensia Ser Bot 46:75–85

Meyers DER, Auchterlonie GJ, Webb RI, Wood B (2008) Uptake and localisation of lead in the root system of *Brassica juncea*. Environ Pollut 153(2):323–332

Miller JE, Hassete JJ, Koppe DE (1975) Interaction of lead and cadmium of electron energy transfer reaction in corn mitochondria. Physiol Plant 28:166–171

Mishra S, Srivastava S, Tripathi R, Kumar R, Seth C, Gupta D (2006) Lead detoxification by coontail (*Ceratophyllum demersum* L.) involves induction of phytochelatins and antioxidant system in response to its accumulation. Chemosphere 65(6):1027–1039

Mishra KK, Rai UN, Prakash O (2007) Bioconcentration and phytotoxicity of Cd in *Eichhornia crassipes*. Environ Monit Assess 130:237–243

Monni S, Salemma M, Millar N (2000) The tolerance of Empetrum nigrum to copper and nickel. Environ Pollut 109:221–229

Morzck E Jr, Funicclli NA (1982) Effect of lead and on germination of *Spartina alterniflora* Losiel seeds at various salinities. Environ Exp Bot 22:23–32

Muhammad S, Shah MT, Khan S (2011) Health risk assessment of heavy metals and their source apportionment in drinking water of Kohistan region, northern Pakistan. Microchem J 98(2):334–343

Mukherji S, Maitra P (1976) Toxic effects of lead growth and metabolism of germinating rice (*Oryza sativa* L.) seeds mitosis of onion (*Allium cepa*) root tip cells. Ind J Exp Biol 14:519–521

Munzuroglu O, Geckil H (2002) Effects of metals on seed germination, root elongation, and coleoptile and hypocotyl growth in *Triticum aestivum* and *Cucumis sativus*. Arch Environ Contam Toxicol 43(2):203–213

Nakos G (1979) Lead pollution: fate of lead in soil and its effects on *Pinus haplenis*. Plant Soil 50:159–161

Navabpour S, Yamchi A, Bagherikia S, Kafi H (2020) Lead-induced oxidative stress and role of antioxidant defense in wheat (*Triticum aestivum* L.). Physiol Mol Biol Plants 26:793–802

Nieboer E, Richardson DHS (1980) The replacement of the nondescript term heavy metals by a biologically and chemistry significant classification of metal ions. Environ Pollut Ser B 1:3–26

Odobasic A, Sestan I, Begic S (2019) Biosensors for determination of heavy metals in waters. In: Biosensors for environmental monitoring. IntechOpen

Osman HE, Fadlallah RS (2023) Impact of lead on seed germination, seedling growth, chemical composition, and forage quality of different varieties of Sorghum. J Umm Al-Qura Univ Appll Sci 9:77–86

Paivoke H (1983) The short-term effect of zinc on growth anatomy and acid phosphate activity of pea seedlings. Ann Bot 20:307–309

Patra M, Bhowmik N, Bandopadhyayand B, Sharma A (2004) Comparison of mercury, lead and arsenic with respect to genotoxic effects on plant systems and the development of genetic tolerance. Environ Exp Bot 52(3):199–223

Patrick L (2006) Lead toxicity, a review of the literature. Part 1: exposure, evaluation, and treatment. Altern Med Rev 11:2–22

Piechalak A, Tomaszewska B, Baralkiewicz D, Malecka A (2002) Accumulation and detoxification of lead ions in legumes. Phytochemistry 60(2):153–162

Pinho S, Ladeiro B (2012) Phytotoxicity by lead as heavy metal focus on oxidative stress. J Bot 2012:1–10

Piotrowska A, Bajguz A, Godlewska-Zylkiewicz B, Czerpak R, Kaminska M (2009) Jasmonic acid as modulator of lead toxicity in aquatic plant *Wolffia arrhiza* (Lemnaceae). Environ Exp Bot 66(3):507–513

Pitzschke A, Forzani C, Hirt H (2006) Reactive oxygen species signaling in plants. Antioxid Redox Signal 8(9–10):1757–1764

Pourrut B, Shahid M, Dumat C, Winterton P, Pinelli E (2011) Lead uptake, toxicity, and detoxification in plants. Rev Environ Contam Toxicol 213:113–136. ISSN 0179-5953

Qufei L, Fashui H (2009) Effects of Pb^{2+} on the structure and function of photosystem II of Spirodelapolyrrhiza. Biol Trace Elem Res 129(1–3):251

Qureshi M, Abdin M, Qadir S, Iqbal M (2007) Lead-induced oxidative stress and metabolic alterations in *Cassia angustifolia* Vahl. Biol Plantarum 51(1):121–128

Rabinowitz D (2017) Resistance and the city. In: De-pathologizing resistance, pp 58–73

Rahimzadeh MR, Kazemi S, Moghadamnia AA (2017) Cadmium toxicity and treatment: an update. Casp J Intern Med 8(3):135

Rahman Z, Singh VP (2019) The relative impact of toxic heavy metals (THMs) (arsenic (As), cadmium (Cd), chromium (Cr)(VI), mercury (Hg), and lead (Pb)) on the total environment: an overview. Environ Monit Assess 191:419

Rao G, Ashraf U, Huang S, Cheng S, Abrar M, Mo Z, Pan S, Tang X (2018) Ultrasonic seed treatment improved physiological and yield traits of rice under lead toxicity. Environ Sci Pollut Res 25:33637–33644

Rascio N, Navari-Izzo F (2011) Heavy metal hyperaccumulating plants: how and why do they do it? And what makes them so interesting? Plant Sci 180:169–181

Rassaei F (2023) The effect of sugarcane bagasse biochar on maize growth factors in lead and cadmium polluted soils. Commun Soil Sci Pl Anal 54(10):1426–1446. https://doi.org/10.1080/00103624.2022.2146704

Reddy AM, Kumar SG, Jyonthsnakumari G, Thimmanaik S, Sudhakar C (2005) Lead induced changes in antioxidant metabolism of horsegram (*Macrotyloma uniflorum* [Lam.] Verdc.) and bengalgram (*Cicer arietinum* L.). Chemosphere 60:97–104

Reeves RD, Baker AJM (2000) Metal-accumulating plants. In: Raskin I, Ensley BD (eds) Phytoremediation of toxic metals: using plants to clean up the environment. Wiley, New York, pp 193–229

Rezania S, Taib SM, Md Din MF, Dahalan FA, Kamyab H (2016) Comprehensive review on phytotechnology: heavy metals removal by diverse aquatic plants species from wastewater. J Hazard Mater 318:587–599

Romanowska E, Igamberdiev AU, Parys E, Gardeström P (2002) Stimulation of respiration by $Pb2^{+}$ in detached leaves and mitochondria of C3 and C4 plants. Physiol Plant 116(2):148–154

Romanowska E, Wróblewska B, Drozak A, Siedlecka M (2006) High light intensity protects photosynthetic apparatus of pea plants against exposure to lead. Plant Physiol Biochem 44(5–6):387–394

Romanowska E, Wróblewska B, Drozak A, Zienkiewicz M, Siedlecka M (2008) Effect of Pb ions on superoxide dismutase and catalase activities in leaves of pea plants grown in high and low irradiance. Biol Plantarum 52(1):80–86

Roth U, Von Roepenack-Lahaye E, Clemens S (2006) Proteome changes in *Arabidopsis thaliana* roots upon exposure to Cd^{2+}. J Exp Bot 57:4003–4013

Rucińska-Sobkowiak R, Nowaczyk G, Krzesłowska M, Rabędaand I, Jurga S (2013) Water status and water diffusion transport in lupine roots exposed to lead. Environ Exp Bot 87:100–109

Ruley AT (2004) Effects of accumulation of lead and synthetic chelators on the physiology and biochemistry of *Sesbania drummondii.* M.S thesis, Western Kentucky University, USA

Saini S, Dhania G (2020) Cadmium as an environmental pollutant: ecotoxicological effects, health hazards, and bioremediation approaches for its detoxification from contaminated sites. In: Bioremediation of industrial waste for environmental safety. Springer, Berlin, pp 357–387

Sajid M, Nazal MK, Ihsanullah, Baig N, Osman AM (2018) Removal of heavy metals and organic pollutants from water using dendritic polymers-based adsorbents: a critical review. Sep Purif Technol 191:400–423

Schützendübel A, Polle A (2002) Plant responses to abiotic stresses: heavy metal-induced oxidative stress and protection by mycorrhization. J Exp Bot 53:1351–1365

Sengar RS, Gautam M, Sengar RS, Sengar RS, Garg SK, Sengar K, Chaudhary R (2009) Lead stress effects on physiobiochemical activities of higher plants. Rev Environ ContamToxicol 196:1–21

Seregin IV, Ivanov VB (2001) Physiological aspects of cadmium and lead toxic effects on higher plants. Russ J Plant Physiol 48(4):523–544

Seregin IV, Kozhevnikova AD (2023) Phytochelatins: sulfur-containing metal(loid)-chelating ligands in plants. Int J Mol Sci 24(3):2430. https://doi.org/10.3390/ijms24032430

Seth CS (2012) A review on mechanisms of plant tolerance and role of transgenic plants in environmental clean-up. Bot Rev 78:32–62

Shahid M, Pinelli E, Pourrut B, Silvestre J, Dumat C (2011) Lead-induced genotoxicity to *Vicia faba* L. roots in relation with metal cell uptake and initial speciation. Ecotoxicol Environ Saf 74(1):78–84

Sharma P, Dubey RS (2005) Lead toxicity in plants. Braz J Plant Physiol 17(1):35–52

Shua WS, Yeb ZH, Lana CY et al (2002) Lead, zinc and copper accumulation and tolerance in populations of *Paspalum distichum* and *Cynodon dactylon.* Environ Pollut 120(2):445–453

Singh R, Tripathi RD, Dwivedi S, Kumar A, Trivedi PK, Chakrabarty D (2010) Lead bioaccumulation potential of an aquatic macrophyte Najas indica are related to antioxidant system. Bioresour Technol 101:3025–3032

Singh R, Kesavan AnupKumar, Landi M, Kaur S, Thakur S, Zheng B, Bhardwaj R, Sharma A (2020) 5-aminolevulinic acid regulates Krebs cycle, antioxidative system and gene expression in *Brassica juncea* L. to confer tolerance against lead toxicity. J Biotechnol 323:283–292

Srivastava V, Sarkar A, Singh S, Singh P, de Araujo AS, Singh RP (2017) Agroecological responses of heavy metal pollution with special emphasis on soil health and plant performances. Front Environ Sci 5:64

Srivastava V, Gupta SK, Singh P, Sharma B, Singh RP (2018) Biochemical, physiological, and yield responses of lady's finger (*Abelmoschus esculentus* L.) grown on varying ratios of municipal solid waste vermicompost. Int J Recycl Org Waste Agric 7(3):241–250

Stiborova M, Pitrichova M, Brezinova A (1987) Effect of heavy metal ions in growth and biochemical characteristic of photosynthesis of barley and maize seedlings. Biol Plant 29:453–467

Sudhakar C, Symalabai L, Veeranjaveyuler K (1992) Lead tolerance of certain legume species grown on lead or tailing. Agri Eco Environ 41:253–261

Sulaiman FR, Hamzah HA (2018) Heavy metals accumulation in suburban roadside plants of a tropical area (Jengka, Malaysia). Ecol Process 7(1):1–11

Suman J, Uhlik O, Viktorova J, Macek T (2018) Phytoextraction of heavy metals: a promising tool for clean-up of polluted environment? Front Plant Sci 9:1476

Tewari RK, Kumar P, Sharma PN, Bisht SS (2002) Modulation of oxidative stress responsive enzymes by excess cobalt. Plant Sci 162:381–388

Tomulescu IM, Radoviciu EM, Merca VV, Tuduce AD (2004) Effect of copper, zinc and lead and their combinations on the germination capacity of two cereals. J Agric Sci 15:39–42

Usman K, Abu-Dieyeh MH, Zouari N, Al-Ghouti MA (2020) Lead (Pb) bioaccumulation and antioxidative responses in *Tetraena qataranse*. Sci Rep 10(1):1–10

Vadas TM, Ahner BA (2009) Cysteine- and glutathione-mediated uptake of lead and cadmium into *Zea mays* and *Brassica napus* roots. Environ Pollut 157(8–9):2558–2563

Valdés J, Guinez M, Castillo A, Vega SE (2014) Cu, Pb, and Zn content in sediments and benthic organisms from San Jorge Bay (northern Chile): accumulation and biotransference in subtidal coastal systems. Cienc Marinas 40(1):45–58

Vamerali T, Bandiera M, Mosca G (2010) Field crops for phytoremediation of metal-contaminated land. A review. Environ Chem Lett 8:1–17

Vangronsveld J, Herzig R, Weyens N, Boulet J, Adriaensen K, Ruttens A, Mench M (2009) Phytoremediation of contaminated soils and groundwater: lessons from the field. Environ Sci Pollut Res 16:765–794

Van der Ent A, Baker AJ, Reeves RD, Pollard AJ, Schat H (2013) Hyperaccumulators of metal and metalloid trace elements: facts and fiction. Plant Soil 362:319–334

Wang S, Shi X (2001) Molecular mechanisms of metal toxicity and carcinogenesis. Mol Cell Biochem 222:3–9

Wang N, Qiu Y, Hu K, Huang C, Xiang J, Li H, Tang J, Wang J, Xiao T (2020) One-step synthesis of cake-like biosorbents from plant biomass for the effective removal and recovery heavy metals: effect of plant species and roles of xanthation. Chemosphere 129129

WHO/FAO/IAEA (1996) Trace elements in human nutrition and health. World Health Organization, Geneva, Switzerland

Wierzbicka M (1987) Lead translocation and localization in *Allium cepa* roots. Can J Bot 65:1851–1860

Wierzbicka MH, Przedpełska E, Ruzik R, Ouerdane L, Połeć-Pawlak K, Jarosz M, Szakiel A (2007) Comparison of the toxicity and distribution of cadmium and lead in plant cells. Protoplasma 231:99–111

Wu G, Kang H, Zhang X, Shao H, Chu L, Ruan C (2010) A critical review on the bio-removal of hazardous heavy metals from contaminated soils: issues, progress, eco-environmental concerns and opportunities. J Hazard Mater 174:1–8

Wuana RA, Okieimen FE (2011) Heavy metals in contaminated soils: a review of sources, chemistry, risks and best available strategies for remediation. Isrn Ecol 2011:402647

Xiong ZT (1997) Bioaccumulation and physiological effects of excess lead in a roadside pioneer species *Sonchus oleraceus* L. Environ Pollut 97(3):275–279

Yadav S (2010) Heavy metals toxicity in plants: an overview on the role of glutathione and phytochelatins in heavy metal stress tolerance of plants. S Afr J Bot 76(2):167–179

Yan ZZ, Keand L, Tam NFY (2010) Lead stress in seedlings of *Avicennia marina*, a common mangrove species in South China, with and without cotyledons. Aquat Bot 92(2):112–118

Yang XE, Long XX, Ni WZ, Fu CX (2002) *Sedum alfredii* H: a new Zn hyperaccumulating plant first found in China. Chin Sci Bull 47:1634–1637

Yingli Y, Yang L, Yuanyuan Y, Zhang Xueling X, Wei J, You J, Wenrui W, Wang L, Ruxia R, Shi A (2011) Comparative antioxidative responses and proline metabolism in two wheat cultivars under short term lead stress. Ecotoxicol Environ Safety 74:4–8

Zaier H, Ghnaya T, Lakhdar A, Baioui R, Ghabriche R, Mnasri M, Sghair S, Lutts S, Abdelly C (2010) Comparative study of Pb-phytoextraction potential in *Sesuvium portulacastrum* and *Brassica juncea*: tolerance and accumulation. J Hazard Mater 183:609–615

Zhang F, Wang Y, Lou Z, Dong J (2007) Effect of heavy metal stress on antioxidative enzymes and lipid peroxidation in leaves and roots of two mangrove plant seedlings (*Kandelia candel* and *Bruguiera gymnorrhiza*). Chemosphere 67(1):44–50

Zheng LJ, Liu XM, Lütz-Meindl U, Peer T (2011) Effects of lead and EDTA-assisted lead on biomass, lead uptake and mineral nutrients in *Lespedeza chinensis* and *Lespedeza davidii*. Water Air Soil Pollut 220:57–68

Index

N. Kumar and A. K. Jha (eds.), *Lead Toxicity Mitigation: Sustainable Nexus Approaches*, Environmental Contamination Remediation and Management,
https://doi.org/10.1007/978-3-031-46146-0

MIX
Papier aus verantwortungsvollen Quellen
Paper from responsible sources
FSC® C105338

If you have any concerns about our products,
you can contact us on
ProductSafety@springernature.com

In case Publisher is established outside the EU,
the EU authorized representative is:
Springer Nature Customer Service Center GmbH
Europaplatz 3, 69115 Heidelberg, Germany

Printed by Libri Plureos GmbH
in Hamburg, Germany